Transition to Analysis with Proof

TEXTBOOKS in MATHEMATICS

Series Editors: Al Boggess and Ken Rosen

PUBLISHED TITLES CONTINUED

PUBLISHED TITLES CONTINUED

TEXTBOOKS in MATHEMATICS

Transition to Analysis with Proof

Steven G. Krantz

CRC Press
Taylor & Francis Group
Boca Raton   London   New York

CRC Press is an imprint of the
Taylor & Francis Group, an **informa** business

A CHAPMAN & HALL BOOK

CRC Press
Taylor & Francis Group
6000 Broken Sound Parkway NW, Suite 300
Boca Raton, FL 33487-2742

© 2018 by Taylor & Francis Group, LLC
CRC Press is an imprint of Taylor & Francis Group, an Informa business

No claim to original U.S. Government works

Printed on acid-free paper
Version Date: 20171011

International Standard Book Number-13: 978-1-1380-6414-0 (Hardback)
International Standard Book Number-13: 978-1-1380-6406-5 (Paperback)

Visit the Taylor & Francis Web site at
http://www.taylorandfrancis.com

and the CRC Press Web site at
http://www.crcpress.com

To the memory of Bernhard Riemann, one of the fathers of modern analysis.

Table of Contents

Preface

In today's world there are more and more students who need to know something about real analysis. Yet real analysis remains one of the most abstruse parts of basic mathematics. Understanding real analysis necessitates a clear understanding of basic logic and set theory, but it also requires considerable practice and determination to master the fundamental ideas. Real analysis is in many ways more about technique than about concept.

Many of the classical texts in real analysis were written for a specialized audience of students planning to go on for an M.S. or Ph.D. in mathematics and then to pursue life as an academic. Yet, in our current situation, there are engineers and computer scientists and many others who need to understand the fundamentals of this subject.

The purpose of this text is to address and teach the latter audience. We begin the book with a quick-and-dirty introduction to set theory and logic. It should be possible to cover that material in about three weeks. Then one can spend the rest of the semester studying real analysis.

Of course by real analysis we mean integration theory and differentiation theory and the theory of sequences and series of functions. That is the nexus of what this book is about. We eschew some of the beautiful but more difficult topics like the Stone–Weierstrass theorem, but we cover all the essential ideas at the heart of the matter. We certainly talk about the Weierstrass nowhere differentiable function and the Weierstrass approximation theorem and the Cantor set, but we do not belabor them.

A student who wants to learn real analysis must do exercises. And we provide plenty of those. There is an exercise set at the end of each section. And the exercises are step-laddered. Both students and teachers can be confident that the first several exercises in each section are basic and accessible. Later exercises in each section are more challenging, or more open-ended. These are marked with an asterisk ∗. Doing an exercise for a student is very much like working an example. And we exploit that connection for didactic purposes.

It is my view that the way to learn real analysis is to analyze examples. So this book has plenty of worked examples. And plenty of figures to illustrate those examples. When practiced properly, real analysis is quite a visual subject.

The book finishes with the Appendix: Elementary Number Systems, a Table of Notation, and a Glossary. Thus the student will find that this text is a self-contained universe of real analysis. It will be a book that he/she will want to refer to in later years, and even as a professional. The index is quite detailed, making the topics of the book particularly accessible.

As always I thank my editor Robert Ross for his support and encouragement. I look forward to hearing from readers of the text.

— Steven G. Krantz
St. Louis, Missouri

Chapter 1

Basic Logic

1.1 Principles of Logic

Preliminary Remarks

Logic is the foundation for all of mathematics. The reason that mathematics is so portable, and that its ideas live forever, is that our methodology is so rigorous and based on solid rules of reasoning. We begin to explore these ideas in the present section.

Strictly speaking, our approach to logic is "intuitive" or "naïve." Whereas in ordinary conversation these emotion-charged words may be used to downgrade the value of that which is being described, our use of these words is more technical. What is meant is that we shall prescribe in this chapter certain rules of logic which are to be followed in the rest of the book. They will be presented to you in such a way that their validity should be intuitively appealing and self-evident. We cannot *prove* these rules. The rules of logic are the point where our learning begins. A more advanced course in logic will explore other logical methods. The ones that we present here are universally accepted in mathematics and in most of science.

We shall begin with sentential logic and elementary connectives. This material is called the *propositional calculus* (to distinguish it from the predicate calculus, which will be treated later). In other words, we shall be discussing *propositions*—which are built up from atomic statements and connectives. The elementary connectives include "and," "or," "not," "if-then," and "if and only if." Each of these will have a precise meaning and will have exact relationships with the other connectives.

An *atomic statement* (or *elementary statement*) is a sentence with a subject and a verb (and sometimes an object) but no connectives (and, or, not, if-then, if-and-only-if). For example,

John is good.

Mary has bread.

Ethel reads books.

are all atomic statements. We build up sentences, or propositions, from atomic statements using connectives.

Next we shall consider the quantifiers "for all" and "there exists" and their relationships with the connectives from the last paragraph. The quantifiers will give rise to the so-called *predicate calculus*. Connectives and quantifiers will prove to be the building blocks of all future statements in this book, indeed in all of mathematics.

1.2 Truth

Preliminary Remarks

What sets mathematics apart from other disciplines is our use of proof to establish the truth of an assertion. This section discusses the concept of truth, and begins to describe what a mathematical proof is.

In everyday conversation, people sometimes argue about whether a statement is true or not. In mathematics there is nothing to argue about. In practice a sensible statement in mathematics is either true or false, and there is no room for opinion about this attribute. How do we determine which statements are true and which are false?

The modern methodology in mathematics works as follows:

- We *define* certain terms.

- We *assume* that these terms have certain properties or truth attributes (these assumptions are called axioms).

- We specify certain rules of logic.

Any statement that can be derived from the axioms, using the rules of logic, is understood to be true (we call such a derivation a *proof*). It is not necessarily the case that every true statement can be derived in this fashion. However, in practice this is our method for verifying that a statement is true.

POINT OF CONFUSION 1.1 About ninety years ago, Kurt Gödel proved that there are true statements in mathematics that cannot be proved. This is a profound result, and one that has had a lasting influence on our subject.

On the other hand, a statement is false if it is inconsistent with the axioms and the rules of logic. That is to say, a statement is false if the assumption that it is true leads to a contradiction. Alternatively, a statement P is false if the negation of P can be established or proved. While it is possible for a statement to be false without our being able to derive a contradiction in this fashion, in practice we establish falsity by the method of contradiction or by giving a counterexample (which is another aspect of the method of contradiction).

The point of view being described here is special to mathematics. While it is indeed true that mathematics is used to model the world around us—in physics, engineering, and in other sciences—the subject of mathematics itself is a man-made system. Its internal coherence is guaranteed by the axiomatic method that we have just described.

It is reasonable to ask whether mathematical truth is a construct of the human mind or an immutable part of nature. For instance, is the assertion that "the area of a circle is π times the radius squared" actually a fact of nature just like Newton's inverse square law of gravitation? Our point of view is that mathematical truth is relative. The formula for the area of a circle is a logical consequence of the axioms of mathematics, nothing more. The fact that the formula seems to describe what is going on in nature is convenient, and is part of what makes mathematics useful. But that aspect is something over which we as mathematicians have no control. Our concern is with the internal coherence of our logical system.

It can be asserted that a proof (a concept to be discussed and developed later in the book) is a psychological device for convincing the reader that an assertion is true. However, our view in this book is more rigid: a proof is a sequence of applications of the rules of logic to derive the assertion from the axioms. There is no room for opinion here. The axioms are plain. The rules are rigid. A proof is like a sequence of moves in a game of chess. If the rules are followed, then the proof is correct; otherwise not.

1.3 "And" and "Or"

Preliminary Remarks

We think of a typical sentence as built up from atomic sentences using connectives. This section discusses the connectives "and" and "or." In particular, we consider their interrelationships and their truth values. We also begin to learn about truth tables.

Let A and B be atomic statements such as "Chelsea is smart" or "The earth is flat." The statement

$$\text{``}A \text{ and } B\text{''}$$

means that both A is true *and* B is true. For instance,

Arvid is old and Arvid is fat.

means both that Arvid is old *and* Arvid is fat. If we meet Arvid and he turns out to be young and fat, then the statement is false. If he is old and thin then the statement is false. Finally, if Arvid is *both* young and thin then the statement is false. The statement is *true* precisely when both properties—oldness and fatness—hold. We may summarize these assertions with a *truth table*. We let

$$A = \text{Arvid is old.}$$

and

$$B = \text{Arvid is fat.}$$

The expression

$$A \wedge B$$

will denote the phrase "A and B." We call this statement the *conjunction* of A and B. The letters "T" and "F" denote "True" and "False," respectively. Then we have

A	B	$A \wedge B$
T	T	T
T	F	F
F	T	F
F	F	F

Notice that we have listed all possible truth values of A and B and the corresponding values of the *conjunction* $A \wedge B$.

POINT OF CONFUSION 1.2 In elementary sentential logic we will find that truth tables are a useful device for establishing the truth value of a sentence. Learn to always lay out the truth values in your truth tables just as we have done in this last example. Doing so will facilitate comparison of truth tables for different statements.

Later on, when we formulate ideas using quantifiers, we shall find that truth tables are no longer of any utility. We will need to learn how to produce mathematical proofs.

In a restaurant, the menu often contains phrases such as

"soup or salad."

This means that we may select soup *or* select salad, but we may not select both. This use of "or" is called the *exclusive* "or"; it is not the meaning of "or" that we use in mathematics and logic. In mathematics we instead say that "A or B" is true provided that A is true or B is true or *both* are true. This is the *inclusive* "or." If we let $A \vee B$ denote "A or B" then the truth table is

A	B	A ∨ B
T	T	T
T	F	T
F	T	T
F	F	F

We call the statement $A \vee B$ the *disjunction* of A and B.

We see from the truth table that the only way that "A or B" can be false is if *both* A is false and B is false. For instance, the statement

Hillary is beautiful or Hillary is poor

means that Hillary is either beautiful or poor or both. In particular, she will not be both ugly and rich. Another way of saying this is that if she is ugly she will compensate by being poor; if she is rich she will compensate by being beautiful. *But she could be both beautiful and poor.*

EXAMPLE 1.3 The statement

$$x > 2 \quad \text{and} \quad x < 5$$

is true for the number $x = 3$ because this value of x is both greater than 2 *and* less than 5. It is false for $x = 6$ because this x value is greater than 2 but not less than 5. It is false for $x = 1$ because this x is less than 5 but not greater than 2. ∎

EXAMPLE 1.4 The statement

$$x \text{ is odd and } x \text{ is a perfect cube}$$

is true for $x = 27$ because both assertions hold. It is false for $x = 7$ because this x, while odd, is not a cube. It is false for $x = 8$ because this x, while a cube, is not odd. It is false for $x = 10$ because this x is neither odd nor is it a cube. ∎

EXAMPLE 1.5 The statement

$$x < 3 \quad \text{or} \quad x > 6$$

is true for $x = 2$ since this x is < 3 (even though it is not > 6). It holds (that is, it is true) for $x = 9$ because this x is > 6 (even though it is not < 3). The statement fails (that is, it is false) for $x = 4$ since this x is neither < 3 nor > 6. ∎

EXAMPLE 1.6 The statement

$$x > 1 \quad \text{or} \quad x < 4$$

is true for every real x. ∎

EXAMPLE 1.7 The statement $(A \vee B) \wedge B$ has the following truth table:

A	B	$A \vee B$	$(A \vee B) \wedge B$
T	T	T	T
T	F	T	F
F	T	T	T
F	F	F	F

Notice in Example 1.7 that the statement $(A \vee B) \wedge B$ has the same truth values as the simpler statement B. In what follows, we shall call such pairs of statements (having the same truth values) *logically equivalent*.

The words "and" and "or" are called *connectives*: their role in sentential logic is to enable us to build up (or to connect together) pairs of statements. The idea is to use very simple statements, like "Jennifer is swift" as building blocks; then we compose more complex statements from these building blocks by using connectives.

POINT OF CONFUSION 1.8 We have seen ample evidence in this section that the connectives "and" and "or" are very closely related. In the next section we shall see that any sentence formulated using the first of these connectives can be reformulated using the second connective. And the opposite is true as well.

In the next two sections, we will become acquainted with the other two basic connectives "not" and "if–then."

A Look Back

1. What is the meaning of the connective "or"?
2. What is the meaning of the connective "and"?
3. From a logical point of view, how are "and" and "or" related?
4. What is the difference between the exclusive use of "or" and the inclusive use of "or"?

Exercises

1. Construct truth tables for each of the following sentences:

(a) $(S \wedge T) \vee (S \vee T)$

(b) $(S \vee T) \wedge (S \wedge T)$

(c) $(\sim S \vee T) \vee (S \wedge T)$

(d) $S \wedge (S \vee (S \vee (S \wedge T)))$

(e) $S \wedge (S \vee (S \wedge (S \vee T)))$

(f) $[S \wedge (S \wedge (S \wedge T) \wedge T)] \vee T$

(g) $S \wedge (T \vee S)$

(h) $(S \wedge T) \vee (T \wedge S)$

2. Let

$$
\begin{aligned}
S &= \text{All fish have eyelids.} \\
T &= \text{There is no justice in the world.} \\
U &= \text{I believe everything that I read.} \\
V &= \text{The moon's a balloon.}
\end{aligned}
$$

Express each of the following sentences using the letters S, T, U, V and the connectives $\vee, \wedge$. *Do not use quantifiers.*

(a) All fish have eyelids and there is no justice in the world.

(b) If I believe everything that I read or either the moon's a balloon or all fish have eyelids.

(c) Either the moon is a balloon if there is no justice in the world or I believe all of the things that I read.

(d) Either fish have eyelids or the moon is a balloon.

(e) Fish have eyelids and there is no justice in the world.

(f) Either there no justice in the world or fish have eyelids.

(g) Fish have eyelids and I believe all of what I read.

(h) The moon is a balloon and either fish have eyelids or there is no justice in the world.

(i) The moon is a balloon and I believe all that I read and all fish have eyelids.

(j) I believe all of what I read or all fish have eyelids.

(k) I believe all of what I read or there is no justice in the world.

3. Let

$$
\begin{aligned}
S &= \text{All politicians are honest.} \\
T &= \text{Some men are fools.} \\
U &= \text{I don't have two brain cells to rub together.} \\
W &= \text{The pie is in the sky.}
\end{aligned}
$$

Translate each of the following into English sentences:

(a) $(S \wedge T) \vee \sim U$

(b) $W \vee (T \wedge U)$

(c) $W \wedge (S \wedge T)$

(d) $S \vee (S \vee U)$

(e) $[(W \wedge T)] \vee [U \wedge S]$

(f) $W \vee (S \wedge T)$

    (g) $S \vee (W \wedge U)$

    (h) $U \wedge (W \vee U)$

    (i) $U \wedge (U \wedge W)$

4. Explain why the sentences $A \vee B$ and $A \wedge B$ are not logically equivalent.

5. Explain why the inclusive use of "or" and the exclusive use of "or" are not logically equivalent.

6. Define a logical connective "nand"—given by the symbol $\triangle$—according to the rule that $A \triangle B$ is true if and only if both A and B are true or both A and B are false. Construct a truth table for nand. How is nand related to "and" and "or"?

1.4 "Not"

Preliminary Remarks

Here we consider the logical connective "not." This connective interacts in interesting and profound ways with "and" and "or" from the preceding section.

The statement "not A," written $\sim A$, is true whenever A is false. For example, the statement

Charles is not happily married

is true provided the statement "Charles is happily married" is false. The truth table for $\sim A$ is as follows:

A	$\sim A$
T	F
F	T

Greater understanding is obtained by combining connectives:

EXAMPLE 1.9 Here is the truth table for $\sim (A \wedge B)$:

A	B	$A \wedge B$	$\sim (A \wedge B)$
T	T	T	F
T	F	F	T
F	T	F	T
F	F	F	T

∎

EXAMPLE 1.10 Now we look at the truth table for $(\sim A) \vee (\sim B)$:

A	B	$\sim A$	$\sim B$	$(\sim A) \vee (\sim B)$
T	T	F	F	F
T	F	F	T	T
F	T	T	F	T
F	F	T	T	T

Notice that the statements $\sim (A \wedge B)$ and $(\sim A) \vee (\sim B)$ have the *same truth table*. As previously noted, such pairs of statements are called *logically equivalent*.

The logical equivalence of $\sim (A \wedge B)$ with $(\sim A) \vee (\sim B)$ makes good intuitive sense: the statement $A \wedge B$ fails precisely when either A is false *or* B is false. Since in mathematics we cannot rely on our intuition to establish facts, it is important to have the truth table technique for establishing logical equivalence. The exercise set will give you further practice with this notion.

One of the main reasons that we use the *inclusive* definition of "or" rather than the exclusive one is so that the connectives "and" and "or" have the nice relationship just discussed. It is also the case that $\sim (A \vee B)$ and $(\sim A) \wedge (\sim B)$ are logically equivalent. These logical equivalences are sometimes referred to as *de Morgan's laws*.

POINT OF CONFUSION 1.11 Augustus de Morgan was an important logician in nineteenth century mathematics. He was the first to formalize the method of mathematical induction. And he was the founder of the London Mathematical Society.

de Morgan was a contentious individual, and frequently lost positions or encomia because of a lack of agreement or cooperation.

A Look Back

1. What is the truth table for "not"?
2. How are "and" and "or" related by way of "not"?
3. Does the phrase "and or" make any sense?
4. Does the phrase "or and" make any sense?

Exercises

1. Construct truth tables for each of the following sentences:

   (a) $(S \wedge T) \vee \sim (S \vee T)$
   (b) $(S \vee T) \vee (S \wedge \sim T)$
   (c) $(\sim S \vee T) \wedge \sim (S \wedge \sim T)$
   (d) $S \wedge (S \vee (S \wedge (\sim S \vee \sim T)))$

(e) $S \vee (\sim S \wedge (S \vee (\sim S \vee T)))$

(f) $[S \wedge (S \wedge (\sim S \wedge T) \wedge T)] \vee T$

(g) $S \wedge (T \vee \sim S)$

(h) $(S \wedge \sim T) \vee (T \wedge \sim S)$

2. Let

$$
\begin{array}{rcl}
S & = & \text{All fish have eyelids.} \\
T & = & \text{There is no justice in the world.} \\
U & = & \text{I believe everything that I read.} \\
V & = & \text{The moon's a balloon.}
\end{array}
$$

Express each of the following sentences using the letters S, T, U, V and the connectives $\vee, \wedge, \sim$.

(a) All fish have eyelids and there is at least justice in the world.

(b) I believe everything that I read and either the moon's a balloon or at least some fish have no eyelids.

(c) Either the moon is not a balloon or if there is some justice in the world and I doubt some of the things that I read.

(d) For fish to have eyelids it is necessary for the moon to be a balloon.

(e) Fish have eyelids and there is at least some justice in the world.

(f) There is some justice in the world or fish have eyelids.

(g) It is not the case that either some fish have no eyelids or that I disbelieve some of what I read.

(h) The moon is a balloon or at least some fish have no eyelids.

(i) The moon is not a balloon or I do not believe all that I read and also at least some fish do not have eyelids.

(j) I do not believe some of what I read or some fish do not have eyelids.

(k) I disbelieve at least some of what I read and there is at least some justice in the world.

3. Let

$$
\begin{array}{rcl}
S & = & \text{All politicians are honest.} \\
T & = & \text{Some men are fools.} \\
U & = & \text{I don't have two brain cells to rub together.} \\
W & = & \text{The pie is in the sky.}
\end{array}
$$

Translate each of the following into English sentences:

(a) $(S \wedge \sim T) \vee \sim U$

(b) $W \vee (T \wedge \sim U)$

     (c) $W \wedge (S \wedge T)$

     (d) $S \vee (S \vee U)$

     (e) $(W \wedge U) \wedge (\sim W \vee U)$

     (f) $[\sim (W \wedge \sim T)] \vee [\sim UW \wedge S]$

     (g) $W \vee (\sim S \wedge T)$

     (h) $S \wedge (W \vee \sim U)$

     (i) $U \wedge (\sim W \wedge U)$

     (j) $U \vee (U \wedge \sim W)$

4. Explain why the sentence $A \vee B$ and $A \wedge \sim B$ are not logically equivalent.

5. Explain why the sentence $A \wedge B$ and $A \vee \sim B$ are not logically equivalent.

6. Explain why the sentences $\sim [(\sim A) \vee (\sim B)]$ and $A \wedge B$ are logically equivalent.

1.5   "If–Then"

Preliminary Remarks

In the present section we study the logical connective "if–then." This connective is essential to logical reasoning. We learn, among other things, that this new connective can be expressed in terms of the earlier logical connectives.

    A statement of the form "If A then B" asserts that, whenever A is true, then B is also true. This assertion (or "promise") is tested when A is true, because it is then claimed that something else (namely B) is true as well. *However*, when A is false, then the statement "If A then B" *claims nothing*. Using the symbols $A \Rightarrow B$ to denote "If A then B," we obtain the following truth table:

A	B	$A \Rightarrow B$
T	T	T
T	F	F
F	T	T
F	F	T

    Notice that we use here an important principle of Aristotelian logic: every sensible statement is either true or false. There is no "in between" status. When A is false, we can hardly assert that

$$A \Rightarrow B$$

is false. For $A \Rightarrow B$ asserts that "whenever A is true then B is true," and A is not true!

    Put in other words, when A is false, then the statement $A \Rightarrow B$ is not tested. It therefore cannot be false. So it must be true.

EXAMPLE 1.12 The statement "If $2 = 4$, then Calvin Coolidge was our greatest president" is true (the antecedent is false and the conclusion may be true or false). This is the case no matter what you think of Calvin Coolidge.

The statement "If fish have hair, then chickens have lips" is true (the antecedent is false, and the conclusion is false).

The statement "If $9 > 5$, then dogs don't fly" is true (the antecedent is true, and the conclusion is true).

[Notice that the "if" part of the sentence and the "then" part of the sentence need not be related in any intuitive sense. The truth or falsity of an "if–then" statement is simply a fact about the logical values of its hypothesis and of its conclusion.]   ∎

EXAMPLE 1.13 The statement $A \Rightarrow B$ is logically equivalent to $(\sim A) \vee B$. The truth table for the latter is

A	B	$\sim A$	$(\sim A) \vee B$
T	T	F	T
T	F	F	F
F	T	T	T
F	F	T	T

which is the same as the truth table for $A \Rightarrow B$.   ∎

You should think for a bit to see that $(\sim A) \vee B$ *says the same thing* as $A \Rightarrow B$. To wit, assume that the statement $(\sim A) \vee B$ is true. Now suppose that A is true. Then the first half of the disjunction is false; so the second half must be true. In other words, B must be true. But that says that $A \Rightarrow B$. For the converse, assume that $A \Rightarrow B$ is true. This means that if A holds, then B must follow. But this may be rephrased as saying that if the first half of the disjunction $(\sim A) \vee B$ is false, then the second half is true. That merely affirms the disjunction. So the two statements are equivalent, i.e., they say the same thing.

Once you believe that assertion, then the truth table for $(\sim A) \vee B$ gives us another way to understand the truth table for $A \Rightarrow B$.

There are in fact infinitely many pairs of logically equivalent statements. But just a few of these equivalences are really important in practice—most others are built up from these few basic ones. Some of the other basic pairs of logically equivalent statements are explored in the exercises.

POINT OF CONFUSION 1.14 Logically equivalent statements are an important part of mathematical thinking. Certainly one of the most effective ways of solving a mathematical problem, or proving a mathematical result, is to reformulate it in a logically equivalent fashion. It is essential to be able to recognize logically equivalent statements.

EXAMPLE 1.15 The statement

$$\text{If } x \text{ is negative, then } -5 \cdot x \text{ is positive}$$

is true. For if $x < 0$, then $-5 \cdot x$ is indeed > 0; if $x \geq 0$, then the statement is unchallenged. ∎

EXAMPLE 1.16 The statement

$$\text{If } (x > 0 \text{ and } x^2 < 0), \quad \text{then } x \geq 10$$

is true since the hypothesis "$(x > 0 \text{ and } x^2 < 0)$" is never true. ∎

EXAMPLE 1.17 The statement

$$\text{If } x > 0, \quad \text{then } (x^2 < 0 \text{ or } 2x < 0)$$

is false since the conclusion "$(x^2 < 0 \text{ or } 2x < 0)$" is false whenever the hypothesis $x > 0$ is true. ∎

EXAMPLE 1.18 Let us construct a truth table for the statement $(A \lor (\sim B)) \Rightarrow ((\sim A) \land B)$.

A	B	$\sim A$	$\sim B$	$(A \lor (\sim B))$	$((\sim A) \land B)$	$(A \lor (\sim B)) \Rightarrow ((\sim A) \land B)$
T	T	F	F	T	F	F
T	F	F	T	T	F	F
F	T	T	F	F	T	T
F	F	T	T	T	F	F

∎

Notice that the statement $(A \lor (\sim B)) \Rightarrow ((\sim A) \land B)$ has the same truth table as $\sim (B \Rightarrow A)$. Can you comment on the logical equivalence of these two statements?

Perhaps the most commonly used logical syllogism is the following. Suppose that we know the truth of A and of $A \Rightarrow B$. We wish to conclude B. Examine the truth table for $A \Rightarrow B$. The only line in which both A is true and $A \Rightarrow B$ is true is the line in which B is true. That justifies our reasoning. In logic texts, the syllogism we are discussing is known as *modus ponendo ponens*.

POINT OF CONFUSION 1.19 Certainly *modus ponendo ponens* is the bedrock of all mathematical reasoning. It is the one and essential rule of logic that is the basis for all proofs. We will use *modus ponendo ponens* frequently and throughout this text.

In fact, *modus ponendo ponens* is quite classical terminology, going back at least to the nineteenth century. It is not commonly used today.

A Look Back

1. Explain the truth table for "if–then."
2. How is "if–then" related to "not" and "or"?
3. How is "if–then" related to "not" and "and"?
4. What does *modus ponendo ponens* mean?

Exercises

1. Construct truth tables for each of the following sentences:

    (a) $(S \wedge T) \vee \sim (S \vee T)$

    (b) $(S \vee T) \Rightarrow (S \wedge T)$

    (c) $(\sim S \vee T) \Leftrightarrow \sim (S \wedge \sim T)$

    (d) $S \Rightarrow \big(S \Rightarrow (S \Rightarrow (S \Rightarrow T))\big)$

    (e) $S \Rightarrow \big(\sim S \Rightarrow (S \Rightarrow (\sim S \Rightarrow T))\big)$

    (f) $\big[S \wedge \big(S \wedge (S \wedge T) \wedge T\big)\big] \vee T$

    (g) $S \wedge (T \vee \sim S)$

    (h) $(S \wedge \sim T) \Rightarrow (T \wedge \sim S)$

2. Let

$$
\begin{aligned}
S &= \text{All fish have eyelids.}\\
T &= \text{There is no justice in the world.}\\
U &= \text{I believe everything that I read.}\\
V &= \text{The moon's a balloon.}
\end{aligned}
$$

Express each of the following sentences using the letters S, T, U, V and the connectives $\vee, \wedge, \sim, \Rightarrow, \Leftrightarrow$. *Do not use quantifiers.*

    (a) If fish have eyelids then there is at least some justice in the world.

    (b) If I believe everything that I read then either the moon's a balloon or at least some fish have no eyelids.

    (c) If either the moon is not a balloon or if there is some justice in the world then I doubt some of the things that I read.

    (d) For fish to have eyelids it is necessary for the moon to be a balloon.

    (e) If fish have eyelids then there is at least some justice in the world.

    (f) For there to be any justice in the world it suffices for fish to have eyelids.

    (g) It is not the case that either some fish have no eyelids or that I disbelieve some of what I read.

(h) In order for the moon to be a balloon it is necessary and sufficient for at least some fish to have no eyelids and for there to be some justice in the world.

(i) If the moon is not a balloon and if I do not believe all that I read then at least some fish do not have eyelids.

(j) If I do not believe some of what I read then some fish do not have eyelids.

(k) For me to disbelieve at least some of what I read it is sufficient for there to be at least some justice in the world.

3. Let

$$S = \text{All politicians are honest.}$$
$$T = \text{Some men are fools.}$$
$$U = \text{I don't have two brain cells to rub together.}$$
$$W = \text{The pie is in the sky.}$$

Translate each of the following into English sentences:

(a) $(S \land \sim T) \Rightarrow \sim U$

(b) $W \lor (T \land \sim U)$

(c) $W \Rightarrow (S \Rightarrow T)$

(d) $S \Rightarrow (S \lor U)$

(e) $(W \land U) \Leftrightarrow (W \lor U)$

(f) $[\sim (W \land \sim T)] \lor [\sim U \land S]$

(g) $W \lor (\sim S \Rightarrow T)$

(h) $S \Leftrightarrow (W \Rightarrow U)$

(i) $U \Leftrightarrow (W \Leftrightarrow U)$

(j) $U \Rightarrow (U \Rightarrow W)$

4. Assume that the universe is the ordinary system $\mathbb{R}$ of real numbers. Which of the following sentences is true? Which is false? Give reasons for your answers.

(a) If π is rational, then the area of a circle is $E = mc^2$.

(b) If $2 + 2 = 4$, then $3/5$ is a rational number.

(c) If $2 + 2 = 5$, then $2 + 3 = 6$.

(d) If both $2 + 3 = 5$ and $2 \cdot 3 = 5$, then the world is flat.

(e) If it is not the case that $3^2 = 9$, then $4^2 = 16$.

(f) If it is not the case that $3^2 = 9$, then $4^2 = 17$.

(g) If it is not the case that $3^2 = 8$, then $4^2 = 16$.

(h) If it is not the case that $3^2 = 8$, then $4^2 = 17$.

(i) If both $3 \cdot 2 = 6$ and $4 + 4 = 8$, then $5 \cdot 5 = 20$.

(j) If both $3 \cdot 2 = 6$ and $4 + 4 = 7$, then $5 \cdot 5 = 20$.

*   **5.** Is it possible to find a statement that is logically equivalent to $S \Rightarrow T$ but that uses only $\wedge$ and $\vee$ (and not $\sim$)? Why or why not?

6. Translate each of the following statements into symbols, connectives, and quantifiers. Your answers should contain no words. State carefully what each of your symbols stands for. [Note: Each statement is true, but you are not required to verify the truth of the statements.]

  (a) The number 5 has a positive square root.

  (b) There is a quadratic polynomial equation with real coefficients that has no real root.

  (c) The sum of two perfect cubes is never itself a perfect cube.

  (d) If $x \cdot y \neq 0$, then $x^2 + y^2 > 0$.

  (e) Every positive real number has two distinct real fourth roots.

  (f) If z and w are complex numbers, then $z \cdot w$ is also a complex number.

  (g) The sum of two irrational numbers need not be irrational.

  (h) The product of two irrational numbers need not be irrational.

  (i) The sum of two rational numbers is always rational.

  (j) The square of a rational number is always rational.

  (k) The square root of a rational number need not be rational.

7. Which of these pairs of statements is logically equivalent? Why?

  (a) $A \vee \sim B$ $\sim A \Rightarrow B$

  (b) $A \wedge \sim B$ $\sim A \Rightarrow \sim B$

  (c) $A \vee (\sim A \wedge B)$ $\sim [\sim A \wedge (A \vee \sim B)]$

  (d) $B \Rightarrow \sim A$ $A \Rightarrow (A \vee B)$

  (e) $A \Leftrightarrow \sim B$ $A \Rightarrow (\sim B \vee \sim A)$

  (f) $\sim (A \vee \sim B)$ $B \wedge \sim A$

  (g) $\sim (A \Rightarrow \sim B)$ $B \Rightarrow A$

8. Formulate, as an English sentence (without symbols), the negation of each of the statements in Exercise 6.

*   **9.** Give a logical demonstration (i.e., a "proof") that any statement in sentential logic can be formulated using just $\sim$ and $\wedge$. That is, given any statement P there is a logically equivalent statement P' such that P' uses only the connectives $\sim$ and $\wedge$.

10. For each of the following statements, formulate a logically equivalent one using only $S, T, \sim$, and $\vee$. (Of course you may use as many parentheses as you need.) *Use a truth table or other means to explain why the statements are logically equivalent.*

  (a) $S \Rightarrow \sim T$

  (b) $\sim S \wedge \sim T$

  (c) $S \Leftrightarrow \sim T$

    (d) $S \wedge (T \vee \sim S)$

    (e) $(S \vee T) \Rightarrow (S \wedge T)$

    (f) $(S \wedge T) \Rightarrow (S \vee T)$

    (g) $(S \Rightarrow T) \vee (T \Rightarrow S)$

    (h) $[\sim (S \vee T)] \Rightarrow S$

11. Redo Exercise 10, this time finding logically equivalent statements that use only $S, T, \sim$, and $\wedge$. *Give reasons for your answers.*

1.6 Contrapositive, Converse, and "Iff"

Preliminary Remarks

The concepts of contrapositive and converse are essential to rigorous mathematical reasoning. We learn here to identify them and to reason with them. The connective "if and only if " is also explored.

The statement

$$\text{If } A \text{ then } B$$

is the same as

$$A \Rightarrow B$$

or

$$A \text{ suffices for } B$$

or as saying

$$A \text{ only if } B$$

All these forms are encountered in practice, and you should think about them long enough to realize that they say the same thing.

    On the other hand,

$$\text{If } B \text{ then } A$$

is the same as saying

$$B \Rightarrow A$$

or

$$A \text{ is necessary for } B$$

or as saying

$$A \text{ if } B.$$

We call the statement $B \Rightarrow A$ the *converse* of $A \Rightarrow B$. The converse of a statement is logically distinct from that original statement; the truth or falsity of one is independent of the truth or falsity of the other. Our examples will illustrate this point.

EXAMPLE 1.20 The converse of the statement

$$\text{If } x \text{ is a healthy horse, then } x \text{ has four legs}$$

is the statement

$$\text{If } x \text{ has four legs, then } x \text{ is a healthy horse.}$$

Notice that these statements have very different meanings: the first statement is true, while the second (its converse) is false. For instance, a chair has four legs, but it is not a healthy horse.   ∎

EXAMPLE 1.21 The converse of the statement

$$\text{If } x > 0, \text{ then } 2x > 0$$

is the statement

$$\text{If } 2x > 0, \text{ then } x > 0.$$

Notice that both statements are true.   ∎

EXAMPLE 1.22 The converse of the statement

$$\text{If } x > 0, \text{ then } x^2 > 0$$

is the statement

$$\text{If } x^2 > 0, \text{ then } x > 0.$$

Notice that the first implication is true, while the second is false.   ∎

POINT OF CONFUSION 1.23 It is quite common in everyday discourse to confuse the converse with the contrapositive (see below). Two people will frequently argue vigorously because they do not understand the difference between the two ideas.

In mathematics we of course keep these two concepts clear and distinct. And we use them both frequently.

The statement

$$A \text{ if and only if } B.$$

is a brief way of saying

$$\text{If } A \text{ then } B. \qquad and \qquad \text{If } B \text{ then } A.$$

We abbreviate A if and only if B as $A \Leftrightarrow B$ or as A iff B. Here is a truth table for $A \Leftrightarrow B$:

A	B	$A \Rightarrow B$	$B \Rightarrow A$	$A \Leftrightarrow B$
T	T	T	T	T
T	F	F	T	F
F	T	T	F	F
F	F	T	T	T

Notice that we can say that $A \Leftrightarrow B$ is true only when both $A \Rightarrow B$ and $B \Rightarrow A$ are true. An examination of the truth table reveals that A $\Leftrightarrow$ B is true precisely when A and B are either both true or both false. Thus, $A \Leftrightarrow B$ means precisely that A and B are logically equivalent. One is true when and *only when* the other is true. One is false when and *only when* the other is false.

EXAMPLE 1.24 The statement

$$x > 0 \Leftrightarrow 2x > 0$$

is true. For if $x > 0$, then $2x > 0$; and if $2x > 0$, then $x > 0$.   ∎

EXAMPLE 1.25 The statement

$$x > 0 \Leftrightarrow x^2 > 0$$

is false. For $x > 0 \Rightarrow x^2 > 0$ is certainly true, while $x^2 > 0 \Rightarrow x > 0$ is false $((-3)^2 > 0$ but $-3 \ngtr 0)$.   ∎

EXAMPLE 1.26 The statement

$$\{\sim (A \vee B)\} \Leftrightarrow \{(\sim A) \wedge (\sim B)\} \qquad (*)$$

is true because the truth table for $\sim (A \vee B)$ and that for $\sim A \wedge \sim B$ are the same. Thus they are logically equivalent: one statement is true precisely when the other is. Another way to see the truth of $(*)$ is to examine the truth table for the full statement:

A	B	$\sim (A \vee B)$	$(\sim A) \wedge (\sim B)$	$\sim (A \vee B) \Leftrightarrow \{(\sim A) \wedge (\sim B)\}$
T	T	F	F	T
T	F	F	F	T
F	T	F	F	T
F	F	T	T	T

∎

POINT OF CONFUSION 1.27 In ordinary conversation we do not generally use the phrase "if and only if." First, it is too subtle for most people. Second, we do not usually reason so rigorously in casual discourse.

But in mathematics this idea is essential, and we use it frequently. Be sure to get it straight.

Given an implication

$$A \Rightarrow B,$$

the *contrapositive* statement is defined to be the implication

$$\sim B \Rightarrow \sim A.$$

The contrapositive is logically equivalent to the original implication, as we see by examining their truth tables:

A	B	$A \Rightarrow B$
T	T	T
T	F	F
F	T	T
F	F	T

and

A	B	$\sim A$	$\sim B$	$\sim B \Rightarrow \sim A$
T	T	F	F	T
T	F	F	T	F
F	T	T	F	T
F	F	T	T	T

EXAMPLE 1.28 The statement

> If it is raining, then it is cloudy

has, as its contrapositive, the statement

> If there are no clouds, then it is not raining.

A moment's thought convinces us that these two statements say the same thing: if there are no clouds, then it could not be raining; for the presence of rain implies the presence of clouds.   ∎

The main point to keep in mind is that, given an implication $A \Rightarrow B$, its *converse* $B \Rightarrow A$ and its *contrapositive* $(B) \Rightarrow (A)$ are entirely different statements. The converse is distinct from, and *logically independent from*, the original statement. The contrapositive is distinct from, but *logically equivalent to*, the original statement.

Some classical treatments augment the concept of *modus ponendo ponens* with the idea of *modus tollendo tollens*. It is in fact logically equivalent to *modus ponendo ponens*. *Modus tollendo tollens* says

If $\sim B$ and $A \Rightarrow B$ then $\sim A$.

Modus tollendo tollens actualizes the fact that $\sim B \Rightarrow \sim A$ is logically equivalent to $A \Rightarrow B$. The first of these implications is of course the *contrapositive* of the second.

A Look Back

1. What is the difference between the converse and the contrapositive?

2. Which of the converse and the contrapositive is logically equivalent to the original implication?

3. What does the statement "*A* iff *B*" say about the logical relationship between *A* and *B*?

4. What is *modus tollendo tollens*?

Exercises

1. State the converse and the contrapositive of each of the following sentences. Be sure to label each.

   (a) In order for it to rain it is necessary that there be clouds.

   (b) In order for it to rain it is sufficient that there be clouds.

   (c) If life is a bowl of cherries, then I am not in the pits.

   (d) If I am not a fool, then mares eat oats.

   (e) A sufficient condition for the liquidity of water is that the temperature exceed 32° Fahrenheit.

   (f) A necessary condition for peace in the world is that all people disarm.

   (g) What's good for the goose is good for the gander.

   (h) If wishes were horses, then beggars would ride.

   (i) If my grandmother had wheels, she'd be a garbage truck.

   (j) If he won't play ball, then he's benched.

   (k) If the Donald won't buy it, then it's not worth a dime.

   (l) If Melania wants it, then the Donald will buy it.

2. Assume that the universe is the ordinary system $\mathbb{R}$ of real numbers. Which of the following sentences is true? Which is false? Give reasons for your answers.

   (a) If π is rational, then the area of a circle is $E = mc^2$.

   (b) If $2 + 2 = 4$, then $3/5$ is a rational number.

   (c) If $2 + 2 = 5$, then $2 + 3 = 6$.

(d) If both $2 + 3 = 5$ and $2 \cdot 3 = 5$, then the world is flat.

(e) If it is not the case that $3^2 = 9$, then $4^2 = 16$.

(f) If it is not the case that $3^2 = 9$, then $4^2 = 17$.

(g) If it is not the case that $3^2 = 8$, then $4^2 = 16$.

(h) If it is not the case that $3^2 = 8$, then $4^2 = 17$.

(i) If both $3 \cdot 2 = 6$ and $4 + 4 = 8$, then $5 \cdot 5 = 20$.

(j) If both $3 \cdot 2 = 6$ and $4 + 4 = 7$, then $5 \cdot 5 = 20$.

3. State the converse of the contrapositive of

> If all men are left-handed then all women are beautiful.

Now state the contrapositive of the converse.

4. Find a statement that is logically equivalent to $A \Leftrightarrow B$ using only $\vee$ and $\sim$ as connectives.

5. Find a statement that is logically equivalent to $A \Leftrightarrow B$ using only $\wedge$ and $\sim$ as connectives.

6. Explain why $A \Leftrightarrow B$ if and only if A is logically equivalent to B.

7. Explain why the contrapositive of the contrapositive of an implication is equivalent to the original implication.

7. Explain why the converse of the converse of an implication is equivalent to the original implication.

1.7   Quantifiers

Preliminary Remarks

The idea of quantifier brings us now to a new level of sophistication. Most nontrivial mathematical statements are formulated using quantifiers. It is essential that we learn to be comfortable with quantifiers.

The mathematical statements that we will encounter in practice will use the *connectives* "and," "or," "not," "if–then," and "iff." They will also use *quantifiers*. The two basic quantifiers are "for all" and "there exists."

EXAMPLE 1.29 Consider the statement

> All automobiles have wheels.

This statement makes an assertion about *all* automobiles. It is true, because every automobile does have wheels.

Compare this statement with the next one:

> There exists a woman who is blonde.

This statement is of a different nature. It does not claim that all women have blonde hair—merely that there exists *at least one* woman who does. Since that is true, the statement is true.   ∎

EXAMPLE 1.30 Consider the statement

> All positive real numbers are integers.

This sentence asserts that something is true for all positive real numbers. It is indeed true for *some* positive numbers, such as 1 and 2 and 193. However, it is false for at least one positive number (such as $1/10$ or π), so the entire statement is false.

Here is a more extreme example:

> The square of any real number is positive.

This assertion is *almost* true—the only exception is the real number 0: $0^2 = 0$ is not positive. But it only takes one exception to falsify a "for all" statement. So the assertion is false.   ∎

POINT OF CONFUSION 1.31 This last example illustrates the principle that the negation of a "for all" statement is a "there exists" statement. It is also true that the negation of a "there exists" statement is a "for all" statement.

EXAMPLE 1.32 To illustrate the last sentence, consider the statement

> There is a man who is at least ten feet tall.

This statement is false. To *verify* that it is false, we must demonstrate that *there does not exist a man who is at least ten feet tall*. In other words, we must show that all men are shorter than ten feet.

The negation of a "there exists" statement is a "for all" statement.

A somewhat different example is the sentence

> There exists a real number x which satisfies the equation
> $$x^3 - 2x^2 + 3x - 6 = 0.$$

There is in fact only one real number that satisfies the equation, and that is $x = 2$. Yet that information is sufficient to show that the statement true.   ∎

EXAMPLE 1.33 Look at the statement

> There exists a real number which is greater than 5.

In fact, there are lots of numbers that are greater than 5; some examples are $7, 42, 2\pi$, and $97/3$. Other numbers, such as 1, 2, and $\pi/6$, are not greater than 5. Since there is *at least one* number satisfying the statement, the assertion is true.   ∎

We often use the symbol $\forall$ to denote "for all" and the symbol $\exists$ to denote "there exists." The assertion

$$\forall x, \ x + 1 < x$$

claims that for every x, the number $x+1$ is less than x. If we take our universe to be the standard real number system, then this statement is false. The assertion

$$\exists x, \ x^2 = x$$

claims that there is a number whose square equals itself. If we take our universe to be the real numbers, then the assertion is satisfied by $x = 0$ and by $x = 1$. Therefore the assertion is true.

In all the examples of quantifiers that we have discussed thus far, we were careful to specify our *universe*. That is, "There is a woman such that ..." or "All positive real numbers are ..." or "All automobiles have ...". The quantified statement makes no sense unless we specify the universe of objects from which we are making our specification. In the discussion that follows, we will always interpret quantified statements in terms of a universe. Sometimes the universe will be explicitly specified, while other times it will be understood from context.

Quite often we will encounter $\forall$ and $\exists$ used together. The following examples are typical:

EXAMPLE 1.34 The statement

$$\forall x \ \exists y, \ y > x$$

claims that for any number x there is a number y that is greater than it. In the realm of the real numbers, this is true. In fact, $y = x + 1$ will always do the trick.

The statement

$$\exists x \ \forall y, \ y > x$$

has quite a different meaning from the first one. It claims that there is an x that is less than *every* y. This is absurd. For instance, x is *not* less than $y = x - 1$. ∎

EXAMPLE 1.35 The statement

$$\forall x \ \forall y, \ x^2 + y^2 \geq 0$$

is true in the realm of the real numbers: it claims that the sum of two squares is always greater than or equal to zero. [This statement happens to be *false* in the realm of the complex numbers. We shall learn about that number system later. When we interpret a logical statement, it will always be important to understand the context, or universe, in which we are working.]

The statement

$$\exists x \ \exists y, \ x + 2y = 7$$

is true in the realm of the real numbers: it claims that there exist x and y such that $x + 2y = 7$. Certainly the numbers $x = 3, y = 2$ will do the job (although there are many other choices that work as well). ∎

We conclude by noting that $\forall$ and $\exists$ are closely related. The statements

$$\forall x,\ B(x) \qquad \text{and} \qquad \sim \exists x,\ \sim B(x)$$

are logically equivalent. The first asserts that the statement $B(x)$ is true for all values of x. The second asserts that there exists no value of x for which $B(x)$ fails, which is the same thing.

Likewise, the statements

$$\exists x, B(x) \qquad \text{and} \qquad \sim \forall x,\ \sim B(x)$$

are logically equivalent. The first asserts that there is some x for which $B(x)$ is true. The second claims that it is not the case that $B(x)$ fails for every x,, which is the same thing. The books [HALM] and [GIH] explore the algebraic structures inspired by these quantifiers.

It is worth noting explicitly that $\forall$ and $\exists$ do *not* commute. That is to say,

$$\forall x \exists y,\ F(x,y) \qquad \text{and} \qquad \exists y \forall x,\ F(x,y)$$

do not say the same thing. We invite you to provide a counterexample.

A "for all" statement is something like the conjunction of a very large number of simpler statements. For example, the statement

$$\text{For every nonzero integer } n, n^2 > 0$$

is actually an efficient way of saying that $1^2 > 0$ and $(-1)^2 > 0$ and $2^2 > 0$, etc. It is not feasible to apply truth tables to "for all" statements, and we usually do not do so.

A "there exists" statement is something like the disjunction of a very large number of statements (the word "disjunction" in the present context means an "or" statement). For example, the statement

$$\text{There exists an integer } n \text{ such that } P(n) = 2n^2 - 5n + 2 = 0$$

is actually an efficient way of saying that $P(1) = 0$ or $P(-1) = 0$ or $P(2) = 0$, etc. It is not feasible to apply truth tables to "there exist" statements, and we usually do not do so.

POINT OF CONFUSION 1.36 We have seen that "there exists" and "for all" are very closely related. But they definitely say very different things and should not be confused. Also *they do not commute*.

We shall use quantifiers frequently in our work, but we must use them carefully.

It is common to say that *first-order logic* consists of the connectives $\wedge$, $\vee$, $\sim$, $\Rightarrow$, $\Longleftrightarrow$, the equality symbol $=$, and the quantifiers $\forall$ and $\exists$, together with an infinite string of variables $x, y, z, \ldots, x', y', z', \ldots$ and, finally, parentheses $(\ ,\ )$ to keep things readable (see [BAR, p. 7]). The word "first" here is used to distinguish the discussion from second-order and higher-order logics.

A Look Back

1. What is the difference between "for all" and "there exists"?

2. How are "for all" and "there exists" related by way of "not"?

3. What is the difference between $\forall\exists$ and $\exists\forall$?

4. Why can we not use a truth table to prove a "for all" statement?

Exercises

1. Translate each of the following statements into symbols, connectives, and quantifiers. Your answers should contain no words. State carefully what each of your symbols stands for. [Note: Each statement is true, but you are not required to verify the truth of the statements.]

    (a) The number 5 has a positive square root.

    (b) There is a quadratic polynomial equation with real coefficients that has no real root.

    (c) The sum of two perfect cubes is never itself a perfect cube.

    (d) If $x \cdot y \neq 0$, then $x^2 + y^2 > 0$.

    (e) Every positive real number has two distinct real fourth roots.

    (f) If z and w are complex numbers, then $z \cdot w$ is also a complex number.

    (g) The sum of two irrational numbers need not be irrational.

    (h) The product of two irrational numbers need not be irrational.

    (i) The sum of two rational numbers is always rational.

    (j) The square of a rational number is always rational.

    (k) The square root of a rational number need not be rational.

2. In each of the following statements, you should treat the real number system $\mathbb{R}$ as your universe. Translate each statement into an English sentence. Your answers should contain no symbols—only words. [Note: Each statement is true, but you are not required to verify the truth of the statements.]

    (a) $\forall x, (x \in \mathbb{R} \wedge x > 0) \Rightarrow (\exists y, y > 0 \wedge y^2 = x)$

    (b) $\exists x \forall y, (y > x) \Rightarrow (y > 5)$

    (c) $\forall x \forall y, x^2 + y^2 \geq 2xy$

    (d) $\exists x, x > 0 \wedge x^3 < x^2$

    (e) $\sim \exists x, x \in \mathbb{R} \wedge (1 < x^2 < x)$

    (f) $\exists x, \sim (x^2 > 0 \Rightarrow x > 0)$

    (g) $\forall y \exists x, x \in \mathbb{R} \wedge x^3 = y + 1$

    (h) $\exists y \forall x, x^2 + y^4 > 2$

3. Use our standard quantifiers $\forall$ and $\exists$ to translate the sentence "There are exactly five solutions to the equation $P(m) = 0$" into symbols. Now translate the sentence "The equation $Q(m) = 0$ is satisfied by all but four integers."

4. An island is populated by truth-tellers and liars. You cannot tell which is which just by looking at them. You meet an inhabitant of the island. What single question (with a yes/no answer) can you ask him/her that will enable you to ascertain whether this person is a truth-teller or a liar?

5. For each of the following statements, formulate an English sentence that is its negation:

    (a) The set S contains at least two integers.

    (b) Mares eat oats and does eat oats.

    (c) I'm rough and I'm tough and I breathe fire.

    (d) This town is not big enough for both of us.

    (e) I will marry Fred and disappoint Irving.

    (f) I cannot marry either Selma or Flo.

    (g) I will pay my taxes and avoid going to jail.

    (h) If I am a good boy, then I will do fine.

    (i) I love everyone and everyone loves me.

    (j) If you study hard, then you will do well in school.

    (k) If you get caught, then you will go to jail.

    (l) If you work hard, then you will succeed.

    (m) If you make more than $100,000, then you pay no income tax.

6. Formulate, as an English sentence (without symbols), the negation of each of the statements in Exercise 2.

7. Explain why $\forall$ is logically equivalent to $\sim \exists \sim$.

8. Explain why $\exists$ is logically equivalent to $\sim \forall \sim$.

9. Explain why $\forall\exists$ and $\exists\forall$ are *not* logically equivalent.

Chapter 2

Methods of Proof

2.1 What Is a Proof?

Preliminary Remarks

As previously noted, proofs are what set mathematics apart from other sciences and other disciplines. The idea of proof dates back to the ancient Greeks, and has kept mathematics rigorous and reliable for 2500 years. There are many different styles and methods of proof, and we begin to learn some of those here.

When a chemist asserts that a substance that is subjected to heat will tend to expand, he/she verifies the assertion through experiment. It is a consequence of the *definition* of heat that heat will excite the atomic particles in the substance; it is plausible that this in turn will necessitate expansion of the substance. However, our knowledge of nature is not such that we may turn these theoretical ingredients into a categorical proof. Additional complications arise from the fact that the word "expand" requires detailed definition. Apply heat to water that is at temperature 40° Fahrenheit or above, and it expands—with enough heat it becomes a gas that surely fills more volume than the original water. But apply heat to a diamond, and there is no apparent "expansion"—at least not to the naked eye.

Mathematics is a less ambitious subject. In particular, it is closed. It does not reach outside itself for verification of its assertions. When we make an assertion in mathematics, we must verify it using the rules that we have laid down. That is, we verify it by applying our rules of logic to our axioms and our definitions; in other words, we construct a *proof*.

In modern mathematics we have discovered that there are perfectly sensible mathematical statements that in fact *cannot* be verified in this fashion, nor can they be proven false. This is a manifestation of Gödel's incompleteness theorem: that any sufficiently complex logical system will contain such unverifiable, indeed untestable, statements. Fortunately, in practice, such statements are the exception rather than the rule. In this book, and in almost all of university-level mathematics, we concentrate on learning about statements whose truth or falsity *is* accessible by way of proof.

This chapter considers the notion of mathematical proof. We shall concentrate on the three principal types of proof: direct proof, proof by contradiction, and proof by induction.

POINT OF CONFUSION 2.1 Proofs are a vital part of what we do in mathematics. We do not reason about mathematics heuristically or casually. We follow very strict standards of rigor, and we prove every statement that we make. This is what sets mathematics apart from other disciplines.

This chapter will make you conversant with the most basic methods of proof. But you should understand that there are literally hundreds of proof techniques. A professional mathematician spends his/her life learning new proof techniques.

In practice, a mathematical proof may contain elements of several or all of these techniques. You will see all the basic elements here. You should be sure to master each of these proof techniques, both so that you can recognize them in your reading and so that they become tools that you can use in your own work.

2.2   Direct Proof

> ### Preliminary Remarks
>
> The method of direct proof is perhaps the simplest and most accessible of all the proof techniques. In this methodology, we prove a statement by following the reasoning suggested by the text of the statement. We learn the method by way of several incisive examples.

In this section, we shall assume that you are familiar with the positive integers, or *natural numbers* (a detailed treatment of the natural numbers appears in the Appendix). This number system $\{1, 2, 3, \dots\}$ is denoted by the symbol $\mathbb{N}$. For now we will take the elementary arithmetic properties of $\mathbb{N}$ for granted.

We shall formulate various statements about natural numbers and prove them. We begin with a definition.

Definition 2.2 A natural number n is said to be *even* if, when it is divided by 2, there is no remainder.

Definition 2.3 A natural number n is said to be *odd* if, when it is divided by 2, the remainder is 1.

You may have never considered the meaning of the terms "odd" or "even" at this level of precision. But your intuition should confirm these definitions. A good definition should be precise, but it should also appeal to your heuristic idea about the concept that is being defined.

Notice that, according to these definitions, any natural number is either even or odd. For if n is any natural number, and if we divide it by 2, then the remainder will be either 0 or 1—there is no other possibility (according to the Division Algorithm—see [HER]). In the first instance, n is even; in the second, n is odd.

In what follows we will find it convenient to think of an even natural number as one having the form $2m$ for some natural number m. We will think of an odd natural number as one having the form $2k + 1$ (or sometimes $2k - 1$) for some natural number k. Check for yourself that, in the first instance, division by 2 will result in a quotient of m and a remainder of 0; in the second instance it will result in a quotient of k and a remainder of 1.

Now let us formulate a statement about the natural numbers and prove it. Following tradition, we refer to formal mathematical statements either as *theorems* or *propositions* or sometimes as *lemmas*. A theorem is supposed to be an important statement that is the culmination of some development of significant ideas. A proposition is a statement of lesser intrinsic importance. Usually a lemma is of no intrinsic interest, but is needed as a step along the way to verifying a theorem or proposition.

Proposition 2.4 *The square of an even natural number is even.*

Proof: Let us begin by using what we learned in Chapter 1. We may reformulate our statement as "If n is even, then $n \cdot n$ is even." This statement makes a promise. Refer to the definition of "even" to see what that promise is:

If n can be written as twice a natural number, then $n \cdot n$ can be written as twice a natural number.

The hypothesis of the assertion is that $n = 2 \cdot m$ for some natural number m. But then

$$n^2 = n \cdot n = (2m) \cdot (2m) = 4m^2 = 2(2m^2).$$

Our calculation shows that n^2 is twice the natural number $2m^2$. So n^2 is also even.

We have shown that the hypothesis that n is twice a natural number entails the conclusion that n^2 is twice a natural number. In other words, if n is even, then n^2 is even. That is the end of our proof. □

Remark 2.5 What is the role of truth tables at this point? Why did we not use a truth table to verify our proposition? One *could* think of the statement that we are proving as the conjunction of infinitely many specific statements about concrete instances of the variable n; and then we could verify each one of those statements. But such a procedure is inelegant and, more importantly, impractical.

For our purposes, the truth table *tells us what we must do to construct a proof.* The truth table for $A \Rightarrow B$ shows that, if A is false, then there is nothing to check; whereas, if A is true, then we must show that B is true. That is just what we did in the proof of Proposition 2.4.

Most of our theorems are "for all" statements or "there exists" statements. In practice, it is not usually possible to verify them directly by use of a truth table.

Proposition 2.6 *The square of an odd natural number is odd.*

Proof: We follow the paradigm laid down in the proof of the previous proposition.

Assume that n is odd. Then $n = 2m + 1$ for some natural number m. But then

$$n^2 = n \cdot n = (2m + 1) \cdot (2m + 1) = 4m^2 + 4m + 1 = 2(2m^2 + 2m) + 1.$$

We see that n^2 is $2m' + 1$, where $m' = 2m^2 + 2m$. In other words, according to our definition, n^2 is odd.   □

Both of the proofs that we have presented are examples of "direct proof." A direct proof proceeds according to the statement being proved; for instance, if we are proving a statement about a square, then we calculate that square. If we are proving a statement about a sum, then we calculate that sum. Here are some additional examples:

EXAMPLE 2.7 Prove that, if n is a positive integer, then the quantity $n^2 + 3n + 2$ is even.

Proof: Denote the quantity $n^2 + 3n + 2$ by K. Observe that

$$K = n^2 + 3n + 2 = (n + 1)(n + 2).$$

Thus K is the product of two successive integers: $n + 1$ and $n + 2$. One of those two integers must be even. So it is a multiple of 2. Therefore K itself is a multiple of 2. Hence K must be even.   ∎

Proposition 2.8 *The sum of two odd natural numbers is even.*

Proof: Suppose that p and q are both odd natural numbers. According to the definition, we may write $p = 2r + 1$ and $q = 2s + 1$ for some natural numbers r and s. Then

$$p + q = (2r + 1) + (2s + 1) = 2r + 2s + 2 = 2(r + s + 1).$$

We have realized $p + q$ as twice the natural number $r + s + 1$. Therefore $p + q$ is even.   □

Remark 2.9 In some subjects, such as literary criticism or philosophy, it is common to reason by analogy, or to present ideas so that they sound good. If we did mathematics solely according to what sounds good, or what appeals intuitively, then we might reason as follows: "If the sum of two odd natural numbers is even then it must be that the sum of two even natural numbers is odd." This is incorrect. For instance, 4 and 6 are each even but their sum $4 + 6 = 10$ is *not* odd.

Intuition definitely plays an important role in the development of mathematics, but all assertions in mathematics must, in the end, be proved by rigorous methods.

EXAMPLE 2.10 Prove that the sum of an even integer and an odd integer is odd.

Proof: An even integer e is divisible by 2, so may be written in the form $e = 2m$, where m is an integer. An odd integer o has remainder 1 when divided by 2, so may be written in the form $o = 2k + 1$, where k is an integer. The sum of these is

$$e + o = 2m + (2k + 1) = 2(m + k) + 1 \,.$$

Thus we see that the sum of an even and an odd integer will have remainder 1 when it is divided by 2. As a result, the sum is odd. ∎

Proposition 2.11 *The sum of two even natural numbers is even.*

Proof: Let $p = 2r$ and $q = 2s$ both be even natural numbers. Then

$$p + q = 2r + 2s = 2(r + s).$$

We have realized $p + q$ as twice a natural number. Therefore we conclude that $p + q$ is even. □

Proposition 2.12 *Let n be a natural number. Then either $n > 6$ or $n < 9$.*

Proof: If you draw a picture of a number line then you will have no trouble convincing yourself of the truth of the assertion. What we want to learn here is to organize our thoughts so that we may write down a rigorous proof.

Our discussion of the connective "or" in Section 1.3 will now come to our aid. Fix a natural number n. If $n > 6$ then the "or" statement is true and there is nothing to prove. If $n \not> 6$, then the truth table teaches us that we must check that $n < 9$. But the statement $n \not> 6$ means that $n \leq 6$ so we have

$$n \leq 6 < 9.$$

That is what we wished to prove. □

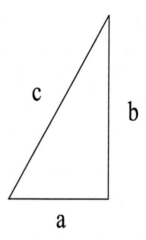

Figure 2.1: A right triangle.

EXAMPLE 2.13 Prove that every even integer may be written as the sum of two odd integers.

Proof: Let the even integer be $K = 2m$, for m an integer. If m is odd then we write

$$K = 2m = m + m$$

and we have written K as the sum of two odd integers. If, instead, m is even, then we write

$$K = 2m = (m - 1) + (m + 1).$$

Since m is even then both $m - 1$ and $m + 1$ are odd. So again we have written K as the sum of two odd integers.   ∎

EXAMPLE 2.14 Prove the Pythagorean theorem.

Proof: The Pythagorean theorem states that $c^2 = a^2 + b^2$, where a and b are the legs of a right triangle and c is its hypotenuse. See Figure 2.1.

Consider now the arrangement of four triangles and a square shown in Figure 2.2. Each of the four triangles is a copy of the original triangle in Figure 2.1. We see that each side of the all-encompassing square is equal to c. So the area of that square is c^2. Now each of the component triangles has base a and height b. So each such triangle has area $ab/2$. And the little square in the middle has side $b - a$. So it has area $(b - a)^2 = b^2 - 2ab + a^2$. We write the total area as the sum of its component areas:

$$c^2 = 4 \cdot \left[\frac{ab}{2} \right] + [b^2 - 2ab + a^2] = a^2 + b^2.$$

That is the desired equality.   ∎

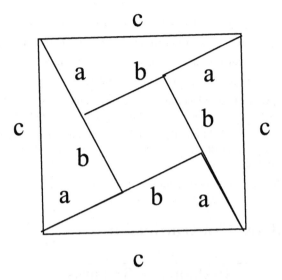

Figure 2.2: The Pythagorean theorem.

POINT OF CONFUSION 2.15 Just because a proof is a "direct proof" does not necessarily mean that it is simple. While the basic logical paradigm is fairly simple, the details of the proof may be quite complex—and tricky too.

The only way to master any proof technique is to practice.

In this section and the next two, we are concerned with form rather than substance. We are not interested in proving anything profound, but rather in showing you what a proof looks like. Later in the book we shall consider some deeper mathematical ideas and correspondingly more profound proofs.

A Look Back

1. What are the characteristics of a direct proof?

2. Give a direct proof that the product of two even numbers is even.

3. Give a direct proof that the sum of two odd numbers is even.

4. Give a direct proof that the square of an odd number is odd.

Exercises

1. Prove that the product of two odd natural numbers must be odd.

2. Prove that if n is an even natural number and if m is *any* natural number, then $n \cdot m$ must be even.

3. Prove that any integer can be written as the sum of at most two odd integers. Is the same true if "odd" is replaced by "even"?

4. Prove that between any two distinct real numbers there is a rational number.

5. Prove that between any two distinct rational numbers there is an irrational number.

6. Prove that, if k is a natural number that is greater than 2, then $2^k > 1 + 2k$.

7. Prove that, if n is an integer greater than 4, then $2^n > n^2 + 1$.

8. Give a direct proof of the pigeonhole principle. This says that if you stuff $k + 1$ letters into k mailboxes, then some mailbox must contains 2 letters.

9. You write 27 letters to 27 different people. Then you address the 27 envelopes. You close your eyes and stuff one letter into each envelope. What is the probability that just one letter is in the wrong envelope?

10. Prove that the sum of two odd numbers is never odd.

2.3   Proof by Contradiction

Preliminary Remarks

The method of proof by contradiction is based on the classical law of the excluded middle. Every sensible statement in mathematics is either true or false. We can prove that a statement is true by eliminating the possibility that it can be false. Obversely, we can prove that a statement is false by eliminating the possibility that it is true.

Aristotelian logic dictates that every sensible statement has a truth value: TRUE or FALSE. If we can demonstrate that a statement A could not possibly be false, then it must be true. On the other hand, if we can demonstrate that A could not be true, then it must be false. Here is a dramatic example of this principle. In order to present it, we shall assume for the moment that you are familiar with the system $\mathbb{Q}$ of rational numbers. These are numbers that may be written as the quotient of two integers (without dividing by zero, of course).

Theorem 2.16 (Pythagoras) *There is no rational number x with the property that $x^2 = 2$.*

Proof: In symbols (refer to Chapter 1), our assertion may be written

$$\sim \exists x, (x \in \mathbb{Q} \wedge x^2 = 2).$$

Seeking a contradiction, we assume the statement to be false. Then what we are assuming is that

$$\exists x, (x \in \mathbb{Q} \wedge x^2 = 2). \tag{$*$}$$

Since x is rational, we may write $x = p/q$, where p and q are integers.

We may as well suppose that both p and q are positive and nonzero. After reducing the fraction, we may assume that it is in lowest terms—so p and q have no common factors.

Now our hypothesis asserts that

$$x^2 = 2$$

or

$$\left(\frac{p}{q}\right)^2 = 2.$$

We may write this out as

$$p^2 = 2q^2. \tag{$**$}$$

Observe that this equation asserts that p^2 is an even number. But then p must be an even number (p cannot be odd, for that would imply that p^2 is odd by Proposition 2.5). So $p = 2r$ for some natural number r.

Substituting this assertion into equation $(**)$ now yields that

$$(2r)^2 = 2q^2.$$

Simplifying, we may rewrite our equation as

$$2r^2 = q^2.$$

This new equation asserts that q^2 is even. But then q itself must be even.

We have proven that both p and q are even. But that means that they have a common factor of 2. This contradicts our starting assumption that p and q have no common factor.

Let us pause to ascertain what we have established: the assumption that a rational square root x of 2 exists, and that it has been written in lowest terms as $x = p/q$, leads to the conclusion that p and q have a common factor and hence are *not* in lowest terms. What does this entail for our logical system?

We cannot allow a statement of the form $C = A \wedge \sim A$ (in the present context the statement A is "$x = p/q$ in lowest terms"). For such a statement C must be false.

But if x exists, then the statement C is true. No statement (such as A) can have two truth values. In other words, the statement C must be false. The only possible conclusion is that x does not exist. That is what we wished to establish.
□

Remark 2.17 In practice, we do not include the last three paragraphs in a proof by contradiction. We provide them now because this is our first exposure to such a proof, and we want to make the reasoning absolutely clear. The point is that the assertions A and $\sim A$ cannot both be true. An assumption that leads to this eventuality cannot be valid. That is the essence of proof by contradiction.

Historically, Theorem 2.16 was extremely important. Prior to Pythagoras ($\sim$300 B.C.E.), the ancient Greeks (following Eudoxus) believed that all numbers (at least all numbers that arise in real life) are rational. However, by the

Pythagorean theorem, the length of the diagonal of a unit square is a number whose square is 2. And our theorem asserts that such a number cannot be rational. We now know that there are many nonrational, or irrational, numbers. In fact, in Section 4.5, we shall learn that, in a certain sense to be made precise, "most" numbers are irrational.

Here is a second example of a proof by contradiction:

Theorem 2.18 (Dirichlet) *Suppose that* $(n + 1)$ *pieces of mail are delivered to* n *mailboxes. Then some mailbox contains at least two pieces of mail.*

Proof: Seeking a contradiction, we suppose that the assertion is false. Then each mailbox contains either zero or one piece of mail. But then the total amount of mail in all the mailboxes cannot exceed

$$\underbrace{1 + 1 + \cdots + 1}_{n \text{ times}}.$$

In other words, there are at most n pieces of mail. That conclusion contradicts the fact that there are $(n + 1)$ pieces of mail. We conclude that some mailbox contains at least two pieces of mail.   □

The last theorem, due to Johann Peter Gustav Lejeune Dirichlet (1805–1859), was classically known as the *Dirichletscher Schubfachschluss*. This German name translates to "Dirichlet's drawer shutting principle." Today, at least in this country, it is more commonly known as "the pigeonhole principle." Since pigeonholes are no longer a common artifact of everyday life, we have illustrated the idea using mailboxes.

EXAMPLE 2.19 Draw the unit interval I in the real line. Now pick 11 points at random from that interval (imagine throwing darts at the interval, or dropping ink drops on the interval). Then some pair of the points has distance not greater than 0.1 inch apart.

To see this, write

$$I = [0, 0.1] \cup [0.1, 0.2] \cup \cdots [0.8, 0.9] \cup [0.9, 1].$$

Here we have used standard interval notation. Think of each of these subintervals as a mailbox. We are delivering 11 letters (that is, the randomly selected points) to these ten mailboxes. By the pigeonhole principle, some mailbox must receive two letters.

We conclude that some subinterval of I, having length 0.1, contains two of the randomly selected points. Thus, their distance does not exceed 0.1 inch. ∎

EXAMPLE 2.20 We shall prove by contradiction that there are infinitely many prime numbers (this is an ancient result of Euclid).

Recall that a prime number is a whole number, or integer, greater than 1 which has no divisors except for 1 and itself. The first several primes are

$2, 3, 5, 7, 11, 13, 17, 19, 23, 29, 31, \ldots$. A natural number which is not prime is called *composite*. A composite number will have nontrivial factors. For example, $18 = 2 \cdot 3 \cdot 3$. In particular, a composite number will always be divisible by a prime.

Now, seeking a contradiction, let us suppose that there are only finitely many primes. Call them $p_1, p_2, \ldots, p_k$. Note that *this is an exhaustive list of all the primes*. Define

$$P = (p_1 \cdot p_2 \cdot p_3 \cdots \cdot p_k) + 1.$$

What can we say about the number P?

If we divide P by p_1, then p_1 goes evenly into the product, and there is a remainder of 1. If we divide P by p_2, then p_2 goes evenly into the product, and there is a remainder of 1. And so it goes for the rest of the p_j. Now P is either prime or composite. But we just checked every known prime—$p_1, p_2, \ldots,$ p_k—and verified that none of them is a divisor of P. So P cannot be composite. We conclude that P is prime.

But this is a contradiction, because P is a prime that is evidently larger than each of the p_j. We had an exhaustive list of the primes, and now we have created one more. That is a contradiction. We conclude that there are infinitely many primes. ∎

EXAMPLE 2.21 We shall show that there are no positive integer solutions to the equation $x^2 - y^2 = 1$. [Such an equation—a polynomial equation for which we seek integer solutions—is called a *diophantine equation*. This in honor of the ancient Greek mathematician Diophantus ($\sim$ 200 C.E.–$\sim$ 284 C.E.).]

Seeking a contradiction, we suppose that our diophantine equation *does* have integer solutions x, y. We write

$$1 = x^2 - y^2 = (x - y) \cdot (x + y).$$

Thus either both $x - y = 1$ and $x + y = 1$ or else $x - y = -1$ and $x + y = -1$. In the first case, we can add the two equations to solve them and find that $x = 1$, $y = 0$. This contradicts the assumption that both x and y are both positive. In the second case, we again can add the two equations and find that $x = -1$, $y = 0$. Again, we contradict the assumption that x and y are both positive.

Either case leads to a contradiction. We conclude that the diophantine equation *cannot* have a solution. ∎

EXAMPLE 2.22 We shall show that the sum of a rational number and an irrational number is always irrational.

Seeking a contradiction, we assume the contrary. So let $a = p/q$ be a rational number (here p and q are integers) and α an irrational number such that $a + \alpha$ is rational. So there is a rational number $b = p'/q'$ (with p', q' integers) with

$$a + \alpha = b.$$

But then we have

$$\alpha = b - a = \frac{p}{q} - \frac{p'}{q'} = \frac{pq' - p'q}{qq'}.$$

We have proved that the irrational number α equals the rational number $(pq' - p'q)/(qq')$. That is a contradiction.   ∎

POINT OF CONFUSION 2.23 People do not usually reason by "proof by contradiction" in ordinary, everyday discourse. It is too subtle for those purposes. The speaker could not easily follow the logic, and neither could the listener.

But proof by contradiction is a powerful tool. Alan Turing used the method to crack the German enigma code during World War II. Proof by contradiction is particularly helpful in establishing existence results, but it can be used for many other purposes as well.

A Look Back

1. What are the characteristics of a proof by contradiction?
2. How does a proof by contradiction differ from a direct proof?
3. Why is the method of proof by contradiction valid?
4. Prove by contradiction that the sum of two even numbers is even.

Exercises

1. Prove that, if n red letters and n blue letters are distributed among n mailboxes, then either some mailbox contains at least two red letters or some mailbox contains at least two blue letters or else some mailbox contains at least one red and one blue letter.

2. Prove that, if m is a power of 3 and n is a power of 3, then $m + n$ is never a power of 3.

3. What is special about the number 3 in Exercise 2? What other natural numbers can be used in its place?

4. Imitate the proof of Pythagoras's theorem to show that the number 5 does not have a rational square root.

5. Prove that, if n is a natural number and if n has a rational square root, then in fact the square root of n is an integer.

6. Complete this sketch to obtain an alternative proof that the number 2 does not have a rational square root:

   (a) Take it for granted that it is known that each positive integer has one and only one factorization into prime factors (a prime number is a positive integer, greater than 1, that can be divided evenly only by 1 and itself).

   (b) Seeking a contradiction, suppose that $\alpha = p/q$ is a rational square root of 2 (we need *not* assume that the rational fraction p/q is reduced to lowest terms).

(c) Then $2 = p^2/q^2$ or $2q^2 = p^2$.

(d) Count the number of prime factors on either side of the last equation in part (c) to arrive at a contradiction.

7. Prove that, if the natural number n is a perfect square, then $n + 1$ will never be a perfect square.

8. Prove that if the product of two integers is even, then one of them must be even.

9. Prove that if the product of two integers is odd, then both of them must be odd.

In each of Exercises 10–19, either prove that the statement is true or give a counterexample. Remember that a counterexample to a "for all" statement consists of a single example; but a counterexample to a "there exists" statement consists in showing that something never occurs.

10. The sum of two perfect squares is a perfect square.

11. Let n be a positive integer. In the list $n, n + 1, n + 2, \ldots, 2n + 2$ there must be a perfect square.

12. There is a positive integer that is the sum of all its divisors that are less than itself (including the divisor 1).

13. The difference of two perfect squares is never a prime (refer to Exercise 2.6 for the definition of "prime").

14. The sum of two perfect squares is never a prime.

15. For x a positive real number we have $1 + x^2 < (1 + x)^2$.

16. For any positive real numbers $a_1, a_2, \ldots, a_n$ we have

$$\frac{a_1 + a_2 + \cdots + a_n}{n} \leq \left(a_1 \cdot a_2 \cdots a_n\right)^{1/n}.$$

17. Between any two distinct real numbers there is a rational number.

18. Between any two distinct rational numbers there is an irrational number.

19. Let m and n be two successive perfect cubes. Then between them must lie a perfect square.

2.4 Proof by Induction

Preliminary Remarks

The method of mathematical induction is intimately bound up with the construction of the natural numbers. It is a means of proving a statement $P(n)$ about the natural numbers. It is a powerful technique that will serve us well in our studies.

The logical validity of the method of proof by induction is intimately bound up with the construction of the natural numbers, with ordinal arithmetic, and with the so-called well-ordering principle (see Section 4.2). However, the topic fits naturally into the present chapter. So we shall present and illustrate the method, and not worry about its logical foundations right now. As with any good idea in mathematics, we shall be able to make it intuitively clear that the method is a valid and useful one. So no confusion should result.

Consider a statement $P(n)$ about the natural numbers. For example, the statement might be "The quantity $n^2 + 5n + 6$ is always even." If we wish to prove this statement, we might proceed as follows:

(1) Prove the statement $P(1)$.

(2) Prove that $P(k) \Rightarrow P(k+1)$ for every $k \in \{1, 2, \dots\}$.

Let us apply the syllogism *modus ponendo ponens* from the end of Section 1.5 to determine what we will have accomplished. We know $P(1)$ and, from **(2)** with $k = 1$, that $P(1) \Rightarrow P(2)$. We may therefore conclude $P(2)$. Now **(2)** with $k = 2$ says that $P(2) \Rightarrow P(3)$. We may then conclude $P(3)$. Continuing in this fashion, we may establish $P(n)$ for every natural number n.

Notice that this reasoning applies to any statement $P(n)$ for which we can establish (1) and (2) above. Thus (1) and (2) taken together constitute a method of proof. It is a method of establishing a statement $P(n)$ for every natural number n. The method is known as *proof by induction*.

It is worth enunciating the steps of the induction process in slightly different language (which should make it easier to remember):

Steps in an Inductive Proof

(a) Enunciate the inductive statement $P(n)$. This should be a simple, declarative sentence about the positive integer n.

(b) Verify the case $n = 1$.

(c) Verify that $P(j)$ implies $P(j+1)$.

EXAMPLE 2.24 Let us use the method of induction to prove that, for every natural number n, the number $n^2 + 5n + 6$ is even.

Solution:

(a) Our statement $P(n)$ is

The number $n^2 + 5n + 6$ is even.

[*Note:* Explicitly identifying $P(n)$ is more than a formality, as Exercise 3 below shows. *Always* record carefully what $P(n)$ is before proceeding.]

We now proceed in two steps:

(b) $P(1)$ **is true.**   When $n = 1$, then

$$n^2 + 5n + 6 = 1^2 + 5 \cdot 1 + 6 = 12,$$

and this is certainly even. We have verified $P(1)$.

(c) $P(j) \Rightarrow P(j+1)$.   We are proving an implication in this step. We *assume* $P(j)$ and *use it* to establish $P(j + 1)$. Thus, we are assuming that

$$j^2 + 5j + 6 = 2m$$

for some natural number m. Then, to check $P(j + 1)$, we calculate

$$
\begin{aligned}
(j+1)^2 + 5(j+1) + 6 &= [j^2 + 2j + 1] + [5j + 5] + 6 \\
&= [j^2 + 5j + 6] + [2j + 6] \\
&= 2m + [2j + 6].
\end{aligned}
$$

Notice that, in the last step, we have *used our hypothesis* that $j^2 + 5j + 6$ is even, that is, that $j^2 + 5j + 6 = 2m$. Now the last line may be rewritten as

$$2(m + j + 3).$$

Thus, we see that $(j + 1)^2 + 5(j + 1) + 6$ is twice the natural number $m + j + 3$. In other words, $(j+1)^2 + 5(j+1) + 6$ is even. But that is the assertion $P(j+1)$.

   In summary, assuming the assertion $P(j)$, we have established the assertion $P(j + 1)$. That completes Step **(c)** of the method of induction. We conclude that $P(n)$ is true for every n.   ∎

   Here is another example to illustrate the method of induction.

Proposition 2.25 *If n is any natural number, then*

$$1 + 2 + \cdots + n = \frac{n(n + 1)}{2}.$$

Proof:
(a) The statement $P(n)$ is

$$1 + 2 + \cdots + n = \frac{n(n + 1)}{2}.$$

Now let us follow the method of induction closely.

(b) $P(1)$ **is true.**   The statement $P(1)$ is

$$1 = \frac{1(1 + 1)}{2}.$$

This is plainly true.

(c) $P(j) \Rightarrow P(j+1)$. We are proving an implication in this step. We *assume* $P(j)$ and *use it* to establish $P(j+1)$. Thus, we are assuming that

$$1 + 2 + \cdots + j = \frac{j(j+1)}{2}. \qquad (*)$$

Let us add the quantity $(j+1)$ to both sides of $(*)$. We obtain

$$1 + 2 + \cdots + j + (j+1) = \frac{j(j+1)}{2} + (j+1).$$

The left side of this last equation is exactly the left side of $P(j+1)$ that we are trying to establish. That is the motivation for our last step.

Now the right-hand side may be rewritten as

$$\frac{j(j+1) + 2(j+1)}{2}.$$

This simplifies to

$$\frac{(j+1)(j+2)}{2}.$$

In conclusion, we have established that

$$1 + 2 + \cdots + j + (j+1) = \frac{(j+1)(j+2)}{2}.$$

This is the statement $P(j+1)$.

Assuming the validity of $P(j)$, we have proved the validity of $P(j+1)$. That completes the third step of the method of induction, and establishes $P(n)$ for all n. □

Some problems are formulated in such a way that it is convenient to begin the induction with some value of n other than $n = 1$. The next example illustrates this notion:

EXAMPLE 2.26 Let us prove that, for $n \geq 4$, we have the inequality

$$3^n > 2n^2 + 3n.$$

Solution:
(a) The statement $P(n)$ is

$$3^n > 2n^2 + 3n.$$

(b) $P(4)$ is true. Observe that the inequality is false for $n = 1, 2, 3$. However, for $n = 4$ it is certainly the case that

$$3^4 > 2 \cdot 4^2 + 3 \cdot 4.$$

(c) $P(j) \Rightarrow P(j+1)$. Now assume that $P(j)$ has been established and let us use it to prove $P(j+1)$. We are hypothesizing that

$$3^j > 2j^2 + 3j.$$

Multiplying both sides by 3 gives

$$3 \cdot 3^j > 3(2j^2 + 3j)$$

or

$$3^{j+1} > 6j^2 + 9j.$$

But now we have

$$
\begin{aligned}
3^{j+1} &> 6j^2 + 9j \\
&= 2(j^2 + 2j + j) + (4j^2 + 3j) \\
&> 2(j^2 + 2j + 1) + (3j + 3) \\
&= 2(j+1)^2 + 3(j+1).
\end{aligned}
$$

This inequality is just $P(j+1)$, as we wished to establish. That completes step three of the induction, and therefore completes the proof.   ∎

EXAMPLE 2.27 Recall that the sequence $\{f_j\}$ of *Fibonacci numbers* is defined by

$$f_0 = 0, f_1 = 1, \ldots, f_j = f_{j-1} + f_{j-2} \qquad \text{for all } j \geq 2.$$

The first several Fibonacci numbers are therefore 0, 1, 1, 2, 3, 5, 8, 13, 21, 34, 55, 89, ... We will prove now that every third Fibonacci number is even (note that the first Fibonacci number is f_0, so we are claiming that $f_3, f_6, f_9, \ldots$ are even). We do so by induction.

(a) The statement $P(n)$ is

$$f_{3n} \text{ is even.}$$

(b) First the case $n = 1$. We may verify by inspection that $f_3 = 2$.

(c) Now suppose that the assertion has been proved for j. We will verify it for $j+1$. In fact we calculate that

$$
\begin{aligned}
f_{3(j+1)} &= f_{3j+3} \\
&= f_{3j+2} + f_{3j+1} \\
&= (f_{3j+1} + f_{3j}) + (f_{3j} + f_{3j-1}) \\
&= \left(\{f_{3j} + f_{3j-1}\} + f_{3j}\right) + (f_{3j} + f_{3j-1}) \\
&= 3f_{3j} + 2f_{3j-1}.
\end{aligned}
$$

Now we simply observe that, by the inductive hypothesis, f_{3j} is even. So the first term in the last line is even. Also, the second term has a factor of 2. So it is even. We conclude that $f_{3(j+1)}$ is even, and that completes the inductive step. ∎

EXAMPLE 2.28 Let us prove, by induction, that the sum of the first n positive, odd integers is equal to n^2.

We begin by writing the sum as

$$S_n = 1 + 3 + 5 + \cdots + \big[2(n-1)+1\big].$$

Notice that when $n = 1$ this just gives 1, with $n = 2$ it gives $1 + 3$, with $n = 3$ it gives $1 + 3 + 5$, and so forth.

(a) The statement $P(n)$ is

$$S_n = n^2.$$

(b) The case $n = 1$ is the assertion $1 = 1^2$. That is certainly true.

(c) Now assume that the identity has been proved for some j. So we are assuming the statement $P(j)$, that

$$1 + 3 + 5 + \cdots + [2(j-1)+1] = j^2.$$

We will check it for $j + 1$. We have that

$$
\begin{aligned}
S_{j+1} &= 1 + 3 + 5 + \cdots + \big[2((j+1)-1)+1\big] \\
&= 1 + 3 + 5 + \cdots + \big[2j+1\big] \\
&= \Big\{ 1 + 3 + 5 + \cdots + [2(j-1)+1] \Big\} + \big[2j+1\big] \\
&= j^2 + \big[2j+1\big] \\
&= (j+1)^2.
\end{aligned}
$$

In the penultimate equality we have used the inductive hypothesis. That completes the inductive step, and hence the proof. ∎

POINT OF CONFUSION 2.29 Of the three proof techniques that we have presented, induction is perhaps the most subtle. But it is an amazingly effective technique for establishing a statement of the form $P(n)$. You could never prove the Pythagorean theorem by induction. But many statements in number theory and especially graph theory are only accessible by induction.

Augustus de Morgan is the father of the use of induction in mathematics. Today everyone uses this technique.

We conclude this section by mentioning an alternative form of the induction paradigm which is sometimes called *complete mathematical induction* or *strong mathematical induction*.

Steps in a Proof by Complete Mathematical Induction:

Let P be a function on the natural numbers.

(a) Enunciate the inductive statement $P(n)$;

(b) prove $P(1)$; and

(c) prove that $[P(k)$ for all $k \leq j] \Rightarrow P(j + 1)$ for every natural number j;

then $P(n)$ is true for every n.

Notice that complete induction modifies Step (c) of the induction principle. Instead of assuming $P(j)$ (in order to prove $P(j + 1)$), we instead assume $P(1), P(2), \ldots, P(j)$. It turns out that the complete induction principle is logically equivalent to the ordinary induction principle enunciated at the outset of this section. But in some instances strong induction is the more useful tool. An alternative terminology for complete induction is "the set formulation of induction."

Complete induction is sometimes more convenient, or more natural, to use than ordinary induction; it finds particular use in abstract algebra.

EXAMPLE 2.30 Let us show, using complete induction, that every integer greater than 1 is either prime or the product of primes. [Here a prime number is an integer whose only factors are 1 and itself.]

For convenience we begin the induction process at the index 2 rather than at 1.

(a) Let $P(n)$ be the assertion "Either n is prime or n is the product of primes."

(b) Then $P(2)$ is plainly true since 2 is the first prime.

(c) Now assume that $P(k)$ is true for $2 \leq k \leq j$ and consider $P(j+1)$. If $j+1$ is prime, then we are done. If $j+1$ is not prime, then $j+1$ factors as $j+1 = p \cdot q$, where p, q are integers less than $j + 1$, but at least 2. By the strong inductive hypothesis, each of p and q factors as a product of primes (or is itself a prime). Thus $j + 1$ factors as a product of primes.

The complete induction is done, and the proof is complete. ∎

Figure 2.3: A balance scale.

A Look Back

1. What are the characteristics of a proof by induction?

2. How does a proof by induction differ from a proof by contradiction?

3. How does a proof by induction differ from a direct proof?

4. Prove by induction that the cube of an odd integer is odd.

Exercises

1. Prove that the sum of the squares of the first n natural numbers is equal to

$$\frac{2n^3 + 3n^2 + n}{6}.$$

2. Prove that the sum of the first k even natural numbers is $k^2 + k$.

* **3.** A popular recreational puzzle hypothesizes that you have nine pearls that are identical in appearance. Eight of these pearls have the same weight, but the ninth is either heavier or lighter—you do not know which. You have a balance scale (see Figure 2.3), and are allowed three weighings to find the odd pearl. How do you proceed?

   Now here is a bogus proof by induction that you can solve the problem in the first paragraph in three weighings not just for nine pearls but for *any number of pearls*. For convenience let us begin the induction with the case $n = 9$ pearls. By the result of the first paragraph, we can handle that case. Now, inductively, suppose that we have an algorithm for handling j pearls. We use this hypothesis to treat $(j + 1)$ pearls. From the $(j + 1)$ pearls, remove one and put it in your pocket There remain j pearls. Apply the j-pearl algorithm to these remaining pearls. If you find the odd pearl, then you are done. If you do not find the odd pearl, then it is the one in your pocket. That completes the case $(j + 1)$ and the proof.

   What is the flaw in this reasoning?

[*Remark*: If you are fiendishly clever, then you can actually handle twelve pearls in the original problem—with just three weighings. However this requires the consideration of 27 cases.]

4. Prove by induction that the sum of the angles interior to a convex polygon with k sides is $(k - 2) \cdot 180°$ (begin with $k = 3$ and you may assume that the result is known for triangles).

* **5.** Prove that complete induction is equivalent to ordinary induction.

6. Give a formal discussion of why the induction process may be begun at any natural number—not just 1.

* **7.** Fix a number $q > 0$. Use induction on the positive integer n to prove the following formula of Ramanujan:

$$1 + \frac{q}{1 - q} + \frac{q^2}{(1 - q)(1 - q^2)} + \frac{q^3}{(1 - q)(1 - q^2)(1 - q^3)}$$
$$+ \cdots + \frac{q^n}{(1 - q)(1 - q^2) \cdots (1 - q^n)}$$
$$= \frac{1}{(1 - q)(1 - q^2) \cdots (1 - q^n)}.$$

8. Use induction on n to prove that the formula in Exercise 7 still holds if the sequence of exponents $1, 2, 3, \ldots, n$ is replaced by the sequence of square exponents $1, 4, 9, \ldots, n^2$.

9. Prove the pigeonhole principle by induction.

10. Let a be a nonzero real number. Prove by induction that, for any positive integer n,

$$\begin{pmatrix} a & 2 \\ 0 & a \end{pmatrix}^n = \begin{pmatrix} a^n & 2na^{n-1} \\ 0 & a^n \end{pmatrix}.$$

11. Use induction to prove that every positive integer of the form $n^3 - n$ is divisible by 6.

12. Use induction to prove the identity

$$\frac{2^2}{1 \cdot 3} \cdot \frac{3^2}{2 \cdot 4} \cdots \frac{n^2}{(n - 1(n + 1)} = \frac{2n}{n + 1}.$$

* **13.** Use induction to prove that

$$\frac{1}{\sqrt{1}} + \frac{1}{\sqrt{2}} + \cdots + \frac{1}{\sqrt{1998}} \geq \sqrt{1998}.$$

[**Hint:** Formulate a statement that depends on n (instead of on 1998) and prove *that* by induction on n.]

Chapter 3

Set Theory

3.1 Undefinable Terms

Preliminary Remarks

In mathematics, we must find a place to begin our studies. Typically a definition in mathematics defines a new term in terms of older terms that have already been treated. But there must be a collection of "first terms" which are *not* defined in terms of earlier terms. These are the undefinables.

Even the most elementary considerations in logic may lead to conundrums. Suppose that we wish to define the notion of "line." We might say that it is the shortest path between two points. This is not completely satisfactory because we have not yet defined "path" or "point." And when we say "the shortest path," do we mean that there is just one unique shortest path? And why does it exist? Every new definition is, perforce, formulated in terms of other ideas. And those ideas in terms of other ones. Where does the regression cease?

The accepted method for dealing with this problem is to begin with certain terms (as few as possible) that are agreed to be "undefinable." These terms should be so simple that there can be little argument as to their meaning. But it is agreed in advance that these undefinable terms simply cannot be defined in terms of ideas that have been previously defined. Our undefined terms are our starting place.

In modern mathematics it is customary to use "set" and "element of" as undefinables. A *set* is declared to be a collection of objects. (Please do not ask what an "object" is or what a "collection" is; when we say that the term "set" is an undefinable, then we mean just that.) If S is a set, then we say that x *is an element of* S, and we write $x \in S$ or $S \ni x$, precisely when x is one of the objects that compose the set S. For example, we write $5 \in \mathbb{N}$ to indicate that the number 5 is an element of the set of natural numbers. We write $-7 \notin \mathbb{N}$ to specify that -7 is *not an element of* the set of natural numbers.

Definition 3.1 We say that two sets S and T are equal precisely when they have the same elements. We write $S = T$.

As an example of equality of sets, if $S = \{x \in \mathbb{N} : x^2 > 3\}$ and $T = \{x \in \mathbb{N} : x \geq 2\}$, then $S = T$.

Incidentally, the method of specifying a set with the notation $\{x : P(x)\}$, where P denotes a property, is the most common method in mathematics of defining a set. This is sometimes called "set-builder notation."

We shall endeavor, in what follows, to formulate all of our set-theoretic notions in a rigorous and logical fashion from the undefinables "set" and "element of."

However, it should be stressed that basic mathematics is *known* to be—indeed has been *proved to be*—logically consistent. [Strictly speaking, all notions of consistency in mathematics are relative to a higher-order system; you learn about these ideas in a course on formal mathematical logic. We shall not give a rigorous treatment of consistency in the present book.] The strict way in which we organize the subject is an important step in establishing this consistency.

3.2　Elements of Set Theory

Preliminary Remarks

All of modern mathematics is formulated in the language of sets and functions. It was Georg Cantor who laid the foundations for modern set theory. In this section we begin to learn the rudiments of set theory.

Beginning in this section, we will be doing mathematics in the way that it is usually done. That is, we shall define terms and we shall state and prove properties that they satisfy. In earlier chapters we were careful, but we were less mathematical. Sometimes we even had to say "This is the way we do it; don't worry." Many of the topics in Chapters 1 and 2 are really only best understood from the advanced perspective of mathematical logic. Now, and for the rest of this book, it is time to show how mathematics is done in practice.

We have already said what theorems, propositions, lemmas, and proofs are. Another formal ingredient of mathematical exposition is the "definition." A definition usually introduces a new piece of terminology or a new idea and *explains what it means in terms of ideas and terminology that have already been presented.* As you read this chapter, pause frequently to check that we are following this paradigm.

Definition 3.2 Let S and T be sets. We say that S is a *subset* of T, and we write $S \subset T$ or $T \supset S$, if

$$x \in S \Rightarrow x \in T.$$

We do not prove our definitions. There is *nothing to prove*. A definition introduces you to a new idea, or piece of terminology, or piece of notation.

EXAMPLE 3.3 Let $S = \{x \in \mathbb{N} : x > 3\}$ and $T = \{x \in \mathbb{N} : x^2 > 4\}$. Determine whether $S \subset T$ or $T \subset S$.

Solution: The key to success and clarity in handling subset questions is to *use the definition*. To see whether $S \subset T$ we must check whether $x \in S$ implies $x \in T$. Now if $x \in S$ then $x > 3$ hence $x^2 > 9$ so certainly $x^2 > 4$. Our syllogism is proved, and we conclude that $S \subset T$.

The reverse inclusion is false. For example, the number 3 is an element of T but is certainly not an element of S. We write $T \not\subset S$. ∎

EXAMPLE 3.4 Let $\mathbb{Z}$ denote the system of integers. Let $S = \{-2, 3\}$. Let $T = \{x \in \mathbb{Z} : x^3 - x^2 - 6x = 0\}$. Determine whether $S \subset T$ or $T \subset S$.

Solution: To see whether $S \subset T$ we must check whether $x \in S$ implies $x \in T$. Let $x \in S$. Then either $x = -2$ or $x = 3$. If $x = -2$ then $x^3 - x^2 - 6x = (-2)^3 - (-2)^2 - 6(-2) = 0$. Also, if $x = 3$ then $x^3 - x^2 - 6x = (3)^3 - (3)^2 - 6(3) = 0$. This verifies the syllogism $x \in S$ implies $x \in T$. Therefore $S \subset T$.

The reverse inclusion fails, for $0 \in T$ but $0 \notin S$. ∎

EXAMPLE 3.5 Let $S = \{x \in \mathbb{N} : x \geq 4\}$ and $T = \{x \in \mathbb{N} : x < 9\}$. Is it true that either $S \subset T$ or $T \subset S$?

Solution: Both inclusions are false. For $10 \in S$ but $10 \notin T$ and $2 \in T$ but $2 \notin S$. ∎

Proposition 3.6 *Let S and T be sets. Then $S = T$ if and only if both $S \subset T$ and $T \subset S$.*

Proof: If $S = T$ then, by definition, S and T have precisely the same elements. In particular, this means that $x \in S$ implies $x \in T$ and also $x \in T$ implies $x \in S$. That is, $S \subset T$ and $T \subset S$.

Now suppose that both $S \subset T$ and $T \subset S$. Seeking a contradiction, suppose that $S \neq T$. Then either there is some element of S that is not an element of T or there is some element of T that is not an element of S. The first eventuality contradicts $S \subset T$, and the second eventuality contradicts $T \subset S$. We conclude that $S = T$. □

Definition 3.7 We let $\emptyset$ denote the set that contains no elements. That is, $\forall x, x \notin \emptyset$. We call $\emptyset$ the *empty set*.

It may seem strange to consider a set with no elements. But this set arises very naturally in many mathematical contexts. For example, consider the set

$$S = \{x \in \mathbb{R} : x^2 < 0\}.$$

There are no real numbers with negative square. So there are no elements in this set. It is useful to be able to write $S = \emptyset$.

EXAMPLE 3.8 If S is any set, then $\emptyset \subset S$. To see this, notice that the statement "If $x \in \emptyset$ then $x \in S$" *must* be true because the hypothesis $x \in \emptyset$ is false. [Check the truth table for "if–then" statements.] This verifies that $\emptyset \subset S$.   ∎

EXAMPLE 3.9 Let $S = \{x \in \mathbb{N} : x + 2 \geq 19 \text{ and } x < 3\}$. Then S is a sensible set. There are no internal contradictions in its definition. But $S = \emptyset$. There are no elements in S.   ∎

Definition 3.10 Let S and T be sets. We say that x is an element of $S \cap T$ if both $x \in S$ *and* $x \in T$. It is useful and enlightening to write

$$x \in S \cap T \Longleftrightarrow x \in S \wedge x \in T.$$

This relates our new set-theoretic idea to basic concepts of logic that we learned in Chapter 1.

We say that x is an element of $S \cup T$ if either $x \in S$ *or* $x \in T$. Again, it is useful to relate the new idea to basic logic by writing

$$x \in S \cup T \Longleftrightarrow x \in S \vee x \in T.$$

We call $S \cap T$ the *intersection* of the sets S and T. We call $S \cup T$ the *union* of the sets S and T.

EXAMPLE 3.11 Let $S = \{x \in \mathbb{N} : 2 < x < 9\}$ and $T = \{x \in \mathbb{N} : 5 \leq x < 14\}$. Then $S \cap T = \{x \in \mathbb{N} : 5 \leq x < 9\}$, for these are the points common to both sets. And $S \cup T = \{x \in \mathbb{N} : 2 < x < 14\}$, for these are the points that are either in S or in T or in both. Draw a diagram on a real line to help you understand this example.   ∎

Remark 3.12 Observe that the use of "or" in the definition of set union justifies our decision to use the "inclusive 'or'" rather than the "exclusive 'or'" in mathematics.

EXAMPLE 3.13 Let $S = \{x \in \mathbb{N} : 1 \leq x \leq 5\}$ and $T = \{x \in \mathbb{N} : 8 < x \leq 12\}$. Then $S \cap T = \emptyset$, for the sets S and T have no elements in common. On the other hand, $S \cup T = \{x \in \mathbb{N} : 1 \leq x \leq 5 \text{ or } 8 < x \leq 12\}$. Draw a diagram on a real line to help you understand this example.   ∎

We may consider the intersection of more than two sets. For example, $x \in S \cap T \cap U$ means that (simultaneously) $x \in S$, $x \in T$, and $x \in U$. In fact one can consider the intersection of any number of sets—even an infinite number.

EXAMPLE 3.14 Let $S = \{1, 2, 3\}$, $T = \{2, 3, 4\}$, and $U = \{3, 4, 1\}$. Then

$$S \cap T = \{2, 3\} \,,\ T \cap U = \{3, 4\} \,,\ U \cap S = \{1, 3\} \ \ S \cap T \cap U = \{3\} \,.$$

∎

EXAMPLE 3.15 Let $S = \{1, 2\}$, $T = \{2, 3\}$, and $U = \{3, 1\}$. Then $S \cap T \cap U = \emptyset$. In other words, the three sets have no elements in common. But $S \cap T = \{2\} \neq \emptyset$, $T \cap U = \{3\} \neq \emptyset$, and $U \cap S = \{1\} \neq \emptyset$. It is common to say that S, T, U are *disjoint sets*.

But S, T, and U are *not pairwise disjoint*. The phrase "pairwise disjoint" means that no two of the sets have any elements in common. For instance, $A = \{1, 2\}$, $B = \{3, 4\}$, and $C = \{5, 7\}$ are pairwise disjoint. ∎

Definition 3.16 Let S and T be sets. We say that $x \in S \setminus T$ if both $x \in S$ and $x \notin T$. We may write this logically as

$$x \in S \setminus T \Longleftrightarrow x \in S \text{ but } x \notin T.$$

We call $S \setminus T$ the *set-theoretic difference* of S and T.

EXAMPLE 3.17 Let $S = \{x \in \mathbb{N} : 2 < x < 7\}$ and $T = \{x \in \mathbb{N} : 5 \leq x < 10\}$. Then $S \setminus T = \{x \in \mathbb{N} : 2 < x < 5\}$ and $T \setminus S = \{x \in \mathbb{N} : 7 \leq x < 10\}$. ∎

Definition 3.18 Suppose that we are studying subsets of a fixed set X. If $S \subset X$, then we use the symbol ^cS to denote $X \setminus S$. In this context, we sometimes refer to X as the *universal set*. We call ^cS the *complement* of S (in X). We may write

$$x \in {}^cS \Longleftrightarrow x \in X \wedge x \notin S.$$

EXAMPLE 3.19 Let $\mathbb{N}$ be the universal set. Let $S = \{x \in \mathbb{N} : 3 < x \leq 20\}$. Then

$$^cS = \{x \in \mathbb{N} : 1 \leq x \leq 3\} \cup \{x \in \mathbb{N} : 20 < x\}.$$

∎

POINT OF CONFUSION 3.20 Now is the time when it is essential to be logical. The set operations $\subset$, $\setminus$, $\cap$, $\cup$, and so forth have very specific meanings which can only be understood by applying basic logic. In the next proposition, when we want to prove the equality of two different sets, we do so by proving two inclusions. This is really the only way to do it.

Proposition 3.21 Let X be the universal set and $S \subset X$, $T \subset X$. Then

(a) $^c(S \cup T) = {}^cS \cap {}^cT$;

(b) $^c(S \cap T) = {}^cS \cup {}^cT$.

Proof: We shall present this proof in detail since it is a good exercise in understanding both our definitions and our method of proof (and also a good exercise with logic).

We begin with the proof of **(a)**. It is often best to treat the proof of the equality of two sets as two separate proofs of containment. [This is why Proposition 3.6 is important.] That is what we now do.

Let $x \in {}^c(S \cup T)$. Then, by definition, $x \notin (S \cup T)$. Thus x is neither an element of S nor an element of T. So both $x \in {}^cS$ and $x \in {}^cT$. Hence $x \in {}^cS \cap {}^cT$. We conclude that ${}^c(S \cup T) \subset {}^cS \cap {}^cT$. Conversely, if $x \in {}^cS \cap {}^cT$, then $x \notin S$ and $x \notin T$. Therefore $x \notin (S \cup T)$. As a result, $x \in {}^c(S \cup T)$. Thus ${}^cS \cap {}^cT \subset {}^c(S \cup T)$. Summarizing, we have ${}^c(S \cup T) = {}^cS \cap {}^cT$.

The proof of part **(b)** is similar, but we include it for practice. Let $x \in {}^c(S \cap T)$. Then, by definition, $x \notin (S \cap T)$. Thus x is not both an element of S and an element of T. So either $x \in {}^cS$ or $x \in {}^cT$. Hence $x \in {}^cS \cup {}^cT$. We conclude that ${}^c(S \cap T) \subset {}^cS \cup {}^cT$. Conversely, if $x \in {}^cS \cup {}^cT$, then either $x \notin S$ or $x \notin T$. Therefore $x \notin (S \cap T)$. As a result, $x \in {}^c(S \cap T)$. Thus ${}^cS \cup {}^cT \subset {}^c(S \cap T)$. Summarizing, we have ${}^c(S \cap T) = {}^cS \cup {}^cT$. $\square$

The two formulas in the last proposition are often referred to as de Morgan's laws. Compare them with de Morgan's laws for $\vee$ and $\wedge$ in Section 1.4.

A Look Back

1. What does $\subset$ mean?

2. What does $\in$ mean? How does it differ from $\subset$?

3. What is the empty set?

4. When are two sets equal?

Exercises

1. Let $S = \{1, 2, 3, 4, 5\}$, $T = \{3, 4, 5, 7, 8, 9\}$, $U = \{1, 2, 3, 4, 9\}$, $V = \{2, 4, 6, 8\}$. [In this exercise let $S \times T$ denote the set of ordered pairs (s, t) of elements of S and T.] Calculate each of the following:

   (a) $S \cap U$

   (b) $(S \cap T) \cup U$

   (c) $(S \cup U) \cap V$

   (d) $(S \cup V) \setminus U$

   (e) $(U \cup V \cup T) \setminus S$

   (f) $(S \cup V) \setminus (T \cap U)$

   (g) $(S \times V) \setminus (T \times U)$

   (h) $(V \setminus T) \times (U \setminus S)$

2. Refer to Exercise 1 for terminology. Let S be any set and let $T = \emptyset$. What can you say about $S \times T$?

3. Prove the following formulas for arbitrary sets S, T, U, and V.

   (a) $S \cap (T \cup U) = (S \cap T) \cup (S \cap U)$

(b) $S \cup (T \cap U) = (S \cup T) \cap (S \cup U)$

(c) $S \cap {}^c T = S \setminus T$

(d) $(S \setminus T) \cup (T \setminus S) = (S \cup T) \setminus (S \cap T)$

(e) $S \setminus (T \cup U) = (S \setminus T) \cap (S \setminus U)$

(f) $S \setminus (T \cap U) = (S \setminus T) \cup (S \setminus U)$

(g) $(S \setminus T) \times (U \setminus V) = (S \times U) \setminus [(S \times V) \cup (T \times U)]$

(h) $(S \cup T) \times V = (S \times V) \cup (T \times V)$

4. Let $A = \{1, 2, 3, 4\}$, $B = \{2, 3, 4, 5\}$, $C = \{3, 4, 5, 6\}$. Calculate $A \setminus (B \setminus C)$.

5. Suppose that $A \subset B \subset C$. What is $A \setminus B$? What is $A \setminus C$? What is $A \cup B$?

In Exercises 6–8, let $\mathbb{N}$ denote the natural numbers, $\mathbb{Z}$ the integers, $\mathbb{Q}$ the rational numbers, and $\mathbb{R}$ the real numbers.

6. Describe the set $\mathbb{Q} \setminus \mathbb{Z}$ in words. Describe $\mathbb{R} \setminus \mathbb{Q}$.

7. Refer to Exercise 1 for terminology. Describe $\mathbb{Q} \times \mathbb{R}$ in words. Describe $\mathbb{Q} \times \mathbb{Z}$.

8. Refer to Exercise 1 for terminology. Describe $(\mathbb{Q} \times \mathbb{R}) \setminus (\mathbb{Z} \times \mathbb{Q})$ in words.

9. TRUE or FALSE: If $S_1, S_2, \ldots$ are sets of integers and if $\cup_{j=1}^{\infty} S_j = \mathbb{Z}$, then one of the sets S_j must have infinitely many elements. Give a proof of your answer.

10. TRUE or FALSE: If $S_1, S_2, \ldots$ are sets of real numbers and if $\cup_{j=1}^{\infty} S_j = \mathbb{R}$, then one of the sets S_j must have infinitely many elements. Give a proof of your answer.

11. Prove that if $A \subset B$ and $B \subset C$ then $A \subset C$.

12. Let $S = \{a, b, c, d\}$, $T = \{1, 2, 3\}$, and $U = \{b, 2\}$. Which of the following statements is true?

(a) $\{a\} \in S$

(b) $a \in S$

(c) $\{a, c\} \subset S$

(d) $\emptyset \in S$

(e) $\{1, 2\} \in T$

(f) $\{a\} \subset S$

(g) $\{a, c, 2, 3\} \subset S \cup T$

(h) $U \subset S \cup T$

(i) $b \in S \cap U$

(j) $\{b\} \subset S \cap U$

(k) $\{1\} \in T$

(l) $\{1, 3\} \subset T$

(m) $\{1,3\} \subset T$

(n) $\emptyset \subset S$

(o) $\emptyset \in S$

(p) $\{\emptyset\} \subset S$

(q) $\{\emptyset\} \in S$

3.3   Venn Diagrams

Preliminary Remarks
The Venn diagram is a pictorial device for depicting relationships among sets. While a Venn diagram is certainly not a rigorous proof, it is a useful means for understanding mathematical ideas.

We sometimes use a *Venn diagram* to aid our understanding of set-theoretic relationships. In a Venn diagram, a set is represented as a region in the plane (for convenience, we use rectangles). The intersection $A \cap B$ of two sets A and B is the region common to the two domains (we have shaded that region with dots in Figure 3.1):

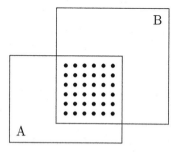

Figure 3.1

Now let A, B, and C be three sets. The Venn diagram in Figure 3.2 makes it easy to see that $A \cap (B \cup C) = (A \cap B) \cup (A \cap C)$.

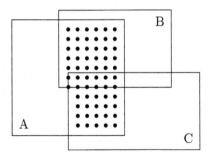

Figure 3.2

The Venn diagram in Figure 3.3 illustrates the fact that

$$A \setminus (B \cup C) = (A \setminus B) \cap (A \setminus C).$$

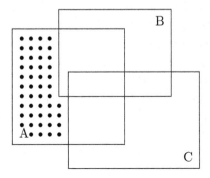

Figure 3.3

A Venn diagram is not a proper substitute for a rigorous mathematical proof. However, it can go a long way toward guiding our intuition.

POINT OF CONFUSION 3.22 A Venn diagram is remarkable because it depicts relationships among sets no matter where those sets may live. The sets could be subsets of Euclidean 3-space, or of space-time in physics, or of a 16-dimensional space in string theory. Nevertheless, a two-dimensional Venn diagram can show clearly what is going on.

A Look Back

1. What is the purpose of a Venn diagram?

2. Draw a Venn diagram to illustrate the equality

$$A \cap B = {}^c({}^cA \cup {}^cB).$$

3. Use a Venn diagram to illustrate that

$$(A \cap B) \setminus (A \cup B) = \emptyset.$$

4. The idea of the Venn diagram is about 100 years old, and was invented by John Venn. Look on the Internet to learn where Venn worked and what his other famous works in logic were.

Exercises

1. Prove the following formulas for arbitrary sets S, T, U, and V. [**Hint:** You may find Venn diagrams useful to guide your thinking, but a Venn diagram is *not* a proof. In this exercise, $S \times T$ denotes the set of ordered pairs (s, t) of elements of S and T.]

   (a) $S \cap (T \cup U) = (S \cap T) \cup (S \cap U)$

   (b) $S \cup (T \cap U) = (S \cup T) \cap (S \cup U)$

   (c) $S \cap {}^cT = S \setminus T$

   (d) $(S \setminus T) \cup (T \setminus S) = (S \cup T) \setminus (S \cap T)$

   (e) $S \setminus (T \cup U) = (S \setminus T) \cap (S \setminus U)$

   (f) $S \setminus (T \cap U) = (S \setminus T) \cup (S \setminus U)$

   (g) $(S \setminus T) \times (U \setminus V) = (S \times U) \setminus [(S \times V) \cup (T \times U)]$

   (h) $(S \cup T) \times V = (S \times V) \cup (T \times V)$

2. Draw Venn diagrams to illustrate parts (a)–(f) of Exercise 1.

3. What is the difference between disjoint and pairwise disjoint? Draw Venn diagrams to illustrate the two ideas.

4. Use a Venn diagram to illustrate the idea that set theoretic difference is related to set complementation.

5. Use a Venn diagram to illustrate the idea that "subset of" and "element of" have different meanings.

6. Illustrate each of de Morgan's set-theoretic laws with a Venn diagram.

3.4   Further Ideas in Elementary Set Theory

Preliminary Remarks

In this section we continue to develop the basic ideas of set theory. This includes set-theoretic product and the power set. Cantor showed us that the power set operation is a useful device for producing arbitrarily large sets.

Now we learn some new ways to combine sets.

Definition 3.23 Let S and T be sets. We define $S \times T$ to be the set of all ordered pairs (s, t) such that $s \in S$ and $t \in T$. The set $S \times T$ is called the *set-theoretic product* (or sometimes just the *product*) of S and T.

EXAMPLE 3.24 Let $S = \{1, 2, 3\}$ and $T = \{a, b\}$. Then

$$S \times T = \{(1, a), (1, b), (2, a), (2, b), (3, a), (3, b)\}.$$

∎

It is no coincidence that, in the last example, the set S has 3 elements, the set T has 2 elements, and the set $S \times T$ has $3 \times 2 = 6$ elements. In fact one can prove that if S has k elements and T has ℓ elements, then $S \times T$ has $k \cdot \ell$ elements. Exercise 15 asks you to prove this assertion by induction on k.

Notice that $S \times T$ is a different set from $T \times S$. With S and T as in the last example,

$$T \times S = \{(a, 1), (b, 1), (a, 2), (b, 2), (a, 3), (b, 3)\}.$$

The phrase "ordered pair" means that the pair $(a, 1)$, for example, is distinct from the pair $(1, a)$.

If S is a set then the *power set* of S is the set of all subsets of S. We denote the power set by $\mathcal{P}(S)$. We may write

$$X \in \mathcal{P}(S) \Longleftrightarrow X \subset S.$$

EXAMPLE 3.25 Let $S = \{1, 2, 3\}$. Then

$$\mathcal{P}(S) = \{\{1\}, \{2\}, \{3\}, \{1, 2\}, \{2, 3\}, \{1, 3\}, \{1, 2, 3\}, \emptyset\}.$$

∎

If the concept of power set is new to you, then you might have been surprised to see $\{1, 2, 3\}$ and $\emptyset$ as elements of the power set. But they are both subsets of S, and they must be listed.

Proposition 3.26 Let $S = \{s_1, \ldots, s_k\}$ be a set. Then $\mathcal{P}(S)$ has 2^k elements.

Proof: We prove the assertion by induction on k.

$P(1)$ is true. In this case, $S = \{s_1\}$ and $\mathcal{P}(S) = \{\{s_1\}, \emptyset\}$. Notice that S has $k = 1$ element and $\mathcal{P}(S)$ has $2^k = 2$ elements.

$P(j) \Rightarrow P(j + 1)$. Assume that any set of the form $S = \{s_1, \ldots, s_j\}$ has power set with 2^j elements. Now let $T = \{t_1, \ldots, t_j, t_{j+1}\}$. Consider the subset $T' = \{t_1, \ldots, t_j\}$ of T. Then $\mathcal{P}(T)$ certainly contains $\mathcal{P}(T')$ (that is, every subset of T' is also a subset of T). But it also contains each of the sets that is obtained by adjoining the element t_{j+1} to each subset of T'. Thus the total number of subsets of T is

$$2^j + 2^j = 2^{j+1}.$$

Notice that we have indeed counted all subsets of T, since any subset either contains t_{j+1} or it does not.

Thus, assuming the validity of our assertion for j, we have proved its validity for $j+1$. That completes our induction and the proof of the proposition. $\square$

We have seen that the operation of set-theoretic product corresponds to the arithmetic product of natural numbers. And now we have seen that the operation of taking the power set corresponds to exponentiation. In Section 4.3 we shall use the concept of function to unify all of these ideas.

POINT OF CONFUSION 3.27 The notion of power set is one of Georg Cantor's great ideas. It is used to generate sets of arbitrarily high cardinality. In higher studies of set theory, it plays a pivotal role. The idea that you can "exponentiate" a set is really quite marvelous.

A Look Back

1. What is the power set of $\{a, \{a\}, \{a, \{a\}\}\}$?

2. What is the power set of $\mathbb{N}$? Can you identify it with all the real numbers between 0 and 1 in a natural way?

3. Describe $\mathbb{Z} \times \mathbb{N}$ in words.

4. Describe $\mathbb{R} \times \mathbb{Z}$ in words.

Exercises

1. Let S be any set and let $T = \emptyset$. What can you say about $S \times T$?

2. Suppose that $A \subset B \subset C$. What is $A \setminus B$? What is $A \setminus C$? What is $A \cup B$?

3. Give an explicit description of the power set of $S = \{a, b, 1, 2\}$ (that is, write out all the elements).

4. Let the set S have k elements. Give a direct proof (different from the one in the text) of the assertion that the number of elements of the power set of S is 2^k. That is, devise an explicit scheme for counting the subsets.

5. Calculate the power set of the power set of $T = \{1, 2\}$.

6. Let S and T be sets. Is it true that $\mathcal{P}(S \times T) = \mathcal{P}(S) \times \mathcal{P}(T)$? Provide either a proof or a counterexample.

7. Prove that $S \subset T$ if and only if $\mathcal{P}(S) \subset \mathcal{P}(T)$.

8. Prove that $S = T$ if and only if $\mathcal{P}(S) = \mathcal{P}(T)$.

9. Let $S = \{a, b, c, d\}$, $T = \{1, 2, 3\}$, and $U = \{b, 2\}$. Which of the following statements is true?

   (a) $\{a\} \in S$

(b) $a \in S$

(c) $\{a, c\} \subset S$

(d) $\emptyset \in S$

(e) $\{a\} \in \mathcal{P}(S)$

(f) $\{\{a\}, \{a, b\}\} \subset \mathcal{P}(S)$

(g) $\{a, c, 2, 3\} \subset S \cup T$

(h) $U \subset S \cup T$

(i) $b \in S \cap U$

(j) $\{b\} \subset S \cap U$

(k) $\{1, 3\} \in T$

(l) $\{1, 3\} \subset T$

(m) $\{1, 3\} \in \mathcal{P}(T)$

(n) $\emptyset \in \mathcal{P}(S)$

(o) $\{\emptyset\} \in \mathcal{P}(S)$

(p) $\emptyset \subset \mathcal{P}(S)$

(q) $\{\emptyset\} \subset \mathcal{P}(S)$

10. Write out the power set of each set:

(a) $\{1, \emptyset, \{a, b\}\}$

(b) $\{\bullet, \triangle, \partial\}$

(c) $\left\{ \emptyset, \{\emptyset\}, \{\{\emptyset\}\} \right\}$

11. Prove or disprove each of the following statements:

(a) $\mathcal{P}(A \cup B) = \mathcal{P}(A) \cup \mathcal{P}(B)$

(b) $\mathcal{P}(A \cap B) = \mathcal{P}(A) \cap \mathcal{P}(B)$

(c) $\mathcal{P}(A \setminus B) = \mathcal{P}(A) \setminus \mathcal{P}(B)$

12. Let S, T, U be finite sets. Verify that the sets $(S \times T) \times U$ and $S \times (T \times U)$ have the same number of elements.

13. Suppose that S is a set and that $S \cup T$ is finite for every choice of finite set T. Prove that S must then be finite.

14. Let $\mathcal{P}$ be the power set of $I = \{1, 2, 3, 4, 5\}$. Let $\mathcal{S}$ be a randomly chosen subset of $\mathcal{P}$. What is the probability that $\mathcal{S}$ is the power set of some subset of I?

15. Prove that if the set S has k elements and the set T has ℓ elements, then $S \times T$ has $k \cdot \ell$ elements.

3.5   Indexing and Extended Set Operations

Preliminary Remarks

It is frequently useful in mathematics to consider very large sets, or sets that are formed using large operations. We consider some of those in the present section.

Frequently we wish to manipulate infinitely many sets. Perhaps we will take their intersection or union. We require suitable notation to perform these operations.

If $S_1, S_2, \ldots$ are sets, then we define

$$\bigcup_{j=1}^{\infty} S_j \equiv \{x : \exists j \in \mathbb{N}, x \in S_j\}.$$

Similarly, we define

$$\bigcap_{j=1}^{\infty} S_j \equiv \{x : \forall j \in \mathbb{N}, x \in S_j\}v.$$

Notice that we employ the common mathematical notation $\equiv$ to mean "is defined to be." Other texts use the notation $\overset{\text{def}}{=}$ or $=:$ or $\doteq$.

EXAMPLE 3.28 Let $\mathbb{Q}$ be the rational numbers and let $S_j = \{x \in \mathbb{Q} : 0 < x < 1 + 1/j\}$, $j = 1, 2, \ldots$. Let us describe $\cup_{j=1}^{\infty} S_j$ and $\cap_{j=1}^{\infty} S_j$.

Notice that $S_1 \supset S_j$, indeed $S_1 \supset S_2 \supset \cdots \supset S_j \supset \cdots$, for every j, hence $\cup_{j=1}^{\infty} S_j = S_1 = \{x \in \mathbb{Q} : 0 < x < 2\}$.

Next, notice that, if $x \in \mathbb{Q}$ and $x > 1$, then if we select $j > 1/(x-1)$ then $x \notin S_j$. It follows that $x \notin \cap_{j=1}^{\infty} S_j$. On the other hand, $\{x \in \mathbb{Q} : 0 < x \leq 1\} \subset S_j$ for every j. It follows that $\cap_{j=1}^{\infty} S_j = \{x \in \mathbb{Q} : 0 < x \leq 1\}$.   ∎

EXAMPLE 3.29 It is entirely possible for *nested*, nonempty sets to have empty intersection. Let $S_j = \{x \in \mathbb{Q} : 0 < x < 1/j\}$. Certainly each S_j is nonempty, for it contains the point $1/(2j)$. Next, $S_1 \supset S_2 \supset \cdots$. Finally, for any positive integer k,

$$\bigcap_{j=1}^{k} S_j = S_k \neq \emptyset.$$

However,

$$\bigcap_{j=1}^{\infty} S_j = \emptyset.$$

To verify this last assertion, notice that, if $x > 0$ and $j > 1/x$, then $x \notin S_j$ hence $x \notin \cap_{j=1}^{\infty} S_j$. However, if $x \leq 0$, then x is not an element of any S_j. As a result, no x lies in the intersection. The intersection is empty.   ∎

In the examples given thus far, the "index set" has been the natural numbers. That is, we let the index j range over $\{1, 2, \dots\}$. It is frequently useful to use a larger index set, such as the real numbers or some unspecified index set. Usually we specify an index set with the letter A and we denote a specific index by $\alpha \in A$.

EXAMPLE 3.30 For each real number α we let $S_\alpha = \{x \in \mathbb{R} : \alpha \le x < \alpha + 1\}$. Thus each S_α is an "interval" of real numbers, and we may speak of

$$\bigcup_{\alpha \in A} S_\alpha \equiv \{x : \exists \alpha \in A, x \in S_\alpha\}$$

and

$$\bigcap_{\alpha \in A} S_\alpha \equiv \{x : \forall \alpha \in A, x \in S_\alpha\}.$$

For the sets S_α that we have specified,

$$\bigcap_{\alpha \in A} S_\alpha = \emptyset.$$

This is so because, if $x \in \mathbb{R}$, then $x \notin S_{x+1}$ hence certainly $x \notin \cap_\alpha S_\alpha$.

On the other hand,

$$\bigcup_{\alpha \in A} S_\alpha = \mathbb{R}$$

since every real x lies in $S_{x-1/2}$. ∎

Proposition 3.31 *Fix a universal set X. Let A be an index set and, for each $\alpha \in A$, let S_α be a subset of X. Then*

(a) $\displaystyle {}^c\!\left(\bigcap_{\alpha \in A} S_\alpha \right) = \bigcup_{\alpha \in A} {}^c S_\alpha;$

(b) $\displaystyle {}^c\!\left(\bigcup_{\alpha \in A} S_\alpha \right) = \bigcap_{\alpha \in A} {}^c S_\alpha.$

Proof: The proof is similar to that of Proposition 3.21 of this chapter. We leave the details to Exercise 1 at the end of the section. □

POINT OF CONFUSION 3.32 It actually happens in functional analysis and other advanced mathematical subjects that one needs to take *really large* products and unions and intersections. Thus one needs to have these ideas under control in order to advance further in mathematics.

Further properties of intersection and union over arbitrary index sets are explored in the exercises. These are some of the most important exercises in the chapter.

A Look Back

1. Why do we want to take the intersection or union or product of a very large number of sets?

2. Calculate the union
$$\bigcup_{\lambda \in \mathbb{R}} \{x \in \mathbb{R} : x > \lambda\}.$$

3. Calculate the intersection
$$\bigcup_{\lambda \in \mathbb{R}} \{x \in \mathbb{R} : x > \lambda\}.$$

4. Describe in words the product
$$\prod_{j \in \mathbb{N}} \{x \in \mathbb{R} : j < x < j+1\}.$$

Exercises

1. Let $S_\alpha \subset X$ be sets indexed over an arbitrary index set A, $\alpha \in A$. Prove each of the following identities:

   (a) $^c\left(\cap_{\alpha \in A} S_\alpha\right) = \cup_{\alpha \in A}{}^c S_\alpha$

   (b) $^c\left(\cup_{\alpha \in A} S_\alpha\right) = \cap_{\alpha \in A}{}^c S_\alpha$

   (c) $T \cap \left(\cup_{\alpha \in A} S_\alpha\right) = \cup_{\alpha \in A} \left(T \cap S_\alpha\right)$

   (d) $T \cup \left(\cap_{\alpha \in A} S_\alpha\right) = \cap_{\alpha \in A} \left(T \cup S_\alpha\right)$

2. TRUE or FALSE: If $S_1, S_2, \ldots$ are sets of integers and if $\cup_{j=1}^\infty S_j = \mathbb{Z}$, then one of the sets S_j must have infinitely many elements. Give a proof of your answer.

3. TRUE or FALSE: If $S_1, S_2, \ldots$ are sets of real numbers and if $\cup_{j=1}^\infty S_j = \mathbb{R}$, then one of the sets S_j must have infinitely many elements. Give a proof of your answer.

4. Let $S_j = \{j, j+1\}$. What is $\cup_{j=1}^\infty S_j$?

5. If $S \subset T_\alpha$ for every $\alpha \in A$ then prove that
$$S \subset \cap_{\alpha \in A} T_\alpha.$$

6. If $S \supset T_\alpha$ for every $\alpha \in A$ then prove that
$$S \supset \cup_{\alpha \in A} T_\alpha.$$

7. Let $\{A_j\}$ be sets. We say that the A_j are *disjoint* if $\cap_j A_j = \emptyset$. On the other hand, the A_j are *pairwise disjoint* if $A_j \cap A_k = \emptyset$ whenever $j \neq k$. Show that these two concepts are different for four sets A_1, A_2, A_3, A_4.

* 8. Let $\mathcal{P}$ be the power set of $I = \{1, 2, 3, 4, 5\}$. Let $\mathcal{S}$ be a randomly chosen subset of $\mathcal{P}$. What is a simple necessary condition on the size of $\mathcal{S}$ for $\mathcal{S}$ to be the power set of a subset of I?

Chapter 4

Relations and Functions

4.1 Relations

> **Preliminary Remarks**
>
> In mathematics, a relation is a quite general device for specifying some inter-action between two sets. The theory of relations is part of the foundation for the theory of functions. Particularly important for us will be the theory of equivalence relations.

Let S and T be sets. A *relation* on S and T is a subset of $S \times T$. If $\mathcal{R}$ is a relation, then we write either $(s, t) \in \mathcal{R}$ or sometimes $s \mathcal{R} t$ to indicate that (s, t) is an element of the relation. We will also write $s \sim t$ when the relation being discussed is understood from context.

EXAMPLE 4.1 Let $S = \mathbb{N}$, the natural numbers; and let $T = \mathbb{R}$, the real numbers. Define a relation $\mathcal{R}$ on S and T by $(s, t) \in \mathcal{R}$ if $s < \sqrt{t} < s + 1$. For instance, $(2, 5) \in \mathcal{R}$ because $\sqrt{5}$ lies between 2 and 3. Also $(4, 17) \in \mathcal{R}$ because $\sqrt{17}$ lies between 4 and 5. However, $(5, 10)$ does not lie in $\mathcal{R}$. Also $(3, \pi)$ does not lie in $\mathcal{R}$. ∎

The *domain* of a relation $\mathcal{R}$ is the set of $s \in S$ such that there exists a $t \in T$ with $(s, t) \in \mathcal{R}$. The *image* of the relation is the set of $t \in T$ such that there exists an $s \in S$ with $(s, t) \in \mathcal{R}$. It is sometimes convenient to refer to the entire set T as the *range* of the relation $\mathcal{R}$. So we see that the image and the range are distinct.

EXAMPLE 4.2 Let $S = \mathbb{N}$ and $T = \mathbb{N}$. Define a relation $\mathcal{R}$ on S and T by the condition $(s, t) \in \mathcal{R}$ if $s^2 < t$. Observe that, for any element $s \in \mathbb{N} = S$, the number $t = s^2 + 1$ satisfies $s^2 < t$. Hence $(s, t) \in \mathcal{R}$. Therefore the domain of the relation is all of $\mathbb{N}$.

Now let us think about the image. The number $1 \in \mathbb{N} = T$ cannot be in the image since there is no element $s \in S = \mathbb{N}$ such that $s^2 < 1$. However, any element $t \in T$ that exceeds 1 satisfies $1^2 < t$. So $(1, t) \in \mathcal{R}$. Thus the image of $\mathcal{R}$ is the set $\{t \in \mathbb{N} : t \geq 2\}$. ∎

EXAMPLE 4.3 Let $S = \mathbb{N}$ and $T = \mathbb{N}$. Define a relation $\mathcal{R}$ on S and T by the condition $(s, t) \in \mathcal{R}$ if $s^2 + t^2$ is itself a perfect square. Then, for instance, $(3, 4) \in \mathcal{R}$, $(4, 3) \in \mathcal{R}$, $(12, 5) \in \mathcal{R}$, and $(5, 12) \in \mathcal{R}$. The number 1 is not in the domain of $\mathcal{R}$ since there is no natural number t such that $1^2 + t^2$ is a perfect square (if there were, this would mean that there are two perfect squares that differ by 1, and that is not the case). The number 2 is not in the domain of $\mathcal{R}$ for a similar reason. Likewise, 1 and 2 are not in the image of $\mathcal{R}$.

In fact, both the domain and image of $\mathcal{R}$ have infinitely many elements. This assertion will be explored in the exercises. ∎

POINT OF CONFUSION 4.4 In everyday conversation, the word "relation" is used to mean a number of different things—ranging from the very personal to the political. But in mathematics the word has a very particular meaning. The idea of "relation" is the basis for our study of functions and a number of other sophisticated ideas. We must get it straight before we can proceed.

Many interesting relations arise for which S and T are the same set. Say that $S = T = A$. Then a relation on S and T is called simply a relation on A.

EXAMPLE 4.5 Let $\mathbb{Z}$ be the integers. Let us define a relation $\mathcal{R}$ on $\mathbb{Z}$ by the condition $(s, t) \in \mathcal{R}$ if $s - t$ is divisible by 2. It is easy to see that both the domain and the image of this relation is $\mathbb{Z}$ itself. It is also worth noting that, if n is any integer, then the set of all elements related to n is either (i) the set of all even integers (if n is even) or (ii) the set of all odd integers (if n is odd). ∎

Notice that the last relation created a division of the domain (=image) into two disjoint sets: the even integers and the odd integers. This was a special instance of an important type of relation that we now define.

Definition 4.6 Let $\mathcal{R}$ be a relation on a set A. We say that $\mathcal{R}$ is an *equivalence relation* if the following properties hold:

$\mathcal{R}$ **is reflexive:** If $x \in A$, then $(x, x) \in \mathcal{R}$.

$\mathcal{R}$ **is symmetric:** If $(x, y) \in \mathcal{R}$, then $(y, x) \in \mathcal{R}$.

$\mathcal{R}$ **is transitive:** If $(x, y) \in \mathcal{R}$ and $(y, z) \in \mathcal{R}$ then $(x, z) \in \mathcal{R}$.

Check for yourself that the relation described in Example 4.5 is in fact an equivalence relation. The most important property of equivalence relations is that which we indicated just before the definition and which we now enunciate formally:

Proposition 4.7 *Let $\mathcal{R}$ be an equivalence relation on a set A. If $x \in A$, then define*

$$E_x \equiv \{a \in A : (a, y) \in \mathcal{R}\}.$$

We call the sets E_x the equivalence classes induced by the relation $\mathcal{R}$. If now s and t are any two elements of A, then either $E_s \cap E_t = \emptyset$ or $E_s = E_t$.

In summary, the set A is the pairwise disjoint union of the equivalence classes induced by the equivalence relation $\mathcal{R}$.

Before we prove this proposition, let us discuss for a moment what it means. Clearly every element $a \in A$ is contained in some equivalence class, for a is contained in E_a itself. The proposition tells us that the set A is in fact the pairwise disjoint union of these equivalence classes. We say that the equivalence classes *partition* the set A.

For instance, in Example 4.5, the equivalence relation gives rise to two equivalence classes: the even integers $\mathcal{E}$ and the odd integers $\mathcal{O}$. Of course $\mathbb{Z} = \mathcal{E} \cup \mathcal{O}$ and $\mathcal{E} \cap \mathcal{O} = \emptyset$. We say that the equivalence relation *partitions* the universal set $\mathbb{Z}$ into two equivalence classes.

Notice that, in Example 4.5, if we pick any element $x \in \mathcal{E}$, then $E_x = \mathcal{E}$. Likewise, if we pick any element $y \in \mathcal{O}$, then $E_y = \mathcal{O}$.

Proof of the Proposition: Let $s, t \in A$ and suppose that $E_s \cap E_t \neq \emptyset$. It is our job to prove that $E_s = E_t$ (think for a moment about the truth table for "or" so that you understand that we are doing the right thing).

Since $E_s \cap E_t \neq \emptyset$, there is an element $x \in E_s \cap E_t$. Then $x \in E_s$. Therefore, by definition, $(s, x) \in \mathcal{R}$. Likewise, $x \in E_t$. Thus $(t, x) \in \mathcal{R}$. By symmetry, it follows that $(x, t) \in \mathcal{R}$. Now transitivity tells us that, since $(s, x) \in \mathcal{R}$ and $(x, t) \in \mathcal{R}$, then $(s, t) \in \mathcal{R}$.

If y is any element of E_t, then $(t, y) \in \mathcal{R}$. Transitivity now implies that since $(s, t) \in \mathcal{R}$ and $(t, y) \in \mathcal{R}$, then $(s, y) \in \mathcal{R}$. Thus $y \in E_s$. We have shown that every element of E_t is an element of E_s. Thus $E_t \subset E_s$.

Reversing the roles of s and t, we find that $E_s \subset E_t$. It follows that $E_s = E_t$. This is what we wished to prove. □

EXAMPLE 4.8 Let A be the set of all people in the United States. If $x, y \in A$, then let us say that $(x, y) \in \mathcal{R}$ if x and y have the same surname (i.e., last name). Then $\mathcal{R}$ is an equivalence relation:

(i) $\mathcal{R}$ is reflexive since any person x has the same surname as his/her self.

(ii) $\mathcal{R}$ is symmetric since if x has the same surname as y, then y has the same surname as x.

(iii) $\mathcal{R}$ is transitive since if x has the same surname as y and y has the same surname as z, then x has the same surname as z.

Thus $\mathcal{R}$ is an equivalence relation. The equivalence classes are all those people with surname Smith, all those people with surname Herkimer, and so forth. ∎

EXAMPLE 4.9 Let S be the set of all residents of the United States. If $x, y \in S$, then let us say that x is related to y (that is, $x \sim y$) if x and y have at least one biological parent in common. It is easy to see that this relation is reflexive and symmetric. It is *not* transitive, as children of divorced parents know too well. What this tells us (mathematically) is that the proliferation of divorce in our society does *not* lead to well-defined families. ∎

EXAMPLE 4.10 Let S be the set of all residents of the United States. If $x, y \in S$, then let us say that x is related to y (that is, $x \sim y$) if x and y have *both* biological parents in common. It is easy to see that this relation is reflexive and symmetric. It is also transitive, since if A has the same Mom and Dad as B and B has the same Mom and Dad as C, then A, B, C are siblings and A has the same Mom and Dad as C. Contrast this situation with that in the last example.

What this tells us (mathematically) is that traditional families are defined by an equivalence relation. ∎

EXAMPLE 4.11 Let S be the set of integers and say that $x \sim y$ if $x \leq y$. This relation is clearly reflexive. It is *not* symmetric, as $3 \leq 5$ but $5 \not\leq 3$. You may check that it is transitive. But the failure of symmetry tells us that this is not an equivalence relation. ∎

EXAMPLE 4.12 For this example let us use the definition of function that you learned in high school: A function from X to Y is a rule that assigns to each element of X an element of Y.

Let f be a function with domain the real numbers and range the real numbers. We say that two numbers $a, b \in \mathbb{R}$ are related if $f(a) = f(b)$. This relation is clearly reflexive and symmetric. Also, if $f(a) = f(b)$ and $f(b) = f(c)$, then $f(a) = f(c)$. So the relation is also transitive, and it is therefore an equivalence relation. The equivalence classes are called *inverse images of points in the range*. For example, the set of all x such that $f(x) = 5$ is an equivalence class. It is the inverse image of 5. ∎

EXAMPLE 4.13 Let
$$S = \{1, 2, 3, 4, 5, 6, 7, 8, 9, 10\}.$$
Define
$$E_1 = \{1, 4, 7, 10\} \quad E_2 = \{2, 5, 8\} \quad E_3 = \{3, 6, 9\}.$$
Then the sets E_1, E_2, E_3 are pairwise disjoint, and their union is S. So these *could be* the equivalence classes for an equivalence relation, and in fact they are. What is that relation?

Say that $a \sim b$ if $b - a$ is divisible by 3. Check for yourself that this relation is reflexive, symmetric, and transitive. And verify that the equivalence class of 1 is E_1, the equivalence class of 2 is E_2 and the equivalence class of 3 is E_3. ∎

This last example is an instance of a general phenomenon. If a set S is partitioned into subsets (pairwise disjoint sets whose union is S), then those subsets will be the equivalence classes for an equivalence relation. In fact if $X = \cup_j S_j$ and the S_j are pairwise disjoint, then we define $x\mathcal{R}y$ if x and y both lie in the same S_j. We ask you in Exercise 17 to check that this defines an equivalence relation.

POINT OF CONFUSION 4.14 The idea of "equivalence relation" is one of the most universal in mathematics. It is used extensively in algebra and other subjects to make new mathematical constructs. As the examples in this section show, this notion formalizes ideas that we commonly use in everyday social intercourse.

A Look Back

1. Give an example of an equivalence relation on the set $\mathbb{Z}$ that has infinitely many equivalence classes and so that each equivalence class is finite.

2. Explain why relations are important in mathematics.

3. Say that two integers m and n are related if $m - n$ is divisible by 7. Explain why this is an equivalence relation.

4. What is an equivalence class?

Exercises

1. Consider the relation on $\mathbb{Z}$ defined by $(m, n) \in \mathcal{R}$ if $m + n$ is even. Prove that this is an equivalence relation. What are the equivalence classes?

2. Consider the relation on $\mathbb{Z} \times (\mathbb{Z} \setminus \{0\})$ defined by $(m, n)\mathcal{R}(m', n')$ provided that $m \cdot n' = m' \cdot n$. Prove that this is an equivalence relation. Can you describe the equivalence classes?

3. Consider the relation on $\mathbb{Z} \times \mathbb{Z}$ defined by $(m, n)\mathcal{R}(m', n')$ provided that $m + n' = m' + n$. Prove that this is an equivalence relation. Can you describe the equivalence classes? Can you pick a representative for each equivalence class that will help to exhibit what the equivalence relation is?

4. Consider the relation defined on the Cartesian plane by $(x, y)\mathcal{R}(x', y')$ if $y = y'$. Prove that this is an equivalence relation. Can you describe the equivalence classes? Can you pick a representative for each equivalence class that will help to exhibit what the equivalence relation is?

5. Consider the relation defined on the Cartesian plane by $(x, y)\mathcal{R}(x', y')$ if $y - y'$ is an integer and $x - x'$ is an integer. Prove that this is an equivalence relation. Can you describe the equivalence classes? Can you pick a representative for each equivalence class that will help to exhibit what the equivalence relation is?

6. Consider the relation defined on the collection of all circles in the Euclidean plane by $C_1\mathcal{R}C_2$ if the circle C_1 and the circle C_2 have the same center. Prove that this is an equivalence relation. Can you describe the equivalence classes? Can you pick a representative for each equivalence class that will help to exhibit what the equivalence relation is?

7. Let S be the set of all living people. Let $x, y \in S$. Say that x is related to y if x and y have some blood relation in common. Is this an equivalence relation? Why or why not?

8. Consider the relation on $\mathbb{Q} \times (\mathbb{Q} \setminus \{0\})$ defined by $(m, n)\mathcal{R}(m', n')$ provided $m \cdot n' = m' \cdot n$. Prove that this is an equivalence relation. Can you describe the equivalence classes? Why is the outcome in this exercise different from that in Exercise 2?

9. Consider a relation defined on all living people defined by $a\mathcal{R}b$ if a and b are of the same sex and a is strictly younger than b. Is this an equivalence relation?

10. Find the domain and image of each of these relations:

   (a) $\{(x, y) \in \mathbb{R} \times \mathbb{R} : x = \sqrt{y + 3}\}$

   (b) $\{(x, y) \in \mathbb{Q} \times \mathbb{Q} : y = 1/(x^2 - 4)\}$

   (c) $\{(\alpha, \beta) : \alpha \text{ is a person}, \beta \text{ is a person}, \text{and } \alpha \text{ is the father of } \beta\}$

   (d) $\{(\alpha, \beta) : \alpha \text{ is a person}, \beta \text{ is a person}, \text{and } \alpha \text{ is a parent of } \beta\}$

   (e) $\{(x, y) \in \mathbb{Q} \times \mathbb{Q} : x^2 + y^2 < 1\}$

   (f) $\{(x, y) \in \mathbb{N} \times \mathbb{Q} : x \cdot y \text{ is an integer}\}$

   (g) $\{(x, y) \in \mathbb{R} \times \mathbb{R} : x \cdot y \text{ is rational}\}$

   (h) $\{(x, y) \in \mathbb{Z} \times \mathbb{Z} : x - y = 2\}$

11. Declare two real numbers to be related if their difference is rational. Prove that this is an equivalence relation. How many elements are in each equivalence class? How many equivalence classes are there?

12. Formulate a notion of composition of two relations. Formulate a notion of the inverse of a relation. Now express the ideas of reflexivity, symmetry, and transitivity for a relation in the language of inverse and composition of a relation.

13. Let S be the set of all living people. Tell which of the following are equivalence relations on S. Give detailed reasons for your answers.

   (a) x is related to y if x and y are siblings.

   (b) x is related to y if y is presently a spouse of x.

   (c) x is related to y if y has at one time or another been a spouse of x.

   (d) x is related to y if y is a parent of x.

   (e) x is related to y if y is a child of x.

   (f) x is related to y if x hates y but y loves x.

   (g) x is related to y if x hates y and y hates x.

   (h) x is related to y if x and y have a common ancestor.

14. Let S be the collection of all polynomials with real coefficients. Say that $p, q \in S$ are related if the number 0 is a root of $p - q$. Is this an equivalence relation on S?

15. Say that two real numbers x and y are related if there is an integer k such that $k < x \le k+1$ and $k < y \le k+1$. Explain why this is an equivalence relation. Draw a figure that shows the equivalence classes in the real line.

16. Say that two real numbers are related if the first five digits of each of their decimal expansions (the five digits to the right of the decimal point) are equal. After giving a precise formulation of this relation, show that it is an equivalence relation. Give a verbal description of each equivalence class.

17. Prove the last assertion in the section.

4.2 Order Relations

Preliminary Remarks

Among the most interesting and useful relations are order relations. There are several different types of orderings, including partial orderings and total orderings. Well ordering will be of particular note for us.

In this section we discuss the concept of ordering a set. There are many different types of orderings; we shall concentrate on just a few of these.

Definition 4.15 Let S be a set and $\mathcal{R}$ a relation on S. We call $\mathcal{R}$ a *partial ordering* on S if it satisfies the following properties:

(a) For all $x \in S$, $(x, x) \in \mathcal{R}$.

(b) If $x, y \in S$ and both $(x, y) \in \mathcal{R}$ and $(y, x) \in \mathcal{R}$, then $x = y$.

(c) If $x, y, z \in S$ and both $(x, y) \in \mathcal{R}$ and $(y, z) \in \mathcal{R}$, then $(x, z) \in \mathcal{R}$.

Not surprisingly, we refer to property **(a)** as *reflexivity*, to property **(b)** as *anti-symmetry*, and to property **(c)** as *transitivity*.

EXAMPLE 4.16 Let S be the power set of $\mathbb{R}$—the set of all sets of real numbers. If $A, B \in S$, then let us say that $(A, B) \in \mathcal{R}$ if $A \subset B$, that is, if every element of A is also an element of B. Check that the three axioms for a partial ordering are satisfied by our relation $\mathcal{R}$. We usually write this relation in the binary form $A \subset B$. ∎

A noteworthy feature of a *partial ordering* (as opposed to a total ordering, to be discussed later) is that not every two elements of the set S need be comparable. The last example illustrates this point: if $A = \{x \in \mathbb{R} : 1 < x < 4\}$ and $B = \{x \in \mathbb{R} : 2 < x < 9\}$, then both $A, B \in S$ yet neither $(A, B) \in \mathcal{R}$ nor $(B, A) \in \mathcal{R}$.

Definition 4.17 Let S be a set and $\mathcal{R}$ a relation on S. We call $\mathcal{R}$ a *simple ordering* on S if it satisfies the following properties:

(a) If $x, y \in S$ and both $(x, y) \in \mathcal{R}$ and $(y, x) \in \mathcal{R}$, then $x = y$.

(b) If $x, y, z \in S$ and both $(x, y) \in \mathcal{R}$ and $(y, z) \in \mathcal{R}$, then $(x, z) \in \mathcal{R}$.

(c) If $x, y \in S$ are distinct, then either $(x, y) \in \mathcal{R}$ or $(y, x) \in \mathcal{R}$.

Observe that property **(c)** distinguishes a simple ordering from a partial ordering.

As before, we refer to **(a)** and **(b)** as the properties of anti-symmetry and transitivity, respectively. We refer to property **(c)** as *strong connectivity*. A simple ordering is sometimes also called a *total ordering*.

EXAMPLE 4.18 Let $S = \mathbb{R}$, the real numbers. Let us say that $(x, y) \in \mathcal{R}$ if $y - x \geq 0$. It is straightforward to verify that properties (a), (b), (c) of a simple ordering hold for this relation. We usually write this relation in the binary form $x \leq y$. ∎

Definition 4.19 Let us say that a set S is *strictly simply ordered* by a relation $\mathcal{R}$ if properties **(b)** and **(c)** of the last definition hold but property **(a)** is replaced by

(a′) If $x, y \in S$ and $(x, y) \in \mathcal{R}$ then $(y, x) \notin \mathcal{R}$.

EXAMPLE 4.20 The real numbers $S = \mathbb{R}$ are strictly simply ordered by the binary relation $<$. ∎

Suppose that S is a set that is equipped with a strict, simple ordering $\mathcal{R}$. Let $A \subset S$. An element $a \in A$ is called *minimal* (for this ordering) in A if $(a, x) \in \mathcal{R}$ for all $x \in A$, $x \neq a$. We also sometimes call a the *least* element of A. It is clear from this definition that the minimal element is unique if it exists.

EXAMPLE 4.21 Let $S = \mathbb{N}$, the natural numbers. If $m, n \in \mathbb{N}$ then we say that $m\mathcal{R}n$ if $m < n$. This is a strict, simple ordering. Then $T = \{8, 4, 9, 17, 3\}$ is a subset of $\mathbb{N}$, and 3 is the minimal element of T. ∎

Definition 4.22 Let us say that a strict, simple ordering $\mathcal{R}$ *well orders* a set S if each nonempty subset $A \subset S$ has a minimal element.

EXAMPLE 4.23 The usual ordering $<$ well orders the natural numbers. That is to say, each nonempty subset of the natural numbers $\{1, 2, 3, \dots\}$ has a minimal element. This statement is intuitively clear, but proving it quickly leads to deep and difficult questions about the foundations of mathematics.

The ordering $<$ does *not* well order the integers $\mathbb{Z}$, nor does it well order the real numbers $\mathbb{R}$. For example $\mathcal{E}$, the even numbers, is a subset of $\mathbb{Z}$ and is also a subset of $\mathbb{R}$; but $\mathcal{E}$ certainly has no minimal element. The subset $S = \{x \in \mathbb{R} : 0 < x < 1\}$ (with the usual ordering on $\mathbb{R}$) has no minimal element.

In fact, one way to well order the integers is to construct a one-to-one correspondence between the integers and the natural numbers and then to pull the natural ordering from the natural numbers back to the integers by way of this correspondence. This gives a well ordering of the integers, but it is certainly not the standard ordering.

It is impossible to explicitly specify a well ordering for the real numbers $\mathbb{R}$, although such a well ordering *does* exist. In fact absolutely any set can be well ordered (although it is often not at all clear how to actually perform the ordering). This matter is intimately connected with the so-called Axiom of Choice. ∎

POINT OF CONFUSION 4.24 The Axiom of Choice is one of the truly profound and mysterious ideas in all of mathematics. Informally, this axiom of set theory says that, given a set S, there is a function that assigns to each subset an element of that set. The Axiom of Choice was first formulated in 1904 by Ernst Zermelo in order to facilitate his study of the well ordering principle. It is a remarkable fact, for instance, that absolutely any set can be well ordered. The proof of this assertion requires the use of the Axiom of Choice.

A Look Back

1. What is a partial ordering?
2. What is a total ordering?
3. What is a well ordering?
4. Explain why the natural numbers $\mathbb{N}$ are well ordered.

Exercises

1. Consider all ordered triples of positive integers. If $\alpha = (a, b, c)$ and $\alpha' = (a', b', c')$ are two such triples, then we say that $\alpha < \alpha'$ if either

   (a) $a < a'$
   (b) $a = a'$ and $b < b'$

   or

   (c) $a = a'$, $b = b'$, and $c < c'$.

   Discuss, in the language of this section, what type of order relation this is. This ordering is called the *lexicographic ordering*. In view of the way that we order words in a dictionary, explain why the ordering just described deserves that name.

2. Explain how the ordering described in Exercise 1 can be generalized to ordered k-tuples $(a_1, \ldots, a_k)$ of positive integers.

3. Consider the set S of all infinite sequences $\{a_1, a_2, \ldots\}$ of real numbers. Say that two such sequences $\alpha = \{a_1, a_2, \ldots\}$ and $\alpha' = \{a_1', a_2', \ldots\}$ satisfy $\alpha \leq \alpha'$ if the terms of α are eventually less than or equal to the terms of α'. This means that there exists a $K > 0$ such that $a_j \leq a_j'$ for all $j \geq K$. Discuss, in the language of this section, what sort of order relation this is.

4. Construct an onto function $f : \mathbb{R} \to \mathbb{R} \times \mathbb{R}$. Use this function to equip $\mathbb{R}$ with an ordering different from the usual one.

5. We have used the phrase "ordered pair" in this book without giving a precise definition of the phrase. We *could* define the ordered pair (a, b) to be the set $\{\{a\}, \{a, b\}\}$. This is clearly distinct from the ordered pair (b, a), which would be $\{\{b\}, \{b, a\}\} = \{\{b\}, \{a, b\}\}$. Using this technical definition, *prove* that $(a, b) = (a', b')$ if and only if $a = a'$ and $b = b'$.

6. Give a rigorous definition of "ordered triple" based on the ideas in Exercise 5.

7. Consider the following relation on $\mathbb{N}$: $(x, y) \in \mathcal{R}$ if $x < y + 2$. What sort of order relation is $\mathcal{R}$?

8. Consider the following relation on $\mathbb{N}$: $(x, y) \in \mathcal{R}$ if $x < y - 2$. What sort of order relation is $\mathcal{R}$?

4.3 Functions

Preliminary Remarks

As noted, functions are part of the bedrock of modern mathematics. It is essential that we have a rigorous definition of function that allows us to consider functions of a very general sort. That is the topic that we treat in the present section.

In more elementary mathematics courses, we define a function as follows: Let S and T be sets. A function f from S to T is a rule that assigns to each element of S a unique element of T.

This definition is problematic. The main difficulty is the use of the words "rule" and "assign." For instance, let $S = T = \mathbb{Z}$. Consider

$$f(x) = \begin{cases} x^2 & \text{if there is life as we know it on Mars} \\ 3x - 5 & \text{if there is not life as we know it on Mars.} \end{cases}$$

Is this a function? Can what we see on the right be considered a rule? Do we have to wait until we have found life on Mars before we can consider this a function?

More significantly, thinking of a function as a rule is extremely limiting. The functions

$$\begin{aligned} f(x) &= x^3 - 3x + 1 \\ g(x) &= \sin x \\ h(x) &= \frac{\ln x}{x^2 + 4} \end{aligned}$$

are inarguably given by rules. But open up your newspaper and look on the financial page at the graph of the Gross National Product. This is certainly the graph of a function, but what "rule" describes it?

It is best in advanced mathematics to have a way to think about function that avoids subjective words like "rule" and "assign." This is the motivation for our next definition.

Definition 4.25 Let S and T be sets. A *function* f from S to T is a relation on S and T such that

(i) every $s \in S$ is in the domain of f;

(ii) if $(s, t) \in f$ and $(s, u) \in f$ then $t = u$.

Of course we refer to S and T as the domain and the range, respectively, of f. Condition (i) mandates that each element s of S is associated to *some* element of T. Condition (ii) mandates, in a formal manner, that each element s of S is associated to *only one* member of T. Notice, however, that the definition neatly sidesteps the notions of "assign" or "rule." Now look back at our "Mars" definition and decide whether it is a function.

We shall frequently speak of the *image* of a given function f from S to T. This just means the set that is the image of f when it is thought of as a relation. It is the set of elements $t \in T$ such that there is an $s \in S$ with $(s, t) \in f$.

EXAMPLE 4.26 Let $S = \{1, 2, 3\}$ and $T = \{a, b, c\}$. Set

$$f = \{(1, a), (2, a), (3, b)\}.$$

This is a function, for it satisfies the properties set down in Definition 4.25. Given the way that you are accustomed to writing functions in earlier courses, you might find it helpful to view this function as

$$\begin{aligned} f(1) &= a \\ f(2) &= a \\ f(3) &= b. \end{aligned}$$

Notice that each element $1, 2, 3$ of the domain is "assigned" to one an only one element of the range. However, the definition of function allows the possibility that two different elements of the domain be assigned to the same range element. Of course the domain of this function f is the set S. The image is $\{a, b\}$ while the range is $\{a, b, c\}$. We did not use all elements of the range, but that is allowed. ∎

EXAMPLE 4.27 Let $S = \{1, 2, 3\}$ and $T = \{a, b, c, d, e\}$. Set

$$f = \{(1, b), (2, c), (3, e)\}.$$

This is a function, for it satisfies the properties set down in Definition 4.25. Notice that each element of the domain S is used once and only once. However, not all elements of the range are used. According to the definition of function, this is allowed. ∎

EXAMPLE 4.28 Let $S = \{1, 2, 3, 4, 5\}$ and let $T = \{a, b, c\}$. This time there are more elements in S than there are in T. Nonetheless,

$$f = \{(1, a), (2, a), (3, b), (4, b), (5, c)\}$$

is a function. It repeats values, but it definitely satisfies Definition 4.25.   ∎

Definition 4.29 Let f be a function with domain S and range T. We often write such a function as $f : S \to T$. We say that f is *one-to-one* or *injective* if, whenever $(s, t) \in f$ and $(s', t) \in f$, then $s = s'$. We sometimes refer to such a mapping as an *injection*. We also refer to such a map as *injective* or *univalent*.

Compare this new definition with Definition 4.25 of function. The new condition is similar to condition **(ii)** for functions. But it is *not* the same. We are now mandating that no two domain elements be associated with the same range element.

EXAMPLE 4.30 Let $S = T = \mathbb{R}$ and let f be the set of all ordered pairs $\{(x, x^2) : x \in \mathbb{R}\}$. We may also write this function as

$$f : \mathbb{R} \quad \to \quad \mathbb{R}$$
$$x \quad \mapsto \quad x^2$$

or as $f(x) = x^2$.

It is easy to verify that f satisfies the definition of function. However, both of the ordered pairs $(-2, 4)$ and $(2, 4)$ are in f (in other words, $f(-2) = 4 = f(2)$) so that f is *not* one-to-one.   ∎

EXAMPLE 4.31 Let $S = T = \mathbb{R}$ and let f be the function $f(x) = x^3 + x - 5$. Then $f'(x) = 3x^2 + 1 > 0$ for every x. Therefore f is a strictly increasing function. In particular, if $s < t$, then $f(s) < f(t)$ so that $f(s) \neq f(t)$. It follows that the function f *is* one-to-one.   ∎

Definition 4.32 Let f be a function with domain S and range T. If, for each $t \in T$ there is an $s \in S$ such that $f(s) = t$, then we say that f is *onto* or *surjective*. We sometimes refer to such a mapping as a *surjection*. Notice that a function is onto precisely when its image equals its range.

EXAMPLE 4.33 Let $f(x) = x^2$ be the function from Example 4.27. Recall from that example that $S = T = \mathbb{R}$. The point $t = -1 \in T$ has the property that there is no $s \in S$ such that $f(s) = t$. As a result, this function f is *not onto*.   ∎

EXAMPLE 4.34 Let $S = \mathbb{R}$, $T = \{x \in \mathbb{R} : 1 \leq x < \infty\}$. Let $g : S \to T$ be given by $g(x) = x^2 + 1$. Then, for each $t \in T$, the number $s = +\sqrt{t-1}$ makes sense and lies in S. Moreover, $g(s) = t$. It follows that this function g is surjective. However, g is not injective.   ∎

EXAMPLE 4.35 Have another look at the function $f : \mathbb{R} \to \mathbb{R}$ given by $f(x) = x^2$. We have already noted, in Example 4.33, that this function is not onto. But if we restrict the range to $T = \{y \in \mathbb{R} : y \geq 0\}$, so that $f : \mathbb{R} \to T$, then it is easy to verify that the function is now onto. In other words, every nonnegative real number has a square root.   ∎

POINT OF CONFUSION 4.36 As indicated earlier, functions are fundamental tools in the study of modern mathematics. Everything we say and everything we do is formulated in the language of functions. For this reason, the informal high school definition of "function" as a "rule that ..." simply will not do. We need a much more rigorous, and more importantly a more general and flexible, definition of the concept. That is what this section is about.

   Most functions that we encounter in real life are *not* given by formulas and *not* given by rules. We need a mathematical description of the function concept that includes all such functions.

A Look Back

1. What are the two key characteristics of a function?

2. Give an example of a function $f : \mathbb{R} \to \mathbb{N}$.

3. Give an example of a function $f : \mathbb{R} \to \mathbb{R} \times \mathbb{R}$.

4. What is the domain of the function

$$f(x) = \frac{1}{\sqrt{1 - x^2}}?$$

Exercises

1. Let $S = \{a, b, c, d\}$ and $T = \{1, 2, 3, 4, 5, 6, 7\}$. Which of the following relations on $S \times T$ is a function?

   (a) $\{(a, 4), (d, 3), (c, 3), (b, 2)\}$

   (b) $\{(a, 5), (c, 4), (d, 3)\}$

   (c) $\{(a, 1), (b, 1), (c, 1), (d, 1)\}$

   (d) $\{(a, 2), (b, 2), (c, 3), (d, 3)\}$

   (e) $\{(d, 1), (c, 2), (b, 3), (a, 4)\}$

   (f) $\{(d, 7), (c, 6), (c, 5), (a, 4), (b, 2)\}$

   (g) $\{(a, 6), (c, 9)\}$

2. Which of the following functions is one-to-one? Which is onto?

   (a) $f : \mathbb{N} \to \mathbb{N}$ $f(m) = m + 2$
   (b) $g : \mathbb{Z} \to \mathbb{Z}$ $g(m) = 2m - 7$
   (c) $h : \mathbb{R} \to \mathbb{R}$ $h(x) = x - x^3$
   (d) $f : \mathbb{Q} \to \mathbb{Q}$ $f(x) = x^2 + 4x$

(e) $g : \mathbb{N} \to \mathbb{N}$ $g(n) = n(n+1)$
(f) $h : \mathbb{R} \to \mathbb{R}$ $h(n) = +\sqrt{n^2 + 1}$
(g) $f : \mathbb{Z} \to \mathbb{N}$ $f(n) = n^2 + n + 1$
(h) $g : \mathbb{N} \to \mathbb{Z}$ $g(k) = k^3 + 2k$
(i) $h : \mathbb{N} \to \mathbb{Q}$ $h(t) = t/(t+1)$
(j) $f : \mathbb{Q} \to \mathbb{Q}$ $f(y) = y^2 - y$

3. Give an example of a function $f : \mathbb{N} \to \mathbb{Z}$ that is onto.

4. Give an example of a function $f : \mathbb{Q} \to \mathbb{N}$ that is onto.

* 5. Prove that there is no function $g : \mathbb{N} \to \mathbb{R}$ that is onto.

* 6. Let S and T be sets and let $f : S \to T$ and $g : T \to S$ be arbitrary functions. Prove that there is a subset $A \subset S$ and a subset $B \subset T$ such that $f(A) = B$ and $g(T \setminus B) = S \setminus A$.

* 7. Find all functions $f : \mathbb{R} \setminus \{0\} \to \mathbb{R} \setminus \{0\}$ that are one-to-one and onto and such that $f^{-1}(x) = 1/f(x)$.

* 8. Let $f : \mathbb{R} \to \mathbb{R}$ be a function with the property that every x in $\mathbb{R}$ is a local minimum. That is, for $x \in \mathbb{R}$ there is an $\epsilon_x > 0$ so that if $t \in (x - \epsilon_x, x + \epsilon_x)$, then $f(t) \geq f(x)$. Then prove that the image of f is countable.

9. Construct an onto function $f : \mathbb{R} \to \mathbb{R} \times \mathbb{R}$. Use this function to equip $\mathbb{R}$ with a new ordering.

10. Let $A = \{1, 2, 3\}$, $B = \{a, b, c\}$, $C = \{s, t, u\}$. Define functions

$$
\begin{aligned}
f &= \{(1, c), (2, c), (3, a)\} \\
g &= \{(a, t), (b, s), (c, u)\}
\end{aligned}
$$

What are the domain and image of f? What are the domain and image of g? Calculate $g \circ f$ and g^{-1}.

11. Give precise meaning to, and prove, the statement that the intersection of two functions is a function. Is it also the case that the union of two functions is a function?

12. Give an explicit example of a function $f : \mathbb{Q} \to \mathbb{Q}$ such that f is one-to-one and onto but such that $f(x) > x^3$ for every x.

13. Let $f : \mathbb{R} \to \mathbb{R}$ be a function. Suppose that, for every surjective function $g : \mathbb{R} \to \mathbb{R}$, it holds that $g \circ f$ is surjective. Then prove that f is surjective.

14. Let X be a set such that there exists a surjective function $f : X \to \mathbb{Z}$. Then prove that X is infinite.

15. Give an example of an onto function that is not one-to-one. Give an example of a one-to-one function that is not onto.

16. Give an explicit example of a function $f : \mathbb{Q} \to \mathbb{Q}$ such that f is one-to-one and onto but such that $f(x) \neq x$ for every x.

17. Let S and T be sets. We let T^S denote the set of all functions from S to T. For the specific example $S = \{1, 2, 3\}$ and $T = \{a, b\}$, write out the set T^S. Also write out the set S^T.

4.4 Combining Functions

Preliminary Remarks

Part of what makes functions useful is that we can combine them in helpful ways. In the present section we consider arithmetic operations on functions and also functional composition. Closely associated with the latter is the idea of inverse function.

There are several elementary operations that allow us to combine functions in useful ways. In this section, and from now on, we shall (whenever possible) write our functions in the form

$$f(x) = (\text{formula})$$

for the sake of clarity. However, we must keep in mind, and we shall frequently see, that many functions *cannot* be expressed with an elegant formula.

Definition 4.37 Let f and g be functions with the same domain S and the same range T. Assume that T is a set in which the indicated arithmetic operation (below) makes sense. Then we define

(a) $(f + g)(x) = f(x) + g(x)$;

(b) $(f - g)(x) = f(x) - g(x)$;

(c) $(f \cdot g)(x) = f(x) \cdot g(x)$;

(d) $(f/g)(x) = f(x)/g(x)$ provided that $g(x) \neq 0$.

Notice that, in each of **(a)**–**(d)**, we are defining a *new function*—either $f + g$ or $f - g$ or $f \cdot g$ or f/g—in terms of the component functions f and g. For practice, we shall express **(a)** in the language of ordered pairs. We ask you to do likewise with **(b)**, **(c)**, **(d)** in Exercise 9.

Let us consider part **(a)** in detail. Now f is a collection of ordered pairs in $S \times T$ that satisfy the conditions for a function, and so is g. The function $f + g$ is given by

$$f + g = \{(s, t + t') : (s, t) \in f, (s, t') \in g\}.$$

Expressing the other combinations of f and g is quite similar, and you should be sure to do the corresponding Exercise 9.

EXAMPLE 4.38 Let $S = T = \mathbb{R}$. Define

$$f(x) = x^3 - x \qquad \text{and} \qquad g(x) = \sin(x^2).$$

Let us calculate $f + g$, $f - g$, $f \cdot g$, f/g.

Now

$$
\begin{aligned}
(f+g)(x) &= (x^3 - x) + \sin(x^2) \\
(f-g)(x) &= (x^3 - x) - \sin(x^2) \\
(f \cdot g)(x) &= (x^3 - x) \cdot [\sin(x^2)] \\
(f/g)(x) &= (x^3 - x)/[\sin(x^2)] \quad \text{provided } x \neq \pm\sqrt{k\pi}, \\
&\qquad\qquad\qquad\qquad k \in \{0, 1, 2, \ldots\}.
\end{aligned}
$$

∎

A more interesting, and more powerful, way to combine functions is through functional composition. Incidentally, in this discussion we will see the value of good mathematical notation.

Definition 4.39 Let $f : S \to T$ be a function, and let $g : T \to U$ be a function. Then we define, for $s \in S$, the composite function

$$(g \circ f)(s) = g(f(s)). \qquad\qquad (*)$$

We call $g \circ f$ the *composition* of the functions g and f.

Notice in this definition that the right-hand side of $(*)$ always makes sense because of the way that we have specified the domain and range of the component functions f and g. In particular, we must have image $f \subset$ domain g in order for the composition to make sense.

EXAMPLE 4.40 Let $f : \mathbb{R} \to \{x \in \mathbb{R} : x \geq 0\}$ be given by $f(x) = x^4 + x^2 + 6$ and $g : \{x \in \mathbb{R} : x \geq 0\} \to \mathbb{R}$ be given by $g(x) = \sqrt{x} - 4$. Notice that f and g fit the paradigm specified in the definition of composition of function. Then

$$
\begin{aligned}
(g \circ f)(x) &= g(f(x)) \\
&= g(x^4 + x^2 + 6) \\
&= \sqrt{x^4 + x^2 + 6} - 4.
\end{aligned}
$$

Notice that $f \circ g$ also makes sense and is given by

$$
\begin{aligned}
(f \circ g)(x) &= f(g(x)) \\
&= f(\sqrt{x} - 4) \\
&= [\sqrt{x} - 4]^4 + [\sqrt{x} - 4]^2 + 6.
\end{aligned}
$$

It is important to understand that $f \circ g$ and $g \circ f$, when both make sense, will generally be different. ∎

It is a good exercise in the ideas of this chapter to express the notion of functional composition in the language of ordered pairs. Thus let $f : S \to T$ be a function and $g : T \to U$ be a function. Then f is a subset of $S \times T$ and g is a subset of $T \times U$, both satisfying the two standard conditions for function. Now $g \circ f$ is a set of ordered pairs specified by

$$g \circ f = \{(s, u) : s \in S, u \in U, \text{ and } \exists t \in T \text{ such that } (s, t) \in f$$
$$\text{and } (t, u) \in g\}.$$

Take a moment to verify that this equation is consistent with the definition of functional composition that we gave earlier. Further note that $g \circ f$ is a set of ordered pairs from $S \times U$.

EXAMPLE 4.41 Let $f : \mathbb{R} \to \mathbb{R}$ be given by $f(x) = \sin x^5$ and let $g : \{x \in \mathbb{R} : x \geq 1\} \to \mathbb{R}$ be given by $g(x) = \sqrt[4]{x - 1}$. We cannot consider $g \circ f$ because the range of f (namely, the set $[-1, 1]$) does not lie in the domain of g. However, $f \circ g$ *does* make sense because the range of g lies in the domain of f. And

$$(f \circ g)(x) = \sin\left[(x - 1)^{5/4}\right].$$

∎

Definition 4.42 Let S and T be sets. Let $f : S \to T$ and $g : T \to S$. We say that f and g are *mutually inverse* provided that both $(f \circ g)(t) = t$ for all $t \in T$ and $(g \circ f)(s) = s$ for all $s \in S$. We write $g = f^{-1}$ or $f = g^{-1}$. We refer to the functions f and g as *invertible*; we call g the *inverse* of f and f the *inverse* of g.

EXAMPLE 4.43 Let $f : \mathbb{R} \to \mathbb{R}$ be given by $f(x) = x^3 - 1$ and $g : \mathbb{R} \to \mathbb{R}$ be given by $g(x) = \sqrt[3]{x + 1}$. Then

$$
\begin{aligned}
(f \circ g)(x) &= \left[\sqrt[3]{x + 1}\right]^3 - 1 \\
&= (x + 1) - 1 \\
&= x
\end{aligned}
$$

and

$$
\begin{aligned}
(g \circ f)(x) &= \sqrt[3]{(x^3 - 1) + 1} \\
&= \sqrt[3]{x^3} \\
&= x
\end{aligned}
$$

for all x. Thus $g = f^{-1}$ (or $f = g^{-1}$). ∎

The idea of inverse function lends itself particularly well to the notation of ordered pairs. For $f : S \to T$ is inverse to $g : T \to S$ (and vice versa) provided that, for every ordered pair $(s, t) \in f$, there is an ordered pair $(t, s) \in g$ and conversely.

Not every function has an inverse. For instance, let $f : S \to T$. Suppose that $f(s) = t$ and also that $f(s') = t$ with $s \neq s'$ (in other words, suppose that f is not one-to-one). If $g : T \to S$, then $g(f(s)) = g(t) = g(f(s'))$ so it cannot be that both $g(f(s)) = s$ and $g(f(s')) = s'$. In other words, f cannot have an inverse. We conclude that a function that *does* have an inverse must be one-to-one.

On the other hand, suppose that $t \in T$ has the property that there is no $s \in S$ with $f(s) = t$ (in other words, suppose that f is not onto). Then, in particular, it could not be that $f(g(t)) = t$ for any function $g : T \to S$. So f could not be invertible. We conclude that a function that *does* have an inverse must be onto.

POINT OF CONFUSION 4.44 One of the many things that we do with functions is that we combine them. The most elementary means of combining functions are arithmetic operations. But we also compose functions, and that can be a profound operation. In particular, composition allows us to understand the idea of inverse function, and of set-theoretic isomorphism (to be treated below).

EXAMPLE 4.45 Let $f : \mathbb{R} \to \{x \in \mathbb{R} : x \geq 0\}$ be given by $f(x) = x^2$. Then f is onto, but f is not one-to-one. It follows that f cannot have an inverse. And indeed it does not, for any attempt to produce an inverse function runs into the ambiguity that every positive number has two square roots.

Let $f : \{x \in \mathbb{R} : x \geq 0\} \to \mathbb{R}$ be given by $f(x) = x^2$. Then f is one-to-one but f is not onto. There certainly is a function $g : \mathbb{R} \to \{x \in \mathbb{R} : x \geq 0\}$ such that $(g \circ f)(x) = x$ for all $x \in \{x \in \mathbb{R} : x \geq 0\}$ (namely $g(x) = \sqrt{|x|}$). But there is no function $g : \mathbb{R} \to \{x \in \mathbb{R} : x \geq 0\}$ such that $(f \circ g)(x) = x$ for all x.   ∎

We have established that *if $f : S \to T$ has an inverse, then f must be one-to-one and onto.* The converse is true too, and we leave the details for you to verify. A function $f : S \to T$ that is one-to-one and onto (and therefore invertible) is sometimes called a *set-theoretic isomorphism* or a *bijection*.

EXAMPLE 4.46 The function $f : \mathbb{R} \to \mathbb{R}$ that is given by $f(x) = x^3$ is a bijection. You should check the details of this assertion for yourself. The inverse of this function f is the function $g : \mathbb{R} \to \mathbb{R}$ given by $g(x) = x^{1/3}$.   ∎

We leave it as an exercise for you to verify that the composition of two bijections (when the composition makes sense) is a bijection.

A Look Back

1. Explain why we only perform arithmetic operations on two functions when they have the same domain.

2. Explain in words what functional composition means.

3. What is the inverse of the function
$$f(x) = \sqrt[3]{x^5 + 1}?$$

4. What is the reciprocal of the function
$$f(x) = \sqrt[3]{x^5 + 1}?$$

Exercises

1. Let $f(x) = x^2 + 2$, $g(x) = \sin(3x)$, $h(x) = xe^x$ Calculate each of the following:

   (a) $f \circ g(x) + h(x)$

   (b) $f \cdot g(x) + h(x)$

   (c) $f \cdot (g \circ h(x))$

   (d) $(g \circ h(x))/f(x)$

   (e) $(g(x) - h(x))/f(x)$

   (f) $g \circ (f + h)(x)$

   (g) $(g - h) \circ f(x)$

2. Formulate a notion of composition of two relations. Formulate a notion of the inverse of a relation. Now express the ideas of reflexivity, symmetry, and transitivity for a relation in the language of inverse and composition of a relation.

3. A function $f : \mathbb{R} \to \mathbb{R}$ is increasing if, whenever $s \leq t$, then $f(s) \leq f(t)$. Show that if $f : \mathbb{R} \to \mathbb{R}$ and $g : \mathbb{R} \to \mathbb{R}$ are both increasing functions then $f \circ g$ is also increasing.

4. Refer to Exercise 3 for terminology. Note that a function f is decreasing if $-f$ is increasing. Show that, if f is increasing and g is decreasing, then $f + g$ may be neither increasing nor decreasing.

5. Show that the function $f(x) = \sin x$ on the interval $[0, 2\pi]$ can be written as the difference of two increasing functions.

6. Show that the product of two increasing functions need not be increasing.

7. Show that the product of two decreasing functions need not be decreasing.

8. What can you say about the quotient of two increasing functions (where the function in the denominator is never vanishing)?

9. Express parts (b), (c), and (d) of Definition 4.37 using the language of ordered pairs.

4.5 Cantor's Notion of Cardinality

Preliminary Remarks

Certainly one of the truly profound ideas in Cantor's set theory was the idea of cardinality. Cantor found a way to compare sizes of even infinite sets. This has turned out to be an extremely important and useful collection of ideas.

One of the most profound ideas of modern mathematics is Georg Cantor's theory of the infinite (Georg Cantor, 1845–1918). Cantor's insight was that infinite sets can be compared by size, just as finite sets can. For instance, we think of the number 2 as *less* than the number 3; so a set with two elements is "smaller" than a set with three elements. We would like to have a similar notion of comparison for infinite sets. In this section we will present Cantor's ideas; we will also give precise definitions of the terms "finite" and "infinite."

Definition 4.47 Let A and B be sets. We say that A and B have the *same cardinality* if there is a function f from A to B which is both one-to-one and onto (that is, f is a bijection from A to B). We write $\mathrm{card}(A) = \mathrm{card}(B)$.

EXAMPLE 4.48 Let $A = \{1, 2, 3, 4, 5\}$, $B = \{\alpha, \beta, \gamma, \delta, \epsilon\}$, and $C = \{a, b, c, d, e, f\}$. Then A and B have the same cardinality because the function

$$f = \{(1, \alpha), (2, \beta), (3, \gamma), (4, \delta), (5, \epsilon)\}$$

is a bijection of A to B. This function is not the *only* bijection of A to B (can you find another?), but we are only required to produce one.

On the other hand, A and C do not have the same cardinality; neither do B and C.   ∎

Notice that if $\mathrm{card}(A) = \mathrm{card}(B)$ via a function f_1 and $\mathrm{card}(B) = \mathrm{card}(C)$ via a function f_2, then $\mathrm{card}(A) = \mathrm{card}(C)$ via the function $f_2 \circ f_1$.

Definition 4.49 Let A and B be sets. If there is a one-to-one function from A to B but no bijection between A and B, then we will write

$$\mathrm{card}(A) < \mathrm{card}(B).$$

This notation is read "A has smaller cardinality than B."

We use the notation

$$\mathrm{card}(A) \leq \mathrm{card}(B)$$

to mean that either $\mathrm{card}(A) < \mathrm{card}(B)$ or $\mathrm{card}(A) = \mathrm{card}(B)$.

Notice that $\mathrm{card}(A) \leq \mathrm{card}(B)$ and $\mathrm{card}(B) \leq \mathrm{card}(C)$ imply that $\mathrm{card}(A) \leq \mathrm{card}(C)$. Moreover, if $A \subset B$, then the inclusion map $i(a) = a$ is a one-to-one function of A into B; therefore $\mathrm{card}(A) \leq \mathrm{card}(B)$.

EXAMPLE 4.50 Let $A = \{1, 2, 3\}$ and $B = \{2, 4, 6, 8, 10\}$. Then the function

$$f(x) = 2x$$

is a one-to-one function from A to B. There is no one-to-one function from B to A (why not?). So we may write

$$\mathrm{card}(A) < \mathrm{card}(B).$$

Now let S be the integers $\mathbb{Z}$ and let T be the rational numbers $\mathbb{Q}$. Certainly the function

$$h(x) = x$$

is one-to-one from S to T. It is not clear (but see below) whether there is a one-to-one function from T to S. We may in any event write

$$\mathrm{card}(S) \leq \mathrm{card}(T).$$

∎

The next theorem gives a useful method for comparing the cardinality of two sets.

Theorem 4.51 (Schroeder-Bernstein) *Let A, B, be sets. If there is a one-to-one function $f : A \to B$ and a one-to-one function $g : B \to A$, then A and B have the same cardinality.*

Remark 4.52 This remarkable theorem says that, if we can find an injection from A to B and an injection from B to A, then in fact there exists a single map that is a *bijection* of A to B. Observe that the two different injections may be completely unrelated. Often it is much easier to construct two separate injections than it is to construct a single bijection.

Proof (optional): It is convenient to assume that A and B are disjoint; we may arrange this if necessary by replacing A by $\{(a, 0) : a \in A\}$ and B by $\{(b, 1) : b \in B\}$. Let D be the image of f and let C be the image of g. Let us define a *chain* to be a sequence of elements of either A or B—that is, a function $\phi : \mathbb{N} \to (A \cup B)$—such that

- $\phi(1) \in B \setminus D$;

- if, for some j, we have $\phi(j) \in B$, then $\phi(j+1) = g(\phi(j))$;

- if, for some j, we have $\phi(j) \in A$, then $\phi(j+1) = f(\phi(j))$.

We see that a chain is a sequence of elements of $A \cup B$ such that the first element is in $B \setminus D$, the second in A, the third in B, and so on. Obviously each element of $B \setminus D$ occurs as the first element of at least one chain.

Define $\mathcal{S} = \{a \in A : a$ is some term of some chain$\}$. It is helpful to note that

$$\mathcal{S} = \{x \in A : x \text{ can be written in the form}$$
$$x = g(f(g(\cdots f(g(y))))) \text{ for some } y \in B \setminus D\}. \tag{4.51.1}$$

Observe that $\mathcal{S} \subset C$.

We set

$$k(x) = \begin{cases} f(x) & \text{if } x \in A \setminus \mathcal{S} \\ g^{-1}(x) & \text{if } x \in \mathcal{S}. \end{cases}$$

Note that the second half of this definition makes sense because $\mathcal{S} \subset C$ and because g is one-to-one. Then $k : A \to B$. We shall show that in fact k is a bijection.

First notice that f and g^{-1} are one-to-one. This is not quite enough to show that k is one-to-one, but we now reason as follows: If $f(x_1) = g^{-1}(x_2)$ for some $x_1 \in A \setminus \mathcal{S}$ and some $x_2 \in \mathcal{S}$, then $x_2 = g(f(x_1))$. But, by (4.51.1), the fact that $x_2 \in \mathcal{S}$ now implies that $x_1 \in \mathcal{S}$. That is a contradiction. Hence k is one-to-one.

It remains to show that k is onto. Fix $b \in B$. We seek an $x \in A$ such that $k(x) = b$.

Case A: If $g(b) \in \mathcal{S}$, then $k(g(b)) \equiv g^{-1}(g(b)) = b$ hence the x that we seek is $g(b)$.

Case B: If $g(b) \notin \mathcal{S}$, then we claim that there is an $x \in A$ such that $f(x) = b$. Assume this claim for the moment.

Now the x that we just found must lie in $A \backslash \mathcal{S}$. For if not, then x would be in some chain. Then $f(x)$ and $g(f(x))$ would also lie in that chain. Hence $g(b) \in \mathcal{S}$, and that is a contradiction. But $x \in A \backslash \mathcal{S}$ tells us that $k(x) = f(x) = b$. That completes the proof that k is onto. Hence k is a bijection.

To prove the claim that we made in Case B, notice that if there is no $x \in A$ with $f(x) = b$ then $b \in B \backslash D$. Thus some chain would begin at b. So $g(b)$ would be a term of that chain. Hence $g(b) \in \mathcal{S}$ and that is a contradiction.

The proof of the Schroeder-Bernstein theorem is complete. $\qquad\qquad\square$

In what follows, we will consistently use some important and universally recognized terminology. An infinite set S is said to be *countable* if it has the same cardinality as the natural number $\mathbb{N}$. If an infinite set is *not* countable, that is if it does *not* have a bijection with the natural numbers $\mathbb{N}$, then it is said to be *uncountable*. Every infinite set is either countable or uncountable. One of our big jobs in this section of the book is to learn to recognize countable and uncountable sets.

Now it is time to look at some specific examples.

EXAMPLE 4.53 Let $\mathcal{E}$ be the set of all even integers and $\mathcal{O}$ the set of all odd integers. Then

$$\mathrm{card}(\mathcal{E}) = \mathrm{card}(\mathcal{O}).$$

Indeed, the function

$$f(j) = j + 1$$

is a bijection from $\mathcal{E}$ to $\mathcal{O}$. ∎

EXAMPLE 4.54 Let $\mathcal{E}$ be the set of even integers. Then

$$\mathrm{card}(\mathcal{E}) = \mathrm{card}(\mathbb{Z}).$$

The function

$$g(j) = j/2$$

gives the bijection. Thus $\mathrm{card}(\mathcal{E}) = \mathrm{card}(\mathbb{Z})$. ∎

This last example is a bit surprising, for it shows that a set (namely, $\mathbb{Z}$, the integers) can be put in one-to-one correspondence with a proper subset (namely $\mathcal{E}$, the even integers) of itself. This phenomenon is impossible for finite sets.

EXAMPLE 4.55 We have
$$\text{card}(\mathbb{Z}) = \text{card}(\mathbb{N}).$$
We define the function f from $\mathbb{Z}$ to $\mathbb{N}$ as follows:

- $f(j) = -(2j + 1)$ if j is negative;

- $f(j) = 2j + 2$ if j is positive or zero.

The values that f takes on the negative integers are $1, 3, 5, \ldots$, on the positive integers are $4, 6, 8, \ldots$, and $f(0) = 2$. Thus f is one-to-one and onto. ∎

By putting together the preceding examples, we see that the set of even integers, the set of odd integers, and the set of all integers are countable sets.

EXAMPLE 4.56 The set of all ordered pairs of positive integers
$$S = \mathbb{N} \times \mathbb{N} = \{(j, k) : j, k \in \mathbb{N}\}$$
is countable.

To see this, we will use the Schroeder-Bernstein theorem. The function
$$f(j) = (j, 1)$$
is a one-to-one function from $\mathbb{N}$ to S. Also, the function
$$g(j, k) = j \cdot 10^{j+k} + k$$
is a function from S to $\mathbb{N}$. Let n be the number of digits in the number k. Notice that $g(j, k)$ is obtained by writing the digits of j, followed by $j + k - n$ zeros, then followed by the digits of k. For instance,
$$g(23, 714) = 23 \underbrace{000 \ldots 000}_{734} 714,$$
where there are $23 + 714 - 3 = 734$ zeros between the 3 and the 7. It is clear that g is one-to-one. By the Schroeder-Bernstein theorem, S and $\mathbb{N}$ have the same cardinality; hence S is countable. ∎

There are other ways to handle the last example, and we shall explore them in the exercises.

Since there is a bijection f of the set of *all* integers $\mathbb{Z}$ with the set $\mathbb{N}$, it follows from the last example that the set $\mathbb{Z} \times \mathbb{Z}$ of all pairs of integers (positive *and* negative) is countable. Indeed the map $(f \times f)(x, y) = (f(x), f(y))$ is a bijection of $\mathbb{Z} \times \mathbb{Z}$ to $\mathbb{N} \times \mathbb{N}$. Let h be the bijection, provided by Example 4.56, from $\mathbb{N} \times \mathbb{N}$ to $\mathbb{N}$. Then $h \circ (f \times f)$ is a bijection of $\mathbb{Z} \times \mathbb{Z}$ to $\mathbb{N}$.

Notice that the word "countable" is a good descriptive word: if S is a countable set, then we can think of S as having a first element s_1 (the one corresponding to $1 \in \mathbb{N}$), a second element s_2 (the one corresponding to $2 \in \mathbb{N}$), and so forth. Thus we write $S = \{s_1, s_2, \ldots\}$.

Definition 4.57 A set S is called *finite* if it is either empty or else there is a bijection of S with a set of the form $I_n \equiv \{1, 2, \ldots, n\}$ for some positive integer n. If S is not empty and if no such bijection exists, then the set is called *infinite*.

Remark 4.58 The empty set is a finite set, but not one of any particular interest. Nevertheless we must account for it, so we include it explicitly in the definition of "finite set."

In some treatments, a different approach is taken to the concepts of "finite" and "infinite" sets. In fact, one defines an infinite set to be one which can be put in one-to-one correspondence with a proper subset of itself. For instance, Example 4.54 shows that the set $\mathbb{Z}$ of all integers can be put in one-to-one correspondence with the set $\mathcal{E}$ of all even integers (and of course $\mathcal{E}$ is a proper subset of $\mathbb{Z}$). By contrast, a finite set *cannot* be put in one-to-one correspondence with a proper subset of itself. This last assertion follows from the pigeonhole principle.

An important property of the natural numbers $\mathbb{N}$ is that any subset $S \subset \mathbb{N}$ has a least element. See the discussion in Section 4.2. This is known as the Well-Ordering Principle, and is studied in a course on logic. In the present chapter, we take the properties of the natural numbers as given (see the Appendix for more on the natural numbers). We use some of these properties in the next proposition.

Proposition 4.59 *If S is a countable set and R is a subset of S, then either R is empty or R is finite or R is countable.*

Proof: Assume that R is not empty.

Write $S = \{s_1, s_2, \ldots\}$. Let j_1 be the least positive integer such that $s_{j_1} \in R$. Let j_2 be the least integer following j_1 such that $s_{j_2} \in R$. Continue in this fashion. If the process terminates at the nth step, then R is finite and has n elements.

If the process does not terminate, then we obtain an enumeration of the elements of R :

$$1 \longleftrightarrow s_{j_1}$$
$$2 \longleftrightarrow s_{j_2}$$
$$\cdots$$

etc.

All elements of R are enumerated in this fashion since $j_\ell \geq \ell$. Therefore R is countable. $\qquad \square$

A set is called *countable* if it is countably infinite, that is, if it can be put in one-to-one correspondence with the natural numbers $\mathbb{N}$. A set is called *denumerable* if it is either empty or finite or countable. In actual practice, mathematicians use the word "countable" to describe sets that are either finite or countable. In other words, they use the word "countable" interchangeably with the word "denumerable."

The set $\mathbb{Q}$ of all rational numbers consists of all expressions

$$\frac{a}{b},$$

where a and b are integers and $b \neq 0$. Thus $\mathbb{Q}$ can be identified with the set of all pairs (a, b) of integers. After discarding duplicates, such as $\frac{2}{4} = \frac{1}{2}$, and using the discussion following Example 4.56 to the effect that $\mathbb{Z} \times \mathbb{Z}$ is countable, we find that the set $\mathbb{Q}$ is countable. We deal with the rational number system in the Appendix.

Theorem 4.60 *Let S_1, S_2 be countable sets. Set $S = S_1 \cup S_2$. Then S is countable.*

Proof: Let us write

$$S_1 = \{s_1^1, s_2^1, \ldots\}$$
$$S_2 = \{s_1^2, s_2^2, \ldots\}.$$

If $S_1 \cap S_2 = \emptyset$, then the function

$$s_j^k \mapsto (j, k)$$

is a bijection of S with a subset of $\{(j, k) : j, k \in \mathbb{N}\}$. We proved earlier (Example 4.56) that the set of ordered pairs of elements of $\mathbb{N}$ is countable. By Proposition 4.59, S is countable as well.

If there exist elements which are common to S_1, S_2, then discard any duplicates. The same argument (use Example 4.56 and Proposition 4.59) shows that S is countable. $\square$

Proposition 4.61 *If S and T are each countable sets, then so is*

$$S \times T = \{(s, t) : s \in S, t \in T\}.$$

Proof: Since S is countable, there is a bijection f from S to $\mathbb{N}$. Likewise there is a bijection g from T to $\mathbb{N}$. Therefore the function

$$(f \times g)(s, t) = (f(s), g(t))$$

is a bijection of $S \times T$ with $\mathbb{N} \times \mathbb{N}$, the set of ordered pairs of positive integers. But we saw in Example 4.56 that the latter is a countable set. Hence so is $S \times T$. $\square$

Corollary 4.62 *If $S_1, S_2, \ldots, S_k$ are each countable sets, then so is the set*

$$S_1 \times S_2 \times \cdots \times S_k = \{(s_1, \ldots, s_k) : s_1 \in S_1, \ldots, s_k \in S_k\}$$

consisting of all ordered k-tuples $(s_1, s_2, \ldots, s_k)$ with $s_j \in S_j$.

Proof: We may think of $S_1 \times S_2 \times S_3$ as $(S_1 \times S_2) \times S_3$. Since $S_1 \times S_2$ is countable (by the Proposition) and S_3 is countable, then so is $(S_1 \times S_2) \times S_3 = S_1 \times S_2 \times S_3$ countable. Continuing in this fashion (i.e., inductively), we can see that any finite product of countable sets is also a countable set.   □

Corollary 4.63 *The countable union of countable sets is countable.*

Proof: Let $A_1, A_2, \ldots$ each be countable sets. If the elements of A_j are enumerated as $\{a_k^j\}_{k=1}^\infty$ and if the sets A_j are pairwise disjoint, then the correspondence

$$a_k^j \longleftrightarrow (j, k)$$

is one-to-one between the union of the sets A_j and the countable set $\mathbb{N} \times \mathbb{N}$. This proves the result when the sets A_j have no common element. If some of the A_j have elements in common, then we discard duplicates in the union and use Proposition 4.56.   □

Proposition 4.64 *The collection $\mathcal{P}$ of all polynomials $p(x)$ with integer coefficients is countable.*

Proof: Let $\mathcal{P}_k$ be the set of polynomials of degree k with integer coefficients. A polynomial p of degree k having integer coefficients has the form

$$p(x) = p_0 + p_1 x + p_2 x^2 + \cdots + p_k x^k,$$

where the p_j are integer constants. The identification

$$p(x) \longleftrightarrow (p_0, p_1, \ldots, p_k)$$

identifies the elements of $\mathcal{P}_k$ with the $(k+1)$-tuples of integers. By Corollary 4.62, it follows that $\mathcal{P}_k$ is countable. But then Corollary 4.63 implies that

$$\mathcal{P} = \bigcup_{j=0}^{\infty} \mathcal{P}_j$$

is countable.   □

Definition 4.65 Let x be a real number. We say that x is *algebraic* if there is a polynomial p with integer coefficients such that $p(x) = 0$.

EXAMPLE 4.66 The number $\sqrt{2}$ is algebraic because it satisfies the polynomial equation $x^2 - 2 = 0$. The number $\sqrt{3} + \sqrt{2}$ is also algebraic. This assertion is less obvious, but in fact the number satisfies the polynomial equation $x^4 - x^2 + 1 = 0$. The numbers π and e are *not* algebraic, but this assertion is extremely difficult to prove. We say that π and e are *transcendental.* In the next proposition, we give an elegant method for showing that "most" real numbers are transcendental without actually saying what any of them are. ∎

Georg Cantor's remarkable discovery is that *not all infinite sets are countable.* We next give an example of this phenomenon.

In what follows, a *sequence* on a set S is a function from $\mathbb{N}$ to S. We usually write such a sequence as $s(1), s(2), s(3), \ldots$ or $s_1, s_2, s_3, \ldots$.

EXAMPLE 4.67 There exists an infinite set which is not countable (we call such a set *uncountable*). Our example will be the set S of all sequences on the set $\{0, 1\}$. In other words, S is the set of all infinite sequences of 0's and 1's.

To see that S is uncountable, assume the contrary—that is, we assume that S is countable. Then there is a first sequence

$$\mathcal{S}^1 = \{s_j^1\}_{j=1}^\infty,$$

a second sequence

$$\mathcal{S}^2 = \{s_j^2\}_{j=1}^\infty,$$

and so forth. This will be a complete enumeration of all the members of S. But now consider the sequence $\mathcal{T} = \{t_j\}_{j=1}^\infty$, which we construct as follows:

- If $s_1^1 = 0$ then set $t_1 = 1$; if $s_1^1 = 1$ then set $t_1 = 0$;

- If $s_2^2 = 0$ then set $t_2 = 1$; if $s_2^2 = 1$ then set $t_2 = 0$;

- If $s_3^3 = 0$ then set $t_3 = 1$; if $s_3^3 = 1$ then set $t_3 = 0$;

$$\cdots$$

- If $s_j^j = 0$ then set $t_j = 1$; if $s_j^j = 1$ then set $t_j = 0$;

etc.

Now the sequence $\mathcal{T} = \{t_j\}$ differs from the first sequence $\mathcal{S}^1$ in the first element: $t_1 \neq s_1^1$.

The sequence $\mathcal{T}$ differs from the second sequence $\mathcal{S}^2$ in the second element: $t_2 \neq s_2^2$.

And so on: the sequence $\mathcal{T}$ differs from the jth sequence $\mathcal{S}^j$ in the jth element: $t_j \neq s_j^j$. So the sequence $\mathcal{T}$ is not in the set S. But $\mathcal{T}$ is *supposed* to be in the set S because it is a sequence of 0's and 1's and all of these are supposed to have been enumerated in our enumeration of S.

This contradicts our assumption, so S must be uncountable. ∎

EXAMPLE 4.68 Consider the set of all decimal representations of numbers strictly between 0 and 1—both terminating and nonterminating. Here a terminating decimal is one of the form

$$0.43926$$

while a nonterminating decimal is one of the form

$$0.14159265\ldots.$$

In the case of the nonterminating decimal, no repetition is implied; the decimal simply continues without cease.

Now the set of all those decimals containing only the digits 0 and 1 can be identified in a natural way with the set of sequences containing only 0 and 1 (just put commas between the digits). And we just saw that the set of such sequences is uncountable.

Since the set of all decimal numbers is an even bigger set, it must be uncountable also. [Put a different way, if the set of all decimal numbers *were* countable, then any of its infinite subsets would be countable—that is the content of Proposition 4.59. Thus the collection of decimal numbers containing only the digits 0 and 1 would be countable, and that is a contradiction.]

As you may know, the set of all decimals identifies with the set of all real numbers. [Many real numbers have two decimal representations—one terminating and one not. Think for a moment about which numbers these are, and why this observation does not invalidate the present discussion.] We find then that the set $\mathbb{R}$ of all real numbers is uncountable. (Contrast this with the situation for the rationals.) In Chapter 5 we will learn more about how the real number system is constructed using just elementary set theory. ∎

Proposition 4.69 *The set of all algebraic real numbers is countable. The set of all transcendental numbers is uncountable.*

Proof: Let $\mathcal{P}$ be the collection of all polynomials with integer coefficients. We have already noted in Proposition 4.64 that $\mathcal{P}$ is a countable set. If $p \in \mathcal{P}$ then let s_p denote the set of real roots of p. Of course s_p is finite, and the number of elements in s_p does not exceed the degree of p. Then the set A of algebraic real numbers may be written as

$$A = \cup_{p \in \mathcal{P}} \, s_p \, .$$

This is the countable union of finite sets, so of course it is countable

Now that we know that the set of algebraic numbers is countable, we can notice that the set T of transcendental numbers must be uncountable. For $\mathbb{R} = A \cup T$. If T were countable then, since A is countable, it would follow that $\mathbb{R}$ is countable. But that is not so. Hence the set T of transcendental numbers is uncountable. □

POINT OF CONFUSION 4.70 Proposition 4.69 is one of Georg Cantor's great triumphs. It is *extremely* difficult to prove that any particular real number (such as π or e) is transcendental. But Cantor proved (by a rather elementary argument that we reproduce here) that *most* real numbers are transcendental. The world had never seen a result like this before. Cantor cannot tell us *which particular* real numbers are transcendental. But he can show that there are a lot of them!

Our last result in this section is a counterpoint to Proposition 4.59 and the discussion leading up to it.

Proposition 4.71 *Let S be any infinite set. Then S has a subset T that is countable.*

Proof: Let $t_1 \in S$ be any element. Now let $t_2 \in S$ be any element that is distinct from t_1. Continue this procedure. It will not terminate, because that would imply that S is finite. And it will produce a countable set T that is a subset of S. $\square$

To repeat the main point of this section, the natural numbers have a cardinality that we call *countable*, and the real numbers have a cardinality that we call *uncountable*. These cardinalities are distinct. In fact the real numbers form a larger set because there is an injective mapping of the natural numbers into the reals but not the other way around. We refer to the cardinality of the natural numbers as "countable" and to that of the real numbers as "the cardinality of the continuum."

It is natural to ask whether there is a set with cardinality strictly between countable and the continuum. Georg Cantor posed this question one hundred years ago, and his failed attempts to resolve the question tormented his final years. Paul Cohen, using very original techniques, solved the problem in 1963.

It is an important result of set theory (due to Cantor) that, given any set S, the set of all subsets of S (called the *power set* of S) has strictly greater cardinality than the set S itself. As a simple example, let $S = \{a, b, c\}$. Then the set of all subsets of S is

$$\{\emptyset, \{a\}, \{b\}, \{c\}, \{a, b\}, \{a, c\}, \{b, c\}, \{a, b, c\}\}.$$

The set of all subsets has eight elements while the original set has three.

We stress that this result is true not just for finite sets but also for infinite sets: if S is an infinite set then the set of all its subsets (the power set) has greater cardinality than S itself. Thus there are infinite sets of arbitrarily large cardinality. In other words, there is no "greatest" cardinal. This fact is so important that we now formulate it as a theorem.

Theorem 4.72 (Cantor) *Let S be any set. Then the power set $\mathcal{P}(S)$, consisting of all subsets of S, has cardinality greater than the cardinality of S. In other words,*

$$\mathrm{card}(S) < \mathrm{card}(\mathcal{P}(S)).$$

Proof: First observe that the function

$$f : S \longrightarrow \mathcal{P}(S)$$
$$s \longmapsto \{s\}$$

is one-to-one. Thus we see that $\mathrm{card}(S) \leq \mathrm{card}(\mathcal{P}(S))$. We need to show that there is no function from S *onto* $\mathcal{P}(S)$. Let $g : S \to \mathcal{P}(S)$. We will produce an element of $\mathcal{P}(S)$ that cannot be in the image of this mapping.

Define $T = \{s \in S : s \notin g(s)\}$. Assume, seeking a contradiction, that $T = g(z)$ for some $z \in S$. By definition of T, the element $z \in T$ if and only if $z \notin g(z)$; thus $z \in T$ if and only if $z \notin T$. That is a contradiction. We see that g cannot map S onto $\mathcal{P}(S)$, therefore $\mathrm{card}(S) < \mathrm{card}(\mathcal{P}(S))$. $\square$

In some of the examples in this section, we constructed a bijection between a given set (such as $\mathbb{Z}$) and a proper subset of that set (such as $\mathcal{E}$, the even integers). It follows from the definitions that this is possible only when the sets involved are infinite. In fact any infinite set can be placed in a set-theoretic isomorphism with a proper subset of itself. We explore this assertion in the exercises.

Put in other words, we have come upon an intrinsic characterization of infinite sets. We state it (without proof) as a proposition:

Proposition 4.73 *Let S be a set. The set S is infinite if and only if it can be put in one-to-one correspondence with a proper subset of itself.*

POINT OF CONFUSION 4.74 This last proposition gives us a truly new way to think about infinite sets. You probably have some intuitive notions of what an infinite set is. But this particular characterization is counterintuitive. And it is profound and important.

The pigeonhole principle guarantees that no finite set can be put in one-to-one correspondence with a proper subset of itself. Cantor taught us that infinite sets are different. For most of the nineteenth century, infinite sets were a forbidden topic among serious mathematicians (because their consideration so frequently led to paradoxes and confusion). Cantor showed us the way out of this thicket, and clarified the entire picture.

Exercise 9 asks you for a proof of Proposition 4.73.

A Look Back

1. What is the cardinality of a set?

2. Do $\mathbb{Z}$ and $\mathbb{N} \times \mathbb{N}$ have the same cardinality?

3. Do $\mathbb{R}$ and $\mathbb{C} \times \mathbb{C}$ have the same cardinality?

4. Give an example of an uncountable set.

Exercises

1. What is the cardinality of each of the following sets?

    (a) $\mathbb{N} \times \mathbb{Q}$

    (b) $\mathbb{N} \times \mathbb{N}$

    (c) $\mathbb{R} \times \mathbb{Q}$

    (d) $\mathcal{P}(\mathbb{Q})$

    (e) $\mathbb{C}$

    (f) $\mathbb{R} \setminus \mathbb{N}$

    (g) $\mathbb{Q} \setminus \mathbb{N}$

    (h) The set of all decimal expansions, terminating or nonterminating, that include only the digits 3 and 7.

    (i) The set of all *terminating* decimal expansions that include only the digits 3 and 7.

    (j) The set of all solutions of all quadratic polynomials with integer coefficients.

    (k) The set of all solutions of all quadratic polynomials with real coefficients.

    (l) The set of all subsets of $\mathbb{N}$ that have at least three and not more than eight elements.

    (m) The set of all subsets of $\mathbb{Z}$ with at least six elements.

2. Let A and B be sets. Let us say that A and B are related if there exists a set theoretic isomorphism from A to B. Prove that this is an equivalence relation. Each equivalence class is called a *cardinal number*.

3. It is a theorem of Cantor that the cardinality of the power set of a given set S is always greater than the cardinality of S. Discuss this result for $S = \{1, 2, 3\}$, for $S = \mathbb{Z}$, and for $S = \mathbb{R}$.

4. We know that the reals $\mathbb{R}$ and the complex numbers $\mathbb{C}$ are both uncountable sets. Give three other examples of uncountable set.

5. Give an explicit example of a set with cardinality that is greater than the cardinality of $\mathbb{R}$.

6. What is the cardinality of the set of polynomials of degree two?

7. What is the cardinality of the set of all functions from $\mathbb{R}$ to $\mathbb{R}$?

8. What is the cardinality of the set of all functions from $\mathbb{Z}$ to $\mathbb{Z}$?

9. Give a proof of Proposition 4.73. [**Hint:** We know that a countable set (such as the integers) is set-theoretically isomorphic to a proper subset of itself. And any infinite set contains a copy of the integers.]

Chapter 5

Essential Number Systems

5.1 The Real Numbers

Preliminary Remarks

Of course real analysis is about the real numbers. The real numbers are a very subtle number system with many deep and mysterious properties. In order to lay a proper foundation for our study of real analysis, we must first say precisely what the real numbers are and what their key properties are. That is our goal in this section.

This is a book about analysis in the real number system. Such a study must be founded on a careful consideration of *what the real numbers are* and *how they are constructed*. In the present section we give a careful treatment of the real number system. In the next we consider the complex numbers.

We know from calculus that, for many purposes, the rational numbers are inadequate. It is important to work in a number system which is closed with respect to the operations we shall perform—meaning that when you perform the operations you stay inside the set. For example, the integers are closed under addition because, when you add two integers, the answer you get is another integer. But the rational numbers are *not* closed under the operation of taking the square root, just because the square root of 2 (the number $\sqrt{2}$) is *not* rational (see Section 2.3).

This idea of closure includes limiting operations. While the rationals are closed under the usual arithmetic operations, they are not closed under the mathematical operation of *taking limits*. For instance, the sequence of rational numbers $3, 3.1, 3.14, 3.141, \ldots$ consists of terms that seem to be getting closer and closer together, *seem* to tend to some limit, and yet there is no rational number which will serve as a limit (of course it turns out that the limit is π—an "irrational" number).

We will now deal with the real number system, a system which contains all limits of sequences of rational numbers (as well as all limits of sequences of real numbers!). In fact our plan will be as follows: in this section we shall discuss all the requisite properties of the reals. The actual construction of the reals is rather complicated, and we shall put that in an optional Appendix to Section 5.1.

Definition 5.1 Let A be an ordered set with ordering $\leq$ and X a subset of A. The set X is called *bounded above* if there is an element $b \in A$ such that $x \leq b$ for all $x \in X$. We call the element b an *upper bound* for the set X.

EXAMPLE 5.2 Let $A = \mathbb{Q}$ (the rational numbers) with the usual ordering. The set $X = \{x \in \mathbb{Q} : 2 < x < 4\}$ is bounded above. For example, 15 is an upper bound for X. So are the numbers 12 and 4. It is interesting to observe that no element of this particular X can actually be an upper bound for X. The number 4 is a good candidate, but 4 is not an element of X. In fact if $x \in X$, then $(x + 4)/2 \in X$ and $x < (x + 4)/2 < 4$, so x could not be an upper bound for X. We think of 4 as a "least upper bound" for X (see below for a more thorough treatment of this idea). ∎

It turns out that the most convenient way to formulate the notion that the real numbers have "no holes" (i.e., that all sequences which seem to be converging actually have something to converge to) is in terms of upper bounds.

Definition 5.3 Let A be an ordered set and X a subset of A. An element $b \in A$ is called a *least upper bound* (or *supremum*) for X if b is an upper bound for X and $b \leq b'$ for every upper bound b' for X. We denote the least upper bound of X by lub X. The least upper bound is also sometimes called the *supremum* and denoted by sup X.

We shall see soon that the characterizing property of the real numbers is that every set with an upper bound has a least upper bound.

POINT OF CONFUSION 5.4 By its very definition, if a least upper bound exists then it is unique. The least upper bound is a tricky idea. Consider the set

$$\{x \in \mathbb{Q} : x^2 < 2\}.$$

This set is clearly bounded above—by 2 for example. So it has a least upper bound *in the real number system*. It does *not* have a least upper bound in the rational number system. Of course that least upper bound is $\sqrt{2}$.

EXAMPLE 5.5 In the last example, we considered the set X of rational numbers strictly between 2 and 4. We observed there that 4 is the least upper bound for X. Note that this least upper bound is not an element of the set X.

The set $Y = \{y \in \mathbb{Q} : -9 \leq y \leq 7\}$ has least upper bound 7. In this case, the least upper bound *is* an element of the set Y. ∎

Notice that we may define a lower bound for a subset of an ordered set in a fashion similar to that for an upper bound: $\ell \in A$ is a lower bound for $X \subset A$ if $\ell \leq x$ for all $x \in X$. An element $\ell \in A$ is called a *greatest lower bound* for X if ℓ is a lower bound for X and $\ell' \leq \ell$ for every lower bound ℓ' for X. We denote the greatest lower bound of X by glb X. The greatest lower bound is also sometimes called the *infimum* and denoted by inf X.

EXAMPLE 5.6 The set $X = \{x \in \mathbb{Q} : 2 < x < 4\}$ in the last two examples has lower bounds $-20, 0, 1, 2$, for instance. The greatest lower bound is 2, which is *not* an element of the set.

The set $Y = \{y \in \mathbb{Q} : -9 \leq y \leq 7\}$ in the last example has lower bounds—among others—given by $-53, -22, -10, -9$. The number -9 is the greatest lower bound. It *is* an element of Y. ∎

The purpose that the real numbers will serve for us is as follows: they will contain the rationals, they will still be an ordered field (that is, a set with operations of multiplication and addition and ordering which have reasonable properties such as commutativity and associativity—see the Appendix to this section for the details. Refer to [KRA1] for a thorough treatment of the concept of ordered field.). Also *every subset which has an upper bound will have a least upper bound*. We formulate this result as a theorem.

Theorem 5.7 *There exists an ordered field $\mathbb{R}$ which* **(i)** *contains $\mathbb{Q}$ and* **(ii)** *has the property that any nonempty subset of $\mathbb{R}$ which has an upper bound has a least upper bound in the number system $\mathbb{R}$.*

The last property described in this theorem is called the Least Upper Bound Property of the real numbers. As mentioned previously, this theorem will be proved in the Appendix to Section 5.1. Now we begin to realize why it is so important to *construct* the number systems that we will use. We are endowing $\mathbb{R}$ with a great many properties. Why do we have any right to suppose that there exists a set with all these properties? We must produce one! We do so in the Appendix to Section 5.1.

Let us begin to explore the richness of the real numbers. The next theorem states a property which is certainly not shared by the rationals. It is fundamental in its importance.

Theorem 5.8 *Let x be a real number such that $x > 0$. Then there is a positive real number y such that $y^2 = y \cdot y = x$.*

Proof: We will use throughout this proof the fact that if $0 < a < b$ then $a^2 < b^2$.
Let
$$S = \{s \in \mathbb{R} : s > 0 \text{ and } s^2 < x\}.$$
Then S is not empty since $x/2 \in S$ if $x < 2$ and $1 \in S$ otherwise. Also S is bounded above since $x + 1$ is an upper bound for S. By Theorem 5.7, the set S has a least upper bound. Call it y. Obviously, $0 < \min\{x/2, 1\} \leq y$ hence

y is positive. We claim that $y^2 = x$. To see this, we eliminate the other two possibilities.

If $y^2 < x$ then set $\epsilon = (x - y^2)/[4(x+1)]$. Then $\epsilon > 0$ and

$$
\begin{aligned}
(y + \epsilon)^2 &= y^2 + 2 \cdot y \cdot \epsilon + \epsilon^2 \\
&= y^2 + 2 \cdot y \cdot \frac{x - y^2}{4(x+1)} + \frac{x - y^2}{4(x+1)} \cdot \frac{x - y^2}{4(x+1)} \\
&< y^2 + 2 \cdot y \cdot \frac{x - y^2}{4y} + \frac{x - y^2}{4(x+1)} \cdot \frac{x - y^2}{4(x+1)} \\
&< y^2 + \frac{x - y^2}{2} + \frac{x - y^2}{4} \cdot \frac{x}{4x} \\
&< y^2 + (x - y^2) \\
&= x.
\end{aligned}
$$

Thus $y + \epsilon \in S$, and y cannot be an upper bound for S. This contradiction tells us that $y^2 \not< x$.

Similarly, if it were the case that $y^2 > x$ then we set $\epsilon = (y^2 - x)/[4(x+1)]$. A calculation like the one we just did (see Exercise 2) then shows that $(y - \epsilon)^2 \geq x$. Hence $y - \epsilon$ is also an upper bound for S, and y is therefore not the *least* upper bound. This contradiction shows that $y^2 \not> x$.

The only remaining possibility is that $y^2 = x$. $\qquad\qquad\square$

POINT OF CONFUSION 5.9 This last proof was fairly tricky. But it proves a very important fact—that every positive real number has a square root. This is in stark contrast to the situation for the rationals. We know, thanks to Pythagoras (see the details in Section 2.3), that the rational number 2 does *not* have a square root *in the rationals*. But it certainly has a square root in the reals.

A similar proof shows that, if n is a positive integer and x a positive real number, then there is a positive real number y such that $y^n = x$. Exercise 15 asks you to provide the details.

We next use the Least Upper Bound Property of the Real Numbers to establish two important qualitative properties of the real numbers:

Theorem 5.10 *The set $\mathbb{R}$ of real numbers satisfies the Archimedean Property:*

> *Let a and b be positive real numbers. Then there is a natural number n such that $na > b$.*

The set $\mathbb{Q}$ of rational numbers satisfies the following Density Property:

> *Let $c < d$ be real numbers. Then there is a rational number q with $c < q < d$.*

Proof: Suppose the Archimedean Property to be false. Then $S = \{na : n \in \mathbb{N}\}$ has b as an upper bound. Therefore S has a finite supremum β. Since $a > 0$, it follows that $\beta - a < \beta$. So $\beta - a$ is not an upper bound for S, and there must be a natural number n' such that $n' \cdot a > \beta - a$. But then $(n' + 1)a > \beta$, and β cannot be the supremum for S. This contradiction proves the first assertion.

For the second property, we assume for simplicity that $d > c > 0$. Let $\lambda = d - c > 0$. By the Archimedean Property, choose a positive integer N such that $N \cdot \lambda > 1$. Again the Archimedean Property gives a natural number P such that $P > N \cdot c$ and another Q such that $Q > -N \cdot c$. Thus we see that Nc falls between the integers $-Q$ and P; therefore there must be an integer M between $-Q$ and P such that

$$M - 1 \leq Nc < M.$$

Thus $c < M/N$. Also

$$M \leq Nc + 1 \quad \text{hence} \quad \frac{M}{N} \leq c + \frac{1}{N} < c + \lambda = d.$$

So M/N is a rational number lying between c and d. $\square$

In Section 4.5 we established that the set of all decimal representations of numbers is uncountable. It follows that the set of all real numbers is uncountable. In fact the same proof shows that the set of all real numbers in the interval $(0, 1)$, or in any nonempty open interval (c, d), is uncountable.

POINT OF CONFUSION 5.11 Between every pair of distinct irrational numbers there is a rational number. And between every pair of distinct rational numbers there is an irrational number. But there are many more irrationals than rationals. So this is a very strange and complicated situation. The rationals and irrationals do *not* alternate. They are arranged in a complex and nonobvious fashion.

The set $\mathbb{R}$ of real numbers is uncountable, yet the set $\mathbb{Q}$ of rational numbers is countable. It follows that the set $\mathbb{R} \setminus \mathbb{Q}$ of *irrational* numbers is uncountable. In particular, it is nonempty. Thus we may see with very little effort that there exist a great many real numbers which cannot be expressed as a quotient of integers. However, it can be quite difficult to see whether any particular real number (such as π or e or $\sqrt[5]{2}$) is irrational.

We conclude by recalling the "absolute value" notation:

Definition 5.12 Let x be a real number. We define

$$|x| = \begin{cases} x & \text{if} \quad x > 0, \\ 0 & \text{if} \quad x = 0, \\ -x & \text{if} \quad x < 0. \end{cases}$$

It is left as an exercise for you to verify the important *triangle inequality*:

$$|x + y| \le |x| + |y|.$$

[**Hint:** It is convenient to verify that the square of the left-hand side is less than or equal to the square of the right-hand side.]

Optional APPENDIX: Construction of the Real Numbers

There are several techniques for constructing the real number system $\mathbb{R}$ from the rational number system $\mathbb{Q}$. We use the method of Dedekind (Julius W. R. Dedekind, 1831–1916) cuts because it uses a minimum of new ideas and is fairly brief.

The number system that we shall be constructing is an instance of a *field* (the complex numbers, in the next section, also form a field). The definition is as follows:

Definition 5.13 A set F is called a *field* if it is equipped with a binary operation (usually called addition and denoted "+") and a second binary operation (called multiplication and denoted "·") such that the following axioms are satisfied. (Here A stands for "addition," M stands for "multiplication," and D stands for "distributive law.")

A1. F is closed under addition: if $x, y \in F$ then $x + y \in F$.

A2. Addition is commutative: if $x, y \in F$ then $x + y = y + x$.

A3. Addition is associative: if $x, y, z \in F$ then $x + (y + z) = (x + y) + z$.

A4. There exists an element, called 0, in F which is an additive identity: if $x \in F$ then $0 + x = x$.

A5. Each element of F has an additive inverse: if $x \in F$ then there is an element $-x \in F$ such that $x + (-x) = 0$.

M1. F is closed under multiplication: if $x, y \in F$ then $x \cdot y \in F$.

M2. Multiplication is commutative: if $x, y \in F$ then $x \cdot y = y \cdot x$.

M3. Multiplication is associative: if $x, y, z \in F$ then $x \cdot (y \cdot z) = (x \cdot y) \cdot z$.

M4. There exists an element, called 1, which is a multiplicative identity: if $x \in F$ then $x \cdot 1 = x$.

M5. Each nonzero element of F has a multiplicative inverse: if $0 \ne x \in F$ then there is an element $x^{-1} \in F$ such that $x \cdot (x^{-1}) = 1$. The element x^{-1} is sometimes denoted $1/x$.

D1. Multiplication distributes over addition: if $x, y, z \in F$ then

$$x \cdot (y + z) = x \cdot y + x \cdot z \,.$$

Definition 5.14 A *cut* is a subset $\mathcal{P}$ of $\mathbb{Q}$ with the following properties:

- $\mathcal{P} \neq \emptyset$.

- If $s \in \mathcal{P}$ and $t < s$ then $t \in \mathcal{P}$.

- If $s \in \mathcal{P}$ then there is a $u \in \mathcal{P}$ such that $u > s$.

- There is a rational number x such that $c < x$ for all $c \in \mathcal{P}$.

You should think of a cut $\mathcal{P}$ as the set of all rational numbers to the left of some point in the real line. Since we have not constructed the real line yet, we cannot define a cut in that simple way; we have to make the construction more indirect. But if you consider the four properties of a cut, they describe a set that looks like a "rational halfline."

Notice that, if $\mathcal{P}$ is a cut and the point $s \notin \mathcal{P}$, then any rational $t > s$ is also not in $\mathcal{P}$. Also, if $r \in \mathcal{P}$ and $s \notin \mathcal{P}$ then it must be that $r < s$.

Definition 5.15 If $\mathcal{P}$ and $\mathcal{Q}$ are cuts then we say that $\mathcal{P} < \mathcal{Q}$ provided that $\mathcal{P}$ is a subset of $\mathcal{Q}$ but $\mathcal{P} \neq \mathcal{Q}$.

Check for yourself that "$<$" is an ordering on the set of all cuts.

Now we introduce operations of addition and multiplication, which will turn the set of all cuts into a field.

Definition 5.16 If $\mathcal{P}$ and $\mathcal{Q}$ are cuts then we define

$$\mathcal{P} + \mathcal{Q} = \{c + d : c \in \mathcal{P}, d \in \mathcal{Q}\}.$$

We define the cut $\widehat{0}$ to be the set of all negative rationals.

The cut $\widehat{0}$ will play the role of the additive identity. We are now required to check that field axioms A1–A5 hold.

For A1, we need to see that $\mathcal{P} + \mathcal{Q}$ is a cut. Obviously $\mathcal{P} + \mathcal{Q}$ is not empty. If s is an element of $\mathcal{P} + \mathcal{Q}$ and t is a rational number less than s, write $s = c + d$, where $c \in \mathcal{P}$ and $d \in \mathcal{Q}$. Then $t - c < s - c = d \in \mathcal{Q}$ so $t - c \in \mathcal{Q}$; and $c \in \mathcal{P}$. Hence $t = c + (t - c) \in \mathcal{P} + \mathcal{Q}$. A similar argument shows that there is an $r > s$ such that $r \in \mathcal{P} + \mathcal{Q}$. Finally, if x is a rational upper bound for $\mathcal{P}$ and y is a rational upper bound for $\mathcal{Q}$, then $x + y$ is a rational upper bound for $\mathcal{P} + \mathcal{Q}$. We conclude that $\mathcal{P} + \mathcal{Q}$ is a cut.

Since addition of rational numbers is commutative, it follows immediately that addition of cuts is commutative. Associativity follows in a similar fashion. Next we show that, if $\mathcal{P}$ is a cut, then $\mathcal{P} + \widehat{0} = \mathcal{P}$. For if $c \in \mathcal{P}$ and $z \in \widehat{0}$ then $c + z < c + 0 = c$ hence $\mathcal{P} + \widehat{0} \subset \mathcal{P}$. Also, if $c' \in \mathcal{P}$ then choose a $d' \in \mathcal{P}$

such that $c' < d'$. Then $c' - d' < 0$ so $c' - d' \in \widehat{0}$. And $c' = d' + (c' - d')$. Hence $\mathcal{P} \subset \mathcal{P} + \widehat{0}$. We conclude that $\mathcal{P} + \widehat{0} = \mathcal{P}$.

Finally, for Axiom A5, we let $\mathcal{P}$ be a cut and set $-\mathcal{P}$ to be equal to $\{d \in \mathbb{Q} : c + d < 0$ for all $c \in \mathcal{P}\}$. If x is a rational upper bound for $\mathcal{P}$ and $c \in \mathcal{P}$ then $-x \in -\mathcal{P}$ so $-\mathcal{P}$ is not empty. By its very definition, $\mathcal{P} + (-\mathcal{P}) \subset \widehat{0}$. Further, if $z \in \widehat{0}$ and $c \in \mathcal{P}$ we set $c' = z - c$. Then $c' \in -\mathcal{P}$ and $z = c + c'$. Hence $\widehat{0} \subset \mathcal{P} + (-\mathcal{P})$. We conclude that $\mathcal{P} + (-\mathcal{P}) = \widehat{0}$.

Having verified the axioms for addition, we turn now to multiplication.

Definition 5.17 If $\mathcal{P}$ and $\mathcal{Q}$ are cuts, then we define the product $\mathcal{P} \cdot \mathcal{Q}$ as follows:

- If $\mathcal{P}, \mathcal{Q} > \widehat{0}$, then $\mathcal{P} \cdot \mathcal{Q} = \{q \in \mathbb{Q} : q < c \cdot d$ for some $c \in \mathcal{P}, d \in \mathcal{Q}$ with $c > 0, d > 0 \}$.

- If $\mathcal{P} > \widehat{0}, \mathcal{Q} < \widehat{0}$, then $\mathcal{P} \cdot \mathcal{Q} = -(\mathcal{P} \cdot (-\mathcal{Q}))$.

- If $\mathcal{P} < \widehat{0}, \mathcal{Q} > \widehat{0}$, then $\mathcal{P} \cdot \mathcal{Q} = -((-\mathcal{P}) \cdot \mathcal{Q})$.

- If $\mathcal{P}, \mathcal{Q} < \widehat{0}$, then $\mathcal{P} \cdot \mathcal{Q} = (-\mathcal{P}) \cdot (-\mathcal{Q})$.

- If either $\mathcal{P} = \widehat{0}$ or $\mathcal{Q} = \widehat{0}$, then $\mathcal{P} \cdot \mathcal{Q} = \widehat{0}$.

Notice that, for convenience, we have defined multiplication of negative numbers just as we did in high school. The reason is that the definition we use for the product of two positive numbers cannot work when one of the two factors is negative (exercise).

It is now a routine exercise to verify that the set of all cuts, with this definition of multiplication, satisfies field axioms M1–M5. The proofs follow those for A1–A5 rather closely.

For the distributive property, one first checks the case when all the cuts are positive, reducing it to the distributive property for the rationals. Then one handles negative cuts on a case-by-case basis.

We now know that the collection of all cuts forms an ordered field. Denote this field by the symbol $\mathbb{R}$. We next verify the crucial property of $\mathbb{R}$ that sets it apart from $\mathbb{Q}$:

Theorem 5.18 The ordered field $\mathbb{R}$ satisfies the Least Upper Bound Property.

Proof: Let S be a subset of $\mathbb{R}$ which is bounded above. Define

$$\mathcal{S}^* = \bigcup_{\mathcal{P} \in S} \mathcal{P}.$$

Then $\mathcal{S}^*$ is clearly nonempty, and it is therefore a cut since it is a union of cuts. It is also clearly an upper bound for S since it contains each element of S. It remains to check that $\mathcal{S}^*$ is the least upper bound for S.

In fact if $\mathcal{T} < \mathcal{S}^*$ then $\mathcal{T} \subset \mathcal{S}^*$ and there is a rational number q in $\mathcal{S}^* \setminus \mathcal{T}$. But, by the definition of $\mathcal{S}^*$, it must be that $q \in \mathcal{P}$ for some $\mathcal{P} \in S$. So $\mathcal{P} > \mathcal{T}$, and $\mathcal{T}$ cannot be an upper bound for S. Therefore $\mathcal{S}^*$ is the least upper bound for S, as desired. $\qquad\qquad\qquad\qquad\qquad\qquad\qquad\qquad\qquad\qquad$ □

We have shown that $\mathbb{R}$ is an ordered field which satisfies the Least Upper Bound Property. It remains to show that $\mathbb{R}$ contains (a copy of) $\mathbb{Q}$ in a natural way. In fact, if $q \in \mathbb{Q}$ we associate to it the element $\varphi(q) = \mathcal{P}_q \equiv \{x \in \mathbb{Q} : x < q\}$. Then $\mathcal{P}_q$ is obviously a cut. It is also routine to check that

$$\varphi(q + q') = \varphi(q) + \varphi(q') \quad \text{and} \quad \varphi(q \cdot q') = \varphi(q) \cdot \varphi(q').$$

Therefore we see that φ represents $\mathbb{Q}$ as a subfield of $\mathbb{R}$.

A Look Back

1. What is a least upper bound?

2. Does the least upper bound of a set X necessarily lie in X?

3. What is the defining property of the real numbers?

4. How do the real numbers differ from the rational numbers?

Exercises

1. Let A be a set of real numbers that is bounded above and set $\alpha = \sup A$. Let $B = \{-a : a \in A\}$. Prove that $\inf B = -\alpha$. Prove the same result with the roles of infimum and supremum reversed.

* 2. Complete the calculation in the proof of Theorem 5.8.

3. What is the least upper bound of the set

$$S = \{x : x^2 < 2\}?$$

Explain why this question has a sensible answer in the real number system but not in the rational number system.

4. Prove that the least upper bound and greatest lower bound for a set of real numbers are unique.

5. Consider the unit circle C. Let

$$S = \{\alpha : 2\alpha < \text{(the circumference of } C)\}.$$

Show that S is bounded above. Let p be the least upper bound of S. Say explicitly what the number p is. This exercise works in the real number system, but not in the rational number system. Why?

6. Give an example of a set that contains its least upper bound but not its greatest lower bound. Give an example of a set that contains its greatest lower bound but not its least upper bound.

7. Give an example of a set of real numbers that does *not* have a least upper bound. Give an example of a set of real numbers that does *not* have a greatest lower bound.

8. Prove the triangle inequality.

9. Prove that addition of the real numbers (as constructed in the Appendix to Section 5.1) is commutative. Now prove that it is associative.

10. Let $\emptyset$ be the empty set (see Section 3.2). Prove that $\sup \emptyset = -\infty$ and $\inf \emptyset = +\infty$.

11. Use the triangle inequality to prove that $|a - b| \leq |a| + |b|$ for any real numbers a and b.

12. Use the triangle inequality to prove that $|a| - |b| \leq |a - b|$ for any real numbers a and b.

* **13.** Let f be a function with domain the reals and range the reals. Assume that f has a local minimum at each point x in its domain. (This means that, for each $x \in \mathbb{R}$, there is an $\epsilon > 0$ such that whenever $|x - t| < \epsilon$ then $f(x) \leq f(t)$). *Do not assume that f is differentiable, or continuous, or anything nice like that.* Prove that the image of f is countable. (**Hint:** When this author solved this problem as a student, his solution was ten pages long; however, there is a one-line solution due to Michael Spivak.)

* **14.** Let λ be a positive irrational real number. If n is a positive integer, choose by the Archimedean Property an integer k such that $k\lambda \leq n < (k + 1)\lambda$. Let $\varphi(n) = n - k\lambda$. Prove that the set of all $\varphi(n)$ is dense in the interval $[0, \lambda]$. [By this we mean that the $\varphi(n)$ get arbitrarily close to each element of $[0, \lambda]$.] (**Hint:** Examine the proof of the density of the rationals in the reals.)

* **15.** Let n be a natural number and x a positive real number. Prove that there is a positive real number y such that $y^n = x$. Is y unique?

5.2   The Complex Numbers

Preliminary Remarks

In this book we do not emphasize the complex numbers. But we do use them occasionally, so this section is provided for your reference. Be sure that, at the least, you understand the basic arithmetic operations on $\mathbb{C}$ and also the special role of the complex number i.

When we first learn about the complex numbers, the most troublesome point is the very beginning: "Let's pretend that the number -1 has a square root. Call it i." What gives us the right to "pretend" in this fashion? The answer is that we have no such right.[1] If -1 has a square root, then we should be able to construct a number system in which that is the case. That is what we shall do in this section.

---

[1]The complex numbers were initially developed so that we would have a number system in which all polynomial equations are solvable. One of the reasons, historically, that mathematicians had trouble accepting the complex numbers is that they did not believe that they really existed—they were just made up. This is, in part, how they came to be called "imaginary." Mathematicians had similar trouble accepting negative numbers; for a time, negative numbers were called "forbidden."

Definition 5.19 The system of *complex numbers*, denoted by the symbol $\mathbb{C}$, consists of all ordered pairs (a, b) of real numbers. We add two complex numbers (a, b) and $(\tilde{a}, \tilde{b})$ by the formula

$$(a, b) + (\tilde{a}, \tilde{b}) = (a + \tilde{a}, b + \tilde{b}).$$

We multiply two complex numbers by the formula

$$(a, b) \cdot (\tilde{a}, \tilde{b}) = (a \cdot \tilde{a} - b \cdot \tilde{b}, a \cdot \tilde{b} + \tilde{a} \cdot b).$$

POINT OF CONFUSION 5.20 If you are puzzled by this definition of multiplication, do not worry. In a few moments you will see that it gives rise to the notion of multiplication of complex numbers that you are accustomed to. Perhaps more importantly, a naive rule for multiplication like $(a, b) \cdot (\tilde{a}, \tilde{b}) = (a\tilde{a}, b\tilde{b})$ gives rise to nonsense like $(1, 0) \cdot (0, 1) = (0, 0)$. It is really necessary for us to use the initially counterintuitive definition of multiplication that is presented here.

EXAMPLE 5.21 Let $z = (3, -2)$ and $w = (4, 7)$ be two complex numbers. Then

$$z + w = (3, -2) + (4, 7) = (3 + 4, -2 + 7) = (7, 5).$$

Also

$$z \cdot w = (3, -2) \cdot (4, 7) = (3 \cdot 4 - (-2) \cdot 7, 3 \cdot 7 + 4 \cdot (-2)) = (26, 13). \qquad \blacksquare$$

As usual, we ought to check that addition and multiplication are commutative and associative, that multiplication distributes over addition, and so forth. We shall leave these tasks to the exercises. Instead we develop some of the crucial, and more interesting, properties of our new number system.

Theorem 5.22 *The following properties hold for the number system $\mathbb{C}$.*

(a) *The number $1 \equiv (1, 0)$ is the multiplicative identity: $1 \cdot z = z$ for any $z \in \mathbb{C}$.*

(b) *The number $0 \equiv (0, 0)$ is the additive identity: $0 + z = z$ for any $z \in \mathbb{C}$.*

(c) *Each complex number $z = (x, y)$ has an additive inverse $-z = (-x, -y)$: it holds that $z + (-z) = 0$.*

(d) *The number $i \equiv (0, 1)$ satisfies $i \cdot i = -1$; in other words, i is a square root of -1.*

Proof: These are direct calculations, but it is important for us to work out these facts.

First, let $z = (x, y)$ be any complex number. Then

$$1 \cdot z = (1, 0) \cdot (x, y) = (1 \cdot x - 0 \cdot y, 1 \cdot y + x \cdot 0) = (x, y) = z.$$

This proves the first assertion.

For the second, we have

$$0 + z = (0,0) + (x,y) = (0+x, 0+y) = (x,y) = z\,.$$

With z as above, set $-z = (-x, -y)$. Then

$$z + (-z) = (x,y) + (-x,-y) = (x+(-x), y+(-y)) = (0,0) = 0\,.$$

Finally, we calculate

$$i \cdot i = (0,1) \cdot (0,1) = (0 \cdot 0 - 1 \cdot 1, 0 \cdot 1 + 0 \cdot 1) = (-1, 0) = -1\,.$$

Thus, as asserted, i is a square root of -1.   □

POINT OF CONFUSION 5.23 The model for the real numbers is the set of all Dedekind cuts. This is a non-obvious construction that guarantees that there really is a number system that satisfies the important and nontrivial properties of the reals.

By contrast, the complex numbers are very easy to construct. The complex number system is simply the two-dimensional Cartesian plane equipped with some interesting algebraic operations.

Proposition 5.24 *If $z \in \mathbb{C}, z \neq 0$, then there is a complex number w such that $z \cdot w = 1$. We call w the multiplicative inverse or sometimes the reciprocal.*

Proof: Write $z = (x,y)$ and set

$$w = \left(\frac{x}{x^2 + y^2}, \frac{-y}{x^2 + y^2} \right)\,.$$

Since $z \neq 0$, this definition makes sense. Then it is straightforward to verify that $z \cdot w = 1$.   □

We see in this last proposition that w is the multiplicative inverse of z. It is often convenient to write w as

$$w = \frac{\overline{z}}{|z|^2}\,.$$

Here $\overline{z} = x - iy$ and $|z|^2 = x^2 + y^2$. We shall say more about these concepts in what follows.

Thus every nonzero complex number has a multiplicative inverse. The other field axioms for $\mathbb{C}$ are easy to check. We conclude that the number system $\mathbb{C}$ forms a field. You will prove in the exercises that it is not possible to order this field. If α is a real number then we associate α with the complex number $(\alpha, 0)$. Thus we have the natural "embedding"

$$\mathbb{R} \ni \alpha \longmapsto (\alpha, 0) \in \mathbb{C}\,.$$

In this way, we can think of the real numbers as a *subset* of the complex numbers. In fact, the real field $\mathbb{R}$ is a *subfield* of the complex field $\mathbb{C}$. This means that if $\alpha, \beta \in \mathbb{R}$ and $(\alpha, 0), (\beta, 0)$ are the corresponding elements in $\mathbb{C}$ then $\alpha + \beta$ corresponds to $(\alpha + \beta, 0)$ and $\alpha \cdot \beta$ corresponds to $(\alpha, 0) \cdot (\beta, 0)$. These assertions are explored more thoroughly in the exercises.

With the remarks in the preceding paragraph we can sometimes ignore the distinction between the real numbers and the complex numbers. For example, we can write

$$5 \cdot i$$

and understand that it means $(5, 0) \cdot (0, 1) = (0, 5)$. Likewise, the expression

$$5 \cdot 1$$

can be interpreted as $5 \cdot 1 = 5$ or as $(5, 0) \cdot (1, 0) = (5, 0)$ without any danger of ambiguity.

Theorem 5.25 *Every complex number can be written in the form* $a + b \cdot i$, *where* a *and* b *are real numbers. In fact, if* $z = (x, y) \in \mathbb{C}$ *then*

$$z = x + y \cdot i.$$

Proof: With the identification of real numbers as a subfield of the complex numbers, we have that

$$x + y \cdot i = (x, 0) + (y, 0) \cdot (0, 1) = (x, 0) + (0, y) = (x, y) = z$$

as claimed. □

Now that we have constructed the complex number field, we will adhere to the usual custom of writing complex numbers as $z = a + b \cdot i$ or, more simply, $a + bi$. We call a the *real part* of z, denoted by Re z, and b the *imaginary part* of z, denoted Im z. We have

$$(a + bi) + (\widetilde{a} + \widetilde{b}i) = (a + \widetilde{a}) + (b + \widetilde{b})i$$

and

$$(a + bi) \cdot (\widetilde{a} + \widetilde{b}i) = (a \cdot \widetilde{a} - b \cdot \widetilde{b}) + (a \cdot \widetilde{b} + \widetilde{a} \cdot b)i.$$

If $z = a + bi$ is a complex number then we define its *complex conjugate* to be the number $\overline{z} = a - bi$. We record some elementary facts about the complex conjugate:

Proposition 5.26 *If* z, w *are complex numbers then*

(1) $\overline{z + w} = \overline{z} + \overline{w}$;

(2) $\overline{z \cdot w} = \overline{z} \cdot \overline{w}$;

(3) $z + \overline{z} = 2 \cdot \text{Re } z$;

(4) $z - \overline{z} = 2 \cdot i \cdot \operatorname{Im} z$;

(5) $z \cdot \overline{z} \geq 0$, with equality holding if and only if $z = 0$.

Proof: Write $z = a + bi, w = c + di$. Then

$$
\begin{aligned}
\overline{z + w} &= \overline{(a + c) + (b + d)i} \\
&= (a + c) - (b + d)i \\
&= (a - bi) + (c - di) \\
&= \overline{z} + \overline{w}.
\end{aligned}
$$

This proves (1). Assertions (2), (3), (4) are proved similarly.
For (5), notice that

$$z \cdot \overline{z} = (a + bi) \cdot (a - bi) = a^2 + b^2 \geq 0.$$

Clearly equality holds if and only if $a = b = 0$.   □

POINT OF CONFUSION 5.27 The concept of complex conjugate was invented for the following reason. If p is a polynomial with real coefficients, and if α is a complex root of p (so $p(\alpha) = 0$), then $\overline{\alpha}$ is also a complex root of p (so $p(\overline{\alpha}) = 0$). It is an exercise for you to confirm this statement.

The expression $|z|$ is defined to be the nonnegative square root of $z \cdot \overline{z}$:

$$|z| = +\sqrt{z \cdot \overline{z}} = \sqrt{x^2 + y^2}$$

when $z = x + iy$. It is called the *modulus* of z and plays the same role for the complex field that absolute value plays for the real field. It is the distance of z to the origin. The modulus has the following properties.

Proposition 5.28 If $z, w \in \mathbb{C}$, then

(1) $|z| = |\overline{z}|$;

(2) $|z \cdot w| = |z| \cdot |w|$;

(3) $|\operatorname{Re} z| \leq |z|$, $|\operatorname{Im} z| \leq |z|$;

(4) $|z + w| \leq |z| + |w|$.

Proof: Write $z = a + bi, w = c + di$. Then (1), (2), (3) are immediate. For (4) we calculate that

$$
\begin{aligned}
|z + w|^2 &= (z + w) \cdot (\overline{z + w}) \\
&= z \cdot \overline{z} + z \cdot \overline{w} + w \cdot \overline{z} + w \cdot \overline{w} \\
&= |z|^2 + 2\operatorname{Re}(z \cdot \overline{w}) + |w|^2 \\
&\leq |z|^2 + 2|z \cdot \overline{w}| + |w|^2 \\
&= |z|^2 + 2|z| \cdot |w| + |w|^2 \\
&= (|z| + |w|)^2.
\end{aligned}
$$

Taking square roots proves (4). □

Observe that, if z is real, then $z = a + 0i$ and the modulus of z equals the absolute value of a. Likewise, if $z = 0 + bi$ is pure imaginary, then the modulus of z equals the absolute value of b. In particular, the fourth part of the proposition reduces, in the real case, to the triangle inequality

$$|a + b| \leq |a| + |b|.$$

If z is any nonzero complex number, then let $r = |z|$. Now define $\xi = z/r$. Certainly ξ is a complex number of modulus 1. Thus ξ lies on the unit circle, so it subtends an angle θ with the positive x-axis. Certainly then $\xi = \cos\theta + i\sin\theta$. It is shown in Section 13.3 that

$$e^{i\theta} = \xi = \cos\theta + i\sin\theta.$$

[**Hint:** You may verify this formula for yourself by writing out the power series for the exponential and writing out the power series for cosine and sine.] As a result, we may write

$$z = re^{i\theta}.$$

We conclude this discussion by recording the most important basic fact about the complex numbers. Carl Friedrich Gauss (1777–1855) gave five proofs of this theorem (the Fundamental Theorem of Algebra) in his doctoral dissertation:

Theorem 5.29 *Let $p(z)$ be any polynomial of degree at least 1. Then p has a root $\alpha \in \mathbb{C}$ such that $p(\alpha) = 0$.*

POINT OF CONFUSION 5.30 Using a little algebra, one can in fact show that a polynomial of degree k has k roots (counting multiplicity).

For example, the polynomial $p(z) = z^4 - 2z^3 + 2z^2 - 2z + 1$ has roots $i, -i, 1$, and 1. Put in other words,

$$p(z) = (z - i)(z + i)(z - 1)(z - 1).$$

A Look Back

1. What is the definition of the complex numbers?

2. How do we multiply two complex numbers?

3. What is the special role of the complex number i?

4. Why is it the case that every nonzero complex number has two distinct square roots?

Exercises

1. Taking the commutative, associative, and distributive laws for the real number system for granted, establish these laws for the complex numbers.

2. Consider the function $\phi : \mathbb{R} \to \mathbb{C}$ given by $\phi(x) = x + i \cdot 0$. Prove that ϕ respects addition and multiplication in the sense that $\phi(x + x') = \phi(x) + \phi(x')$ and $\phi(x \cdot x') = \phi(x) \cdot \phi(x')$.

3. If $z, w \in \mathbb{C}$ then prove that $\overline{z/w} = \overline{z}/\overline{w}$.

4. Prove that the set of all complex numbers is uncountable.

5. Prove that the set of all complex numbers with rational real part is uncountable.

6. Prove that the set of all complex numbers with both real and imaginary parts rational is countable.

7. Prove that the set $\{z \in \mathbb{C} : |z| = 1\}$ is uncountable.

8. Prove that the field of complex numbers cannot be made into an *ordered* field. (**Hint:** Since $i \neq 0$ then either $i > 0$ or $i < 0$. Both lead to a contradiction.)

9. Find all cube roots of the complex number $1 + i$.

10. Use the Fundamental Theorem of Algebra, together with the Division Algorithm, to prove that any polynomial of degree k has k (not necessarily distinct) roots.

11. Prove that the complex roots of a polynomial with real coefficients occur in complex conjugate pairs.

12. Calculate the square roots of i.

*   13. In the complex plane, draw a picture of

$$S = \{z \in \mathbb{C} : |z - 1| + |z + 1| = 2\}.$$

*   14. In the complex plane, draw a picture of

$$T = \{z \in \mathbb{C} : |z + \overline{z}| - |z - \overline{z}| = 2\}.$$

*   15. Prove that any nonzero complex number has kth roots $w_1, w_2, \ldots, w_k$. That is, prove that there are k of them.

Chapter 6

Sequences

6.1 Convergence of Sequences

Preliminary Remarks

Sequences are the nuts and bolts of real analysis. All the basic ideas are formulated in terms of sequences. In particular, we care about *limits* of sequences. We need to be able to determine when a sequence has a limit, and we often need to calculate that limit. This section introduces you to these ideas.

A *sequence* of real numbers is a function $\varphi : \mathbb{N} \to \mathbb{R}$. We often write the sequence as $\varphi(1), \varphi(2), \ldots$ or, more simply, as $\varphi_1, \varphi_2, \ldots$.

EXAMPLE 6.1 The function $\varphi(j) = 1/j$ is a sequence of real numbers. We will often write such a sequence as $\varphi_j = 1/j$ or as $\{1, 1/2, 1/3, \ldots\}$ or as $\{1/j\}_{j=1}^{\infty}$. ∎

POINT OF CONFUSION 6.2 Do not be misled into thinking that a sequence must form a pattern, or be given by a formula. Obviously the ones which are given by formulas are easy to write down, but they are certainly not typical. For example, the coefficients in the decimal expansion of π, $\{3, 1, 4, 1, 5, 9, 2, \ldots\}$, fit our definition of sequence—but they are not given by any obvious pattern.

The most important question about a sequence is whether it converges. We define this notion as follows.

Definition 6.3 A sequence $\{a_j\}$ of real numbers is said to *converge* to a real number α if, for each $\epsilon > 0$, there is an integer $N > 0$ such that if $j > N$ then $|a_j - \alpha| < \epsilon$. We call α the *limit* of the sequence $\{a_j\}$. We write $\lim_{j \to \infty} a_j = \alpha$. We also sometimes write $a_j \to \alpha$.

If a sequence $\{a_j\}$ does not converge, then we frequently say that it *diverges*.

EXAMPLE 6.4 Let $a_j = 1/j, j = 1, 2, \ldots$. Then the sequence converges to 0. For let $\epsilon > 0$. Choose N to be the next integer after $1/\epsilon$. If $j > N$, then

$$|a_j - 0| = |a_j| = \frac{1}{j} < \frac{1}{N} < \epsilon,$$

proving the claim.

Let $b_j = (-1)^j, j = 1, 2, \ldots$. Then the sequence *does not converge*. To prove this assertion, suppose to the contrary that it does. Say that the sequence converges to a number α. Let $\epsilon = 1/2$. By definition of convergence, there is an integer $N > 0$ such that, if $j > N$, then $|b_j - \alpha| < \epsilon = 1/2$. For such j we have

$$2 = |(-1)^j - (-1)^{j+1}| = |b_j - b_{j+1}| \leq |b_j - \alpha| + |\alpha - b_{j+1}|$$

(by the triangle inequality—see the end of Section 5.1). But this last is

$$< \epsilon + \epsilon = 1 \,.$$

We have proved that $2 < 1$, a clear contradiction. So the sequence $\{b_j\}$ has no limit. ∎

POINT OF CONFUSION 6.5 Given any sequence, it either converges or it diverges. There is no in-between status, and no undecided status.

We begin with a few intuitively appealing properties of convergent sequences which will be needed later. First, a definition.

Definition 6.6 A sequence $\{a_j\}$ is said to be *bounded* if there is a number $M > 0$ such that $|a_j| \leq M$ for every j.

Now we have

Proposition 6.7 Let $\{a_j\}$ be a convergent sequence. Then we have:

- *The limit of the sequence is unique.*

- *The sequence is bounded.*

Proof: Suppose that the sequence has two limits α and $\widetilde{\alpha}$. Let $\epsilon > 0$. Then there is an integer $N > 0$ such that for $j > N$ we have the inequality $|a_j - \alpha| < \epsilon/2$. Likewise, there is an integer $\widetilde{N} > 0$ such that for $j > \widetilde{N}$ we have $|a_j - \widetilde{\alpha}| < \epsilon/2$. Let $N_0 = \max\{N, \widetilde{N}\}$. Then, for $j > N_0$, we have

$$|\alpha - \widetilde{\alpha}| \leq |\alpha - a_j| + |a_j - \widetilde{\alpha}| < \epsilon/2 + \epsilon/2 = \epsilon \,.$$

Since this inequality holds for any $\epsilon > 0$ we have that $\alpha = \widetilde{\alpha}$. So the limit of the sequence is unique.

Next, with α the limit of the sequence and $\epsilon = 1$, we choose an integer $N > 0$ such that $j > N$ implies that $|a_j - \alpha| < \epsilon = 1$. For such j we have that

$$|a_j| \leq |a_j - \alpha| + |\alpha| < 1 + |\alpha| \equiv P.$$

Let $Q = \max\{|a_1|, |a_2|, \ldots, |a_N|\}$. If j is any natural number then either $1 \leq j \leq N$ (in which case $|a_j| \leq Q$) or else $j > N$ (in which case $|a_j| \leq P$). Set $M = \max\{P, Q\}$. Then $|a_j| \leq M$ for all j, as desired. So the sequence is bounded. □

The next proposition records some elementary properties of limits of sequences.

Proposition 6.8 *Let $\{a_j\}$ be a sequence of real numbers with limit α and $\{b_j\}$ be a sequence of real numbers with limit β. Then we have*

(1) *If c is a constant then the sequence $\{c \cdot a_j\}$ converges to $c \cdot \alpha$.*

(2) *The sequence $\{a_j + b_j\}$ converges to $\alpha + \beta$.*

(3) *The sequence $\{a_j \cdot b_j\}$ converges to $\alpha \cdot \beta$.*

(4) *If $b_j \neq 0$ for all j and $\beta \neq 0$ then the sequence a_j/b_j converges to α/β.*

Proof: For the first part, we may assume that $c \neq 0$ (for when $c = 0$ there is nothing to prove). Let $\epsilon > 0$. Choose an integer $N > 0$ such that for $j > N$ it holds that

$$|a_j - \alpha| < \frac{\epsilon}{|c|}.$$

For such j we have that

$$|c \cdot a_j - c \cdot \alpha| = |c| \cdot |a_j - \alpha| < |c| \cdot \frac{\epsilon}{|c|} = \epsilon.$$

This proves the first assertion.

The proof of the second assertion is similar, and we leave it as an exercise.

For the third assertion, notice that the sequence $\{a_j\}$ is bounded (by the second part of Proposition 6.7): say that $|a_j| \leq M$ for every j. Let $\epsilon > 0$. Choose an integer $N > 0$ so that $|a_j - \alpha| < \epsilon/(2M + 2|\beta|)$ when $j > N$. Also choose an integer $\widetilde{N} > 0$ such that $|b_j - \beta| < \epsilon/(2M + 2|\beta|)$ when $j > \widetilde{N}$. Then, for $j > \max\{N, \widetilde{N}\}$, we have that

$$\begin{aligned}
|a_j b_j - \alpha\beta| &= |a_j(b_j - \beta) + \beta(a_j - \alpha)| \\
&\leq |a_j(b_j - \beta)| + |\beta(a_j - \alpha)| \\
&< M \cdot \frac{\epsilon}{2M + 2|\beta|} + |\beta| \cdot \frac{\epsilon}{2M + 2|\beta|} \\
&\leq \frac{\epsilon}{2} + \frac{\epsilon}{2} \\
&= \epsilon.
\end{aligned}$$

So the sequence $\{a_j b_j\}$ converges to $\alpha\beta$.

Part (4) is proved in a similar fashion and we leave the details as an exercise.
$\square$

POINT OF CONFUSION 6.9 You were probably puzzled by the choice of N and $\widetilde{N}$ in the proof of part (3) of Proposition 6.8—where did the number $\epsilon/(2M + 2|\beta|)$ come from? The answer of course becomes obvious when we read on further in the proof. So the lesson here is that a proof is constructed backward: you look to the end of the proof to see what you need to specify earlier on. Skill in these matters can come only with practice.

When discussing the convergence of a sequence, we often find it inconvenient to deal with the definition of convergence as given. For this definition makes reference to the number to which the sequence is supposed to converge, and we often do not know this number in advance. Would it not be useful to be able to decide whether a sequence converges *without knowing to what limit it converges*?

Definition 6.10 Let $\{a_j\}$ be a sequence of real numbers. We say that the sequence satisfies the *Cauchy criterion* (A. L. Cauchy, 1789–1857)—more briefly, that the sequence is *Cauchy*—if, for each $\epsilon > 0$, there is an integer $N > 0$ such that if $j, k > N$ then $|a_j - a_k| < \epsilon$.

Notice that the concept of a sequence being Cauchy simply makes precise the notion of the elements of the sequence (i) *getting* close together and (ii) *staying* close together.

Lemma 6.11 *Every Cauchy sequence is bounded.*

Proof: Let $\epsilon = 1 > 0$. There is an integer $N > 0$ such that $|a_j - a_k| < \epsilon = 1$ whenever $j, k > N$. Thus if $j \geq N + 1$ we have

$$
\begin{aligned}
|a_j| &= |a_{N+1} + (a_j - a_{N+1})| \\
&\leq |a_{N+1}| + |a_j - a_{N+1}| \\
&\leq |a_{N+1}| + 1 \equiv K.
\end{aligned}
$$

Let $L = \max\{|a_1|, |a_2|, \ldots, |a_N|\}$. If j is any natural number, then either $1 \leq j \leq N$, in which case $|a_j| \leq L$, or else $j \geq N + 1$, in which case $|a_j| \leq K$. Set $M = \max\{K, L\}$. Then, for any j, $|a_j| \leq M$ as required. $\square$

Theorem 6.12 *Let $\{a_j\}$ be a sequence of real numbers. The sequence is Cauchy if and only if it converges to some limit α.*

Proof: First assume that the sequence converges to a limit α. Let $\epsilon > 0$. Choose, by definition of convergence, an integer $N > 0$ such that if $j > N$ then $|a_j - \alpha| < \epsilon/2$. If $j, k > N$ then

$$
|a_j - a_k| \leq |a_j - \alpha| + |\alpha - a_k| < \frac{\epsilon}{2} + \frac{\epsilon}{2} = \epsilon.
$$

So the sequence is Cauchy.

Conversely, suppose that the sequence is Cauchy. Define

$$S = \{x \in \mathbb{R} : x < a_j \text{ for all but finitely many } j\}.$$

[**Hint:** You might find it helpful to think of this set as

$$S = \{x \in \mathbb{R} : \text{there is a positive integer } k \text{ such that } x < a_j \text{ for all } j \geq k\}.]$$

By the lemma, the sequence $\{|a_j|\}$ is bounded by some number M. If x is a real number less than $-M$, then $x \in S$, so S is nonempty. Also S is bounded above by M. Let $\alpha = \sup S$. Then α is a well-defined real number, and we claim that α is the limit of the sequence $\{a_j\}$.

To see this, let $\epsilon > 0$. Choose an integer $N > 0$ such that $|a_j - a_k| < \epsilon/2$ whenever $j, k > N$. Notice that this last inequality implies that

$$|a_j - a_{N+1}| < \epsilon/2 \quad \text{when} \quad j \geq N+1, \tag{6.12.1}$$

hence

$$a_j > a_{N+1} - \epsilon/2 \quad \text{when} \quad j \geq N+1.$$

Thus $a_{N+1} - \epsilon/2 \in S$ and it follows that

$$\alpha \geq a_{N+1} - \epsilon/2. \tag{6.12.2}$$

Line (6.12.1) also shows that

$$a_j < a_{N+1} + \epsilon/2 \quad \text{when} \quad j \geq N+1.$$

Thus $a_{N+1} + \epsilon/2 \notin S$ and

$$\alpha \leq a_{N+1} + \epsilon/2. \tag{6.12.3}$$

Combining lines (6.12.2) and (6.12.3) gives

$$|\alpha - a_{N+1}| \leq \epsilon/2. \tag{6.12.4}$$

But then line (6.12.4) yields, for $j > N$, that

$$|\alpha - a_j| \leq |\alpha - a_{N+1}| + |a_{N+1} - a_j| < \epsilon/2 + \epsilon/2 = \epsilon.$$

This proves that the sequence $\{a_j\}$ converges to α, as claimed. □

POINT OF CONFUSION 6.13 Any convergent sequence is Cauchy. And, in the real number system, any Cauchy sequence is convergent. So why do we have both concepts? The answer is easy. Convergent sequences are what we are really interested in. But the concept of Cauchy sequence helps us to identify them.

Definition 6.14 Let $\{a_j\}$ be a sequence of real numbers. The sequence is said to be *increasing* if $a_1 \leq a_2 \leq \ldots$. It is *decreasing* if $a_1 \geq a_2 \geq \ldots$.

A sequence is said to be *monotone* if it is either increasing or decreasing.

Proposition 6.15 *If $\{a_j\}$ is an increasing sequence which is bounded above—$a_j \leq M < \infty$ for all j—then $\{a_j\}$ is convergent. If $\{b_j\}$ is a decreasing sequence which is bounded below—$b_j \geq K > -\infty$ for all j—then $\{b_j\}$ is convergent.*

Proof: Let $\epsilon > 0$. Let $\alpha = \sup a_j < \infty$. By definition of supremum, there is an integer N so that $|a_N - \alpha| < \epsilon$. Then, if $\ell \geq N + 1$, we have $a_N \leq a_\ell \leq \alpha$ hence $|a_\ell - \alpha| < \epsilon$. Thus the sequence converges to α.

The proof for decreasing sequences is similar and we omit it.   □

Remark 6.16 Let $a_1 = \sqrt{2}$ and set $a_{j+1} = \sqrt{2 + a_j}$ for $j \geq 1$. You can verify that $\{a_j\}$ is increasing and bounded above (by 4 for example). What is its limit (which is guaranteed to exist by the proposition)?   ∎

A proof very similar to that of the proposition gives the following useful fact:

Corollary 6.17 *Let S be a nonempty set of real numbers which is bounded above and below. Let β be its supremum and α its infimum. If $\epsilon > 0$ then there are $s, t \in S$ such that $|s - \beta| < \epsilon$ and $|t - \alpha| < \epsilon$.*

Proof: This is essentially a restatement of the proof of the proposition.   □

We conclude the section by recording one of the most useful results for calculating the limit of a sequence:

Proposition 6.18 (The Pinching Principle) *Let $\{a_j\}, \{b_j\}$, and $\{c_j\}$ be sequences of real numbers satisfying*

$$a_j \leq b_j \leq c_j$$

for every j sufficiently large. If

$$\lim_{j \to \infty} a_j = \lim_{j \to \infty} c_j = \alpha$$

for some real number α, then

$$\lim_{j \to \infty} b_j = \alpha.$$

Proof: This proof is requested of you in the exercises.   □

POINT OF CONFUSION 6.19 The Pinching Principle gives us a way to compare an unknown sequence with two known sequences. It is a powerful tool. But it again drives home the point that we need to have knowledge of a library of reference sequences.

EXAMPLE 6.20 Define
$$a_j = \frac{\sin j \cos 2j}{j^2} .$$

Then
$$0 \le |a_j| \le \frac{1}{j^2} .$$

It is clear that
$$\lim_{j \to \infty} 0 = 0$$

and
$$\lim_{j \to \infty} \frac{1}{j^2} = 0 .$$

Therefore
$$\lim_{j \to \infty} |a_j| = 0$$

so that
$$\lim_{j \to \infty} a_j = 0 . \qquad \blacksquare$$

A Look Back

1. Speaking intuitively, what does it mean for a sequence to converge?
2. Speaking intuitively, what does it mean for a sequence to be Cauchy?
3. What is a bounded sequence?
4. What is a monotone sequence?
5. Why does any given sequence have at most one limit?

Exercises

1. Prove parts (2) and (4) of Proposition 6.8.
2. Prove the following result, which we have used without comment in the text: Let S be a set of real numbers which is bounded above and let $t = \sup S$. For any $\epsilon > 0$ there is an element $s \in S$ such that $t - \epsilon < s \le t$. [**Hint:** Notice that this result makes good intuitive sense: the elements of S should become arbitrarily close to the supremum t, otherwise there would be enough room to decrease the value of t and make the supremum even smaller.] Formulate and prove a similar result for the infimum.
3. Prove Proposition 6.18.
4. Let $a_1, a_2 > 0$ and for $j \ge 3$ define $a_j = a_{j-1} + a_{j-2}$. Show that this sequence cannot converge to a finite limit.
5. Suppose a sequence $\{a_j\}$ has the property that, for every natural number N, there is a j_N such that $a_{j_N} = a_{j_N+1} = \cdots = a_{j_N+N}$. In other words, the sequence has arbitrarily long repetitive strings. Does it follow that the sequence converges?
6. Let γ be an irrational real number and let a_j be a sequence of rational numbers converging to γ. Suppose that each a_j is a fraction expressed in lowest terms: $a_j = \alpha_j / \beta_j$. Prove that the β_j tend to ∞.

7. Use the integral of $1/(1+t^2)$, together with Riemann sums (ideas which you know from calculus, and which we shall treat rigorously later in the book), to develop a scheme for calculating the digits of π.

8. Does the sequence $a_j = (-1)^j j^2/(j^2 + 1)$ converge? If so, to what limit?

9. Does the sequence $a_j = e^j/(e^{2j} + j^2)$ converge? If so, to what limit?

10. Define $a_1 = 1$, $a_2 = 1$, $a_3 = a_1 + a_2$, and $a_j = a_{j-2} + a_{j-1}$ for $j \geq 4$. Does this sequence converge? If so, to what limit?

11. Give an example of a sequence of rational numbers that converges to π. Give an example of a sequence of irrational numbers that converges to 2.

12. Prove that the sum of two irrational numbers can be either rational or irrational. Prove that the product of two irrational numbers can be either rational or irrational.

13. Give an example of a sequence $\{a_j\}$ that converges to $+\infty$ but so that it is *not* the case that $a_1 \leq a_2 \leq a_3 \leq \cdots$.

6.2   Subsequences

Preliminary Remarks

A subsequence of a given sequence is a "junior" sequence that lives inside the parent sequence. Many of the most important ideas in analysis, including closure and compactness, are formulated in terms of subsequences. It is a rather slippery idea, but one that you need to master.

Let $\{a_j\}$ be a given sequence. If

$$0 < j_1 < j_2 < \cdots$$

are positive integers then the function

$$k \mapsto a_{j_k}$$

is called a *subsequence* of the given sequence. We usually write the subsequence as

$$\{a_{j_k}\}_{k=1}^{\infty} \quad \text{or} \quad \{a_{j_k}\}.$$

EXAMPLE 6.21 Consider the sequence

$$\{2^j\} = \{2, 4, 8, \ldots\}.$$

Then the sequence

$$\{2^{2k}\} = \{4, 16, 64, \ldots\} \tag{6.21.1}$$

is a subsequence. Notice that the subsequence contains a subcollection of elements of the original sequence *in the same order*. In this example, $j_k = 2k$.

Another subsequence is

$$\{2^{(2^k)}\} = \{4, 16, 256, \ldots\}.$$

In this instance, it holds that $j_k = 2^k$. Notice that this new subsequence is in fact a subsequence of the first subsequence (6.21.1). That is, it is a sub-subsequence of the original sequence $\{2^j\}$. ∎

Proposition 6.22 *If $\{a_j\}$ is a convergent sequence with limit α, then every subsequence converges to the limit α.*

Conversely, if a sequence $\{b_j\}$ has the property that each of its subsequences is convergent then $\{b_j\}$ itself is convergent.

Proof: Assume $\{a_j\}$ is convergent to a limit α, and let $\{a_{j_k}\}$ be a subsequence. Let $\epsilon > 0$ and choose $N > 0$ such that $|a_j - \alpha| < \epsilon$ whenever $j > N$. Now if $k > N$ then $j_k > N$ hence $|a_{j_k} - \alpha| < \epsilon$. Therefore, by definition, the subsequence $\{a_{j_k}\}$ also converges to α.

The converse is trivial, simply because the entire sequence is a subsequence of itself. □

POINT OF CONFUSION 6.23 A subsequence of a given sequence is also a *subset* of that sequence. But it is much more than that, because it maintains the same order of terms.

Now we present one of the most fundamental theorems of basic real analysis (due to B. Bolzano, 1781–1848, and K. Weierstrass, 1815–1897).

Theorem 6.24 (Bolzano-Weierstrass) *Let $\{a_j\}$ be a bounded sequence in $\mathbb{R}$. Then there is a subsequence which converges.*

Proof: Say that $|a_j| \leq M$ for every j. We may assume that $M > 0$.

One of the two intervals $[-M, 0]$ and $[0, M]$ must contain infinitely many elements of the sequence. Say that $[0, M]$ does. Choose a_{j_1} to be one of the infinitely many sequence elements in $[0, M]$.

Next, one of the intervals $[0, M/2]$ and $[M/2, M]$ must contain infinitely many elements of the sequence. Say that it is $[0, M/2]$. Choose an element $a_{j_2} \in [0, M/2]$ with $j_2 > j_1$. Continue in this fashion, halving the interval, choosing a half with infinitely many sequence elements, and selecting the next subsequential element from that half.

Let us analyze the resulting subsequence. Notice that $|a_{j_1} - a_{j_2}| \leq M$ since both elements belong to the interval $[0, M]$. Likewise, $|a_{j_2} - a_{j_3}| \leq M/2$ since both elements belong to $[0, M/2]$. In general, $|a_{j_k} - a_{j_{k+1}}| \leq 2^{-k+1} \cdot M$ for each $k \in \mathbb{N}$. Now let $\epsilon > 0$. Choose an integer $N > 0$ such that $2^{-N} < \epsilon/(4M)$.

Then, for any $m > l > N$ we have

$$
\begin{aligned}
|a_{j_l} - a_{j_m}| &= |(a_{j_l} - a_{j_{l+1}}) + (a_{j_{l+1}} - a_{j_{l+2}}) + \cdots + (a_{j_{m-1}} - a_{j_m})| \\
&\leq |a_{j_l} - a_{j_{l+1}}| + |a_{j_{l+1}} - a_{j_{l+2}}| + \cdots + |a_{j_{m-1}} - a_{j_m}| \\
&\leq 2^{-l+1} \cdot M + 2^{-l} \cdot M + \cdots + 2^{-m+2} \cdot M \\
&= \left(2^{-l+1} + 2^{-l} + 2^{-l-1} + \cdots + 2^{-m+2}\right) \cdot M \\
&= \left((2^{-l+2} - 2^{-l+1}) + (2^{-l+1} - 2^{-l}) + \cdots \right. \\
&\qquad \left. + (2^{-m+3} - 2^{-m+2})\right) \cdot M \\
&= \left(2^{-l+2} - 2^{-m+2}\right) \cdot M \\
&< 2^{-l+2} \cdot M \\
&< 4 \cdot \frac{\epsilon}{4M} \cdot M \\
&= \epsilon.
\end{aligned}
$$

We see that the subsequence $\{a_{j_k}\}$ is Cauchy, so it converges.   □

Remark 6.25 Of course it is not true that every bounded sequence converges. But the Bolzano-Weierstrass theorem is a good substitute result. Often, in practice, all that we need is a convergent subsequence.

The Bolzano-Weierstrass theorem is a generalization of our result from the last section about increasing sequences which are bounded above (resp. decreasing sequences which are bounded below). For such a sequence is surely bounded above *and* below (why?). So it has a convergent subsequence. And thus it follows easily that the entire sequence converges. Details are left as an exercise.

It is a fact—which you can verify for yourself—that *any* real sequence has a monotone subsequence. This fact implies Bolzano-Weierstrass.

POINT OF CONFUSION 6.26 In this text we have not yet given a rigorous definition of the function $\sin x$ (see Section 13.3). However, just for the moment, use the definition you learned in calculus class and consider the sequence $\{\sin j\}_{j=1}^{\infty}$. Notice that the sequence is bounded in absolute value by 1. The Bolzano-Weierstrass theorem guarantees that there is a convergent subsequence, even though it would be very difficult to say precisely what that convergent subsequence is.   □

A Look Back

1. What is a subsequence?

2. How can it happen that a sequence does not converge but one of its subsequences does converge?

3. Is it possible for two different subsequences to converge to two different limits?

4. Is it possible for two different subsequences to converge to the same limit?

Exercises

1. Use the Bolzano-Weierstrass theorem to show that every increasing sequence that is bounded above converges.

2. Give an example of a sequence of rational numbers with the property that, for any real number α, there is a subsequence converging to α.

3. Let $x_1 = 2$. For $j \geq 1$, set

$$x_{j+1} = x_j - \frac{x_j^2 - 2}{2x_j} \, .$$

   Show that the sequence $\{x_j\}$ is decreasing and bounded below. What is its limit?

4. The sequence $\{\cos j + \sin j\}$ has a convergent subsequence. Explain why. Can you say what that subsequence is?

5. Give an example of a sequence that is bounded below but does not have a convergent subsequence. Give an example of a sequence that is bounded above but does not have a convergent subsequence.

* 6. Provide the details of the assertion that the sequence $\{\cos j\}$ is dense in the interval $[-1, 1]$.

* 7. Let $S = \{0, 1, 1/2, 1/3, 1/4, \dots \}$. Give an example of a sequence $\{a_j\}$ with the property that, for each $s \in S$, there is a subsequence converging to s, but no subsequence converges to any limit not in S.

* 8. Give another proof of the Bolzano-Weierstrass theorem as follows. If $\{a_j\}$ is a bounded sequence let $b_j = \inf\{a_j, a_{j+1}, \dots \}$. Then each b_j is finite, $b_1 \leq b_2 \leq \dots$, and $\{b_j\}$ is bounded above. Now use Proposition 6.15.

* 9. Give an example of a sequence with infinitely many distinct subsequences that converge to π.

10. Give an example of a sequence which does not converge, but which has infinitely many different subsequences that do converge.

11. Prove that a sequence $\{a_j\}$ converges if and only if every subsequence has a subsequence that converges.

6.3   Lim sup and Lim inf

Preliminary Remarks

While our interest in sequences is in their limits, it is a fact that most sequences do not have a limit. So what can we do? Is there a substitute idea?

In fact there is, and that is the idea of lim sup (limit superior) and lim inf (limit inferior). These are, in effect, the greatest limit of any subsequence and the least limit of any subsequence.

You can see that this new set of ideas combines many of the earlier ideas. But it leads to greater depth and insight, and it is important.

Convergent sequences are useful objects, but the unfortunate truth is that most sequences do not converge. Nevertheless, we would like to have a language for discussing the asymptotic behavior of *any* real sequence $\{a_j\}$ as $j \to \infty$. That is the purpose of the concepts of "limit superior" (or "upper limit") and "limit inferior" (or "lower limit"). It should be stressed that *any sequence whatever* has a lim sup and a lim inf.

Definition 6.27 Let $\{a_j\}$ be a sequence of real numbers. For each j let

$$A_j = \inf\{a_j, a_{j+1}, a_{j+2}, \dots\}.$$

Then $\{A_j\}$ is an increasing sequence (since, as j increases, we are taking the infimum of a smaller and smaller set of numbers), so it has a limit. We define the *limit infimum* of $\{a_j\}$ to be

$$\liminf a_j = \lim_{j\to\infty} A_j.$$

It is common to refer to this number as the lim inf of the sequence. Note that the lim inf could be $\pm\infty$, or it could be a finite real number.

Likewise, let

$$B_j = \sup\{a_j, a_{j+1}, a_{j+2}, \dots\}.$$

Then $\{B_j\}$ is a decreasing sequence (since, as j increases, we are taking the supremum of a smaller and smaller set of numbers), so it has a limit. We define the *limit supremum* of $\{a_j\}$ to be

$$\limsup a_j = \lim_{j\to\infty} B_j.$$

It is common to refer to this number as the lim sup of the sequence. Note that the lim sup could be $\pm\infty$, or it could be a finite real number.

POINT OF CONFUSION 6.28 Notice that the lim sup or lim inf of a sequence can be $\pm\infty$. For instance, the sequence $a_j = j^2 - j$ has lim sup equal to $+\infty$. The sequence $-2j + 6$ has lim inf equal to $-\infty$.

Remark 6.29 What is the intuitive content of this definition? For each j, A_j picks out the greatest lower bound of the sequence in the jth position or later. So the sequence $\{A_j\}$ should tend to the *smallest* possible limit of any subsequence of $\{a_j\}$.

Likewise, for each j, B_j picks out the least upper bound of the sequence in the jth position or later. So the sequence $\{B_j\}$ should tend to the *greatest* possible limit of any subsequence of $\{a_j\}$. We shall make these remarks more precise in Proposition 6.32 below.

Notice that it is implicit in the definition that *every* real sequence has a limit supremum and a limit infimum.

POINT OF CONFUSION 6.30 It is important to keep in mind that the lim sup of a given sequence is in fact associated to a subsequence. And so is the lim inf. In practice we think of the lim sup as the limit of the "greatest" subsequence and the lim inf as the limit of the "least" subsequence.

A further comment is that we can talk about the limit infimum of a sequence even when the sequence is *not* bounded below. But then some or all of the A_j may be $-\infty$ and the limit infimum may be $-\infty$. Likewise we may discuss the limit supremum of a sequence that is not bounded above.

EXAMPLE 6.31 Consider the sequence $\{(-1)^j\}$. Of course this sequence does not converge. Let us calculate its lim sup and lim inf.

Referring to the definition, we have that $A_j = -1$ for every j. So

$$\liminf (-1)^j = \lim (-1) = -1.$$

Similarly, $B_j = +1$ for every j. Therefore

$$\limsup (-1)^j = \lim (+1) = +1.$$

As we predicted in the remark, the lim inf is the least subsequential limit, and the lim sup is the greatest subsequential limit. ∎

Now let us prove the characterizing property of lim sup and lim inf to which we have been alluding.

Proposition 6.32 *Let $\{a_j\}$ be a sequence of real numbers. Let us set $\beta = \limsup_{j\to\infty} a_j$ and $\alpha = \liminf_{j\to\infty} a_j$. If $\{a_{j_\ell}\}$ is any subsequence of the given sequence then*

$$\alpha \le \liminf_{\ell\to\infty} a_{j_\ell} \le \limsup_{\ell\to\infty} a_{j_\ell} \le \beta.$$

Moreover, there is a subsequence $\{a_{j_k}\}$ such that

$$\lim_{k\to\infty} a_{j_k} = \alpha$$

and another sequence $\{a_{n_k}\}$ such that

$$\lim_{k\to\infty} a_{n_k} = \beta.$$

Proof: For simplicity in this proof we assume that the lim sup and lim inf are finite.

We begin by considering the lim inf. There is a $j_1 \geq 1$ such that $|A_1 - a_{j_1}| < 2^{-1}$. We choose j_1 to be as small as possible. Next, we choose j_2, necessarily greater than j_1, such that j_2 is as small as possible and $|a_{j_2} - A_2| < 2^{-2}$. Continuing in this fashion, we select $j_k > j_{k-1}$ such that $|a_{j_k} - A_k| < 2^{-k}$, etc.

Recall that $A_k \to \alpha = \liminf_{j \to \infty} a_j$. Now fix $\epsilon > 0$. If N is an integer so large that $k > N$ implies that $|A_k - \alpha| < \epsilon/2$ and also that $2^{-N} < \epsilon/2$ then, for such k, we have

$$
\begin{aligned}
|a_{j_k} - \alpha| &\leq |a_{j_k} - A_k| + |A_k - \alpha| \\
&< 2^{-k} + \frac{\epsilon}{2} \\
&< \frac{\epsilon}{2} + \frac{\epsilon}{2} \\
&= \epsilon.
\end{aligned}
$$

Thus the subsequence $\{a_{j_k}\}$ converges to α, the lim inf of the given sequence. A similar construction gives a (different) subsequence $\{a_{n_k}\}$ converging to β, the lim sup of the given sequence.

Now let $\{a_{j_\ell}\}$ be *any* subsequence of the sequence $\{a_j\}$. Let β^* be the lim sup of this subsequence. Then, by the first part of the proof, there is a subsequence $\{a_{j_{\ell_m}}\}$ such that

$$
\lim_{m \to \infty} a_{j_{\ell_m}} = \beta^*.
$$

But $a_{j_{\ell_m}} \leq B_{j_{\ell_m}}$ by the very definition of the Bs. Thus

$$
\beta^* = \lim_{m \to \infty} a_{j_{\ell_m}} \leq \lim_{m \to \infty} B_{j_{\ell_m}} = \beta
$$

or

$$
\limsup_{\ell \to \infty} a_{j_\ell} \leq \beta,
$$

as claimed. A similar argument shows that

$$
\liminf_{l \to \infty} a_{j_l} \geq \alpha.
$$

This completes the proof of the proposition. □

Corollary 6.33 *If $\{a_j\}$ is a sequence and $\{a_{j_k}\}$ is a convergent subsequence then*

$$
\liminf_{j \to \infty} a_j \leq \lim_{k \to \infty} a_{j_k} \leq \limsup_{j \to \infty} a_j.
$$

We close this section with a fact that is analogous to one for the supremum and infimum. Its proof is analogous to arguments we have seen before.

Proposition 6.34 *Let $\{a_j\}$ be a sequence. We write $\limsup a_j = \beta$ and also $\liminf a_j = \alpha$. We also assume that α, β are finite real numbers. Let $\epsilon > 0$. Then there are arbitrarily large j such that $a_j > \beta - \epsilon$. Also there are arbitrarily large k such that $a_k < \alpha + \epsilon$.*

POINT OF CONFUSION 6.35 Given a sequence, there are always elements of that sequence which are arbitrarily close to the lim sup, and there are elements of the sequence which are arbitrarily close to the lim inf. It is often this particular property of the concepts which is most useful.

A Look Back

1. Intuitively speaking, what is a lim sup?

2. Intuitively speaking, what is a lim inf?

3. In what sense is the lim sup of a sequence greater than or equal to the lim inf of that sequence?

4. What does it mean when the lim sup and lim inf are equal?

Exercises

1. Consider $\{a_j\}$ both as a sequence and as a set. How are the lim sup and the sup related? How are the lim inf and the inf related? Give examples.

2. Prove the last proposition in this section.

3. How are the lim sup and lim inf of $\{a_j\}$ related to the lim sup and lim inf of $\{-a_j\}$?

4. Let $\{a_j\}$ be a real sequence. Prove that if

$$\liminf a_j = \limsup a_j$$

then the sequence $\{a_j\}$ converges. Prove the converse as well.

5. Let $a < b$ be real numbers. Give an example of a real sequence whose lim sup is b and whose lim inf is a.

6. How is $\limsup(a_j + b_j)$ related to $\limsup a_j$ and $\limsup b_j$?

7. How is $\limsup(a_j - b_j)$ related to $\limsup a_j$ and $\limsup b_j$?

8. What is $\limsup_{j \to \infty} \cos j$? What is $\limsup_{j \to \infty} \sin j$?

9. What is $\limsup_{j \to \infty} \ln j$?

10. Prove that, if $\liminf a_j < \limsup a_j$, then the sequence $\{a_j\}$ cannot converge.

11. Explain why we do not consider lim sup and lim inf for complex numbers.

12. Give an example of a sequence whose lim inf is 0 and whose lim sup is $+\infty$.

* 13. Find the lim sup and lim inf of the sequences

$$\{|\sin j|^{\sin j}\} \quad \text{and} \quad \{|\cos j|^{\cos j}\}.$$

6.4   Some Special Sequences

Preliminary Remarks

One of the ways that we understand a new sequence is to compare it with a known sequence. Thus we need a library of "known" sequences that we can use as the basis for our studies. This section begins to assemble such a library.

We often obtain information about a new sequence by comparison with a sequence that we already know. Thus it is well to have a catalogue of fundamental sequences which provide a basis for comparison.

EXAMPLE 6.36 Fix a real number a. The sequence $\{a^j\}$ is called a *power sequence*. If $-1 < a < 1$ then the sequence converges to 0. If $a = 1$ then the sequence is a constant sequence and converges to 1. If $a > 1$ then the sequence diverges to $+\infty$. Finally, if $a \leq -1$, then the sequence diverges. ∎

Recall that, in Section 5.1, we discussed the existence of nth roots of positive real numbers. If $\alpha > 0, m \in \mathbb{Z}$, and $n \in \mathbb{N}$, then we may define

$$\alpha^{m/n} = (\alpha^m)^{1/n}.$$

Thus we may talk about rational powers of a positive number. Next, if $\beta \in \mathbb{R}$, then we may define

$$\alpha^\beta = \sup\{\alpha^q : q \in \mathbb{Q}, q < \beta\}.$$

Thus we can define *any real power* of a positive real number. Some of the exercises ask you to verify several basic properties of these exponentials.

Lemma 6.37 *If $\alpha > 1$ is a real number and $\beta > 0$ then $\alpha^\beta > 1$.*

Proof: Let q be a positive rational number which is less than β. Say that $q = m/n$, with m, n integers. It is obvious that $\alpha^m > 1$ and hence that $(\alpha^m)^{1/n} > 1$. Since α^β majorizes this last quantity, we are done. □

EXAMPLE 6.38 Fix a real number β and consider the sequence $\{j^\beta\}$. If $\beta > 0$, then it is easy to see that $j^\beta \to +\infty$: to verify this assertion fix $M > 0$ and take the number N to be the first integer after $M^{1/\alpha}$.

If $\beta = 0$ then j^β is a constant sequence, identically equal to 1.

If $\beta < 0$ then $j^\beta = 1/j^{-\beta}$. The denominator of this last expression tends to $+\infty$, hence the sequence j^β tends to 0. ∎

EXAMPLE 6.39 The sequence $\{j^{1/j}\}$ converges to 1. In fact, consider the expressions $\alpha_j = j^{1/j} - 1 > 0$. We have that

$$j = (\alpha_j + 1)^j \geq \frac{j(j-1)}{2}(\alpha_j)^2$$

(the latter being just one term from the Binomial expansion). Thus

$$0 < \alpha_j \le \sqrt{2/(j-1)}$$

as long as $j \ge 2$. It follows that $\alpha_j \to 0$ or $j^{1/j} \to 1$. ∎

EXAMPLE 6.40 Let α be a positive real number. Then the sequence $\alpha^{1/j}$ converges to 1. To see this, first note that the case $\alpha = 1$ is trivial, and the case $\alpha > 1$ implies the case $\alpha < 1$ (by taking reciprocals). So we concentrate on $\alpha > 1$. But then we have

$$1 < \alpha^{1/j} < j^{1/j}$$

when $j > \alpha$. Since $j^{1/j}$ tends to 1, Proposition 6.18 applies and the proof is complete. ∎

EXAMPLE 6.41 Let $\lambda > 1$ and let α be real. Then the sequence

$$\left\{ \frac{j^\alpha}{\lambda^j} \right\}_{j=1}^\infty$$

converges to 0.

To see this, fix an integer $k > \alpha$ and consider $j > 2k$. [Notice that k is fixed once and for all but j will be allowed to tend to $+\infty$ at the appropriate moment.] Writing $\lambda = 1 + \mu, \mu > 0$, we have that

$$\lambda^j = (\mu+1)^j > \frac{j(j-1)(j-2)\cdots(j-k+1)}{k(k-1)(k-2)\cdots 2 \cdot 1} \mu^k \cdot 1^{j-k}.$$

Of course this comes from picking out the kth term of the Binomial expansion for $(\mu + 1)^j$. Notice that, since $j > 2k$, then each of the expressions $j, (j-1), \ldots (j-k+1)$ in the numerator on the right exceeds $j/2$. Thus

$$\lambda^j > \frac{j^k}{2^k \cdot k!} \cdot \mu^k$$

and

$$0 < \frac{j^\alpha}{\lambda^j} < j^\alpha \cdot \frac{2^k \cdot k!}{j^k \cdot \mu^k} = \frac{j^{\alpha-k} \cdot 2^k \cdot k!}{\mu^k}.$$

Since $\alpha - k < 0$, the right side tends to 0 as $j \to \infty$. ∎

POINT OF CONFUSION 6.42 We should always bear in mind that expressions of the form α^j grow *much faster* than expressions of the form j^β (when $\alpha > 1$, $\beta > 1$, for instance). That's because exponentials grow faster than polynomials. Keeping this fact in focus helps in evaluating the limits of particular sequences.

EXAMPLE 6.43 The sequence

$$\left\{ \left(1 + \frac{1}{j}\right)^j \right\}$$

converges. In fact it is increasing and bounded above. Use the binomial Expansion to prove this assertion. The limit of the sequence is the number that we shall later call e (in honor of Leonhard Euler, 1707–1783, who first studied it in detail). We shall study this sequence in detail in Chapter 13.   ∎

EXAMPLE 6.44 The sequence

$$\left(1 - \frac{1}{j}\right)^j$$

converges to $1/e$, where the definition of e is given in the last example. More generally, the sequence

$$\left(1 + \frac{x}{j}\right)^j$$

converges to e^x (here e^x is defined as in the discussion following Example 6.36 above).   ∎

A Look Back

1. What is a sequence that converges to Euler's number e?

2. What is a sequence that converges to π?

3. Is there a power sequence that converges to 2?

4. Is there a sequence of rational numbers that converges to Euler's number e?

Exercises

1. Let α be a positive real number and let $p/q = m/n$ be two different representations of the same rational number r. Prove that

$$(\alpha^m)^{1/n} = (\alpha^p)^{1/q} .$$

Also prove that

$$(\alpha^{1/n})^m = (\alpha^m)^{1/n}.$$

If β is another positive real and γ is any real then prove that

$$(\alpha \cdot \beta)^\gamma = \alpha^\gamma \cdot \beta^\gamma .$$

2. Discuss the convergence of the sequence $\{(1/j)^{1/j}\}_{j=1}^\infty$.

3. Discuss the convergence of the sequence $\{(j^j)/(2j)!\}_{j=1}^\infty$.

4. Prove that the exponential, as defined in this section, satisfies

$$(a^b)^c = a^{bc} \qquad \text{and} \qquad a^b a^c = a^{b+c} .$$

5. Discuss the convergence of the sequence

$$1 \, , 1 + 1/2! \, , \ 1 + 1/2! + 1/3! \, , \ \ldots .$$

What is the limit?

6. For which values of $\alpha > 0$ and $\beta > 0$ does the sequence

$$a_j = \frac{\alpha^j}{j^\beta}$$

converge? What about the sequence

$$b_j = \frac{j^\beta}{\alpha^j} \; ?$$

7. It is not known whether $\pi + e$ or $\pi - e$ is irrational. But one of them must be. Explain.

8. Discuss convergence of the sequence $\sqrt{j}^{1/j}$.

9. Discuss convergence of the sequence $(1 + 1/j^2)^j$.

* 10. Consider the sequence given by

$$a_j = \left[1 + \frac{1}{2} + \frac{1}{3} + \cdots + \frac{1}{j} \right] - \log j \, .$$

Use a picture (remember that log is the antiderivative of $1/x$) to give a convincing argument that the sequence $\{a_j\}$ converges. The limit number is called γ. This number was first studied by Euler. It arises in many different contexts in analysis and number theory.

As a challenge problem, show that

$$|a_j - \gamma| \leq \frac{C}{j}$$

for some universal constant $C > 0$.

It is not known whether γ is rational or irrational.

* 11. A sequence is defined by the rule $a_0 = 2$, $a_1 = 1$, and $a_j = 3a_{j-1} - a_{j-2}$. Find a formula for a_j.

Chapter 7

Series of Numbers

7.1 Convergence of Series

Preliminary Remarks

A series is an infinite sum. We think of the series as the limit of the sequence of its partial sums. While at first a bit confusing, this approach avoids many conundrums and redundancies that have plagued the history of series.

Of course we care about whether a series converges, and what it converges to. This chapter will be devoted to the study of these questions.

In this section we will use standard summation notation:

$$\sum_{j=m}^{n} a_j \equiv a_m + a_{m+1} + \cdots + a_n \,.$$

EXAMPLE 7.1 We calculate two sample sums:

$$\sum_{j=2}^{5} j^2 = 2^2 + 3^2 + 4^2 + 5^2 = 54\,,$$

$$\sum_{j=4}^{8} 2^{-j} = 2^{-4} + 2^{-5} + 2^{-6} + 2^{-7} + 2^{-8} = \frac{2^5 - 1}{2^8}\,. \qquad \blacksquare$$

A series is an infinite sum. One of the most effective ways to handle an infinite process in mathematics is with a limit. This consideration leads to the following definition:

Definition 7.2 The formal expression

$$\sum_{j=1}^{\infty} a_j ,$$

where the a_j are real numbers, is called a *series*. For $N = 1, 2, 3, \ldots$, the expression

$$S_N = \sum_{j=1}^{N} a_j = a_1 + a_2 + \ldots a_N$$

is called the Nth *partial sum* of the series. In case

$$\lim_{N \to \infty} S_N$$

exists and is finite, we say that the series *converges*. The limit of the partial sums is called the *sum* of the series. If the series does not converge, then we say that the series *diverges*.

POINT OF CONFUSION 7.3 Notice that the question of convergence of a series, which should be thought of as an *addition process*, reduces to a question about the *sequence* of partial sums.

An obvious way to be misled here is to confuse the roles of sequences and series. The way that we analyze a series is that we think of it as a sequence of partial sums. The series converges if and only if the sequence of partial sums converges.

EXAMPLE 7.4 Consider the series

$$\sum_{j=1}^{\infty} 2^{-j} .$$

The Nth partial sum for this series is

$$S_N = 2^{-1} + 2^{-2} + \cdots + 2^{-N} .$$

In order to determine whether the sequence $\{S_N\}$ has a limit, we rewrite S_N as

$$\begin{aligned} S_N &= \left(2^{-0} - 2^{-1}\right) + \left(2^{-1} - 2^{-2}\right) + \ldots \\ &\quad \left(2^{-N+1} - 2^{-N}\right) . \end{aligned}$$

The expression on the right of the last equation telescopes (i.e., successive pairs of terms cancel) and we find that

$$S_N = 2^{-0} - 2^{-N} .$$

Thus

$$\lim_{N \to \infty} S_N = 2^{-0} = 1 .$$

We conclude that the series converges to 1.   ∎

EXAMPLE 7.5 Let us examine the series

$$\sum_{j=1}^{\infty} \frac{1}{j}$$

for convergence or divergence. (This series is commonly called the *harmonic series* because it describes the harmonics in music.) Now

$$S_1 = 1 = \frac{2}{2}$$

$$S_2 = 1 + \frac{1}{2} = \frac{3}{2}$$

$$S_4 = 1 + \frac{1}{2} + \left(\frac{1}{3} + \frac{1}{4}\right)$$

$$\geq 1 + \frac{1}{2} + \left(\frac{1}{4} + \frac{1}{4}\right) \geq 1 + \frac{1}{2} + \frac{1}{2} = \frac{4}{2}$$

$$S_8 = 1 + \frac{1}{2} + \left(\frac{1}{3} + \frac{1}{4}\right) + \left(\frac{1}{5} + \frac{1}{6} + \frac{1}{7} + \frac{1}{8}\right)$$

$$\geq 1 + \frac{1}{2} + \left(\frac{1}{4} + \frac{1}{4}\right) + \left(\frac{1}{8} + \frac{1}{8} + \frac{1}{8} + \frac{1}{8}\right)$$

$$= \frac{5}{2}.$$

In general this argument shows that

$$S_{2^k} \geq \frac{k+2}{2}.$$

The sequence $\{S_N\}$ is increasing since the series contains only positive summands. The fact that the partial sums $S_1, S_2, S_4, S_8, \ldots$ increase without bound shows that the entire sequence of partial sums must increase without bound. We conclude that the series diverges. ∎

POINT OF CONFUSION 7.6 The harmonic series diverges, but it diverges *very* slowly. For example, the sum of the first million terms of the harmonic series is less than 13.82.

Just as with sequences, we have a Cauchy criterion for series:

Proposition 7.7 *The series $\sum_{j=1}^{\infty} a_j$ converges if and only if, for every $\epsilon > 0$, there is an integer $N > 0$ such that, if $n \geq m > N$, then*

$$\left| \sum_{j=m}^{n} a_j \right| < \epsilon. \tag{7.7.1}$$

The condition (7.7.1) is called the Cauchy criterion for series.

Proof: Suppose that the Cauchy criterion holds. Pick $\epsilon > 0$ and choose N so large that (7.7.1) holds for $n \geq m > N$. If $n \geq m > N$, then

$$|S_n - S_m| = \left| \sum_{j=m+1}^{n} a_j \right| < \epsilon$$

by hypothesis. Thus the sequence $\{S_N\}$ is Cauchy in the sense discussed for sequences in Section 6.1. We conclude that the sequence $\{S_N\}$ converges; by definition, therefore, the series converges.

Conversely, if the series converges then, by definition, the sequence $\{S_N\}$ of partial sums converges. In particular, the sequence $\{S_N\}$ must be Cauchy. Thus, for any $\epsilon > 0$, there is a number $N > 0$ such that if $n \geq m > N$ then

$$|S_n - S_m| < \epsilon.$$

This just says that

$$\left| \sum_{j=m+1}^{n} a_j \right| < \epsilon,$$

and this last inequality is the Cauchy criterion for series.   □

POINT OF CONFUSION 7.8 The Cauchy criterion for series simply says that

$$|S_n - S_m| < \epsilon.$$

So the partial sums are getting closer and closer together.

EXAMPLE 7.9 Let us use the Cauchy criterion to verify that the series

$$\sum_{j=1}^{\infty} \frac{1}{j \cdot (j+1)}$$

converges.

Notice that, if $n \geq m > 1$, then

$$\left| \sum_{j=m}^{n} \frac{1}{j \cdot (j+1)} \right| = \left(\frac{1}{m} - \frac{1}{m+1} \right) + \left(\frac{1}{m+1} - \frac{1}{m+2} \right) + \ldots$$
$$+ \left(\frac{1}{n} - \frac{1}{n+1} \right).$$

The sum on the right plainly telescopes and we have

$$\left| \sum_{j=m}^{n} \frac{1}{j \cdot (j+1)} \right| = \frac{1}{m} - \frac{1}{n+1}.$$

Let $\epsilon > 0$. Let us choose N to be the next integer after $1/\epsilon$. Then, for $n \geq m > N$, we may conclude that

$$\left| \sum_{j=m}^{n} \frac{1}{j \cdot (j+1)} \right| = \frac{1}{m} - \frac{1}{n+1} < \frac{1}{m} < \frac{1}{N} < \epsilon .$$

This is the desired conclusion. The series satisfies the Cauchy criterion for series, so it converges. ∎

The next result gives a necessary condition for a series to converge. It is a useful device for detecting divergent series, although it can never tell us that a series converges.

Proposition 7.10 (The Zero Test) *If the series*

$$\sum_{j=1}^{\infty} a_j$$

converges, then the terms a_j tend to 0 as $j \to \infty$.

Proof: Since we are assuming that the series converges, then it must satisfy the Cauchy criterion for series. Let $\epsilon > 0$. Then there is an integer $N > 0$ such that, if $n \geq m > N$, then

$$\left| \sum_{j=m}^{n} a_j \right| < \epsilon . \tag{7.10.1}$$

We take $n = m$ and $m > N$. Then (7.10.1) becomes

$$|a_m| < \epsilon .$$

But this is precisely the conclusion that we desire. □

EXAMPLE 7.11 The series $\sum_{j=1}^{\infty} (-1)^j$ must diverge, *even though its terms appear to be cancelling each other out.* The reason is that the summands do not tend to zero; hence the preceding proposition applies.

Write out several partial sums of this series to see more explicitly that the partial sums are $-1, 0, -1, 0, \ldots$ and hence that the series diverges. ∎

POINT OF CONFUSION 7.12 Series with nonnegative summands are generally much easier to understand, and to analyze, than series with both positive and negative summands. This is because series of the first type converge because of *size of the terms* alone. But series of the second type can and do converge because of cancellation. Cancellation is subtle.

We conclude this section with a necessary and sufficient condition for convergence of a series of nonnegative terms. As with some of our other results on series, it amounts to little more than a restatement of a result on sequences.

Proposition 7.13 *A series*

$$\sum_{j=1}^{\infty} a_j$$

with all $a_j \geq 0$ is convergent if and only if the sequence of partial sums is bounded above.

Proof: Notice that, because the summands are nonnegative, we have

$$S_1 = a_1 \leq a_1 + a_2 = S_2,$$

$$S_2 = a_1 + a_2 \leq a_1 + a_2 + a_3 = S_3,$$

and in general

$$S_N \leq S_N + a_{N+1} = S_{N+1}.$$

Thus the sequence $\{S_N\}$ of partial sums forms an increasing sequence. We know that such a sequence is convergent to a finite limit if and only if it is bounded above (see Section 6.1). This completes the proof.   □

EXAMPLE 7.14 The series $\sum_{j=1}^{\infty} 1$ is divergent since the summands are nonnegative and the sequence of partial sums $\{S_N\} = \{N\}$ is unbounded.

Referring back to Example 7.5, we see that the series $\sum_{j=1}^{\infty} \frac{1}{j}$ diverges because its partial sums are unbounded.

We see from Example 7.4 that the series $\sum_{j=1}^{\infty} 2^{-j}$ converges because its partial sums are all bounded above by 1.   ∎

It is frequently convenient to begin a series with summation at $j = 0$ or some other term instead of $j = 1$. All of our convergence results still apply to such a series because of the Cauchy criterion. In other words, the convergence or divergence of a series will depend only on the behavior of its "tail."

A Look Back

1. Intuitively speaking, what does it mean for a series to converge?

2. Describe in words what the Cauchy criterion for series says.

3. Give an example of a convergent series.

4. Give an example of a divergent series.

Exercises

1. Discuss convergence or divergence for each of the following series:

(a) $\displaystyle\sum_{j=1}^{\infty} \frac{(2^j)^2}{j!}$

(b) $\displaystyle\sum_{j=1}^{\infty} \frac{(2j)!}{(3j)!}$

(c) $\displaystyle\sum_{j=1}^{\infty} \frac{j!}{j^j}$

(d) $\displaystyle\sum_{j=1}^{\infty} \frac{(-1)^j}{3j^2 - 5j + 6}$

(e) $\displaystyle\sum_{j=1}^{\infty} \frac{2j-1}{3j^2-2}$

(f) $\displaystyle\sum_{j=1}^{\infty} \frac{2j-1}{3j^3-2}$

(g) $\displaystyle\sum_{j=1}^{\infty} \frac{\log(j+1)}{[1+\log j]^j}$

(h) $\displaystyle\sum_{j=12}^{\infty} \frac{1}{j \log^3 j}$

(i) $\displaystyle\sum_{j=2}^{\infty} \frac{\log(2)}{\log j}$

(j) $\displaystyle\sum_{j=2}^{\infty} \frac{1}{j \log^{1.1} j}$

2. If $b_j > 0$ for every j and if $\sum_{j=1}^{\infty} b_j$ converges then prove that $\sum_{j=1}^{\infty} (b_j)^2$ converges. Prove that the assertion is false if the positivity hypothesis is omitted. How about third powers?

3. If $b_j > 0$ for every j and if $\sum_{j=1}^{\infty} b_j$ converges then prove that $\sum_{j=1}^{\infty} \frac{1}{1+b_j}$ diverges.

4. Let $\sum_{j=1}^{\infty} a_j$ be a divergent series of positive terms. Prove that there exist numbers $b_j, 0 < b_j < a_j$, such that $\sum_{j=1}^{\infty} b_j$ diverges.

Similarly, let $\sum_{j=1}^{\infty} c_j$ be a convergent series of positive terms. Prove that there exist numbers $d_j, 0 < c_j < d_j$, such that $\sum_{j=1}^{\infty} d_j$ converges.

Thus we see that there is no "smallest" divergent series and no "largest" convergent series.

5. TRUE or FALSE: If $a_j > c > 0$ and $\sum 1/a_j$ converges, then $\sum a_j$ converges.

6. If $b_j > 0$ and $\sum_j b_j$ converges, then what can you say about $\sum_j b_j/(1 + b_j)$?

7. If $b_j > 0$ and $\sum_j b_j$ diverges, then what can you say about $\sum_j 2^{-j} b_j$?

8. If $b_j > 0$ and $\sum_j b_j$ converges, then what can you say about $\sum_j b_j/j^2$?

9. If $a_j > 0$ and $\sum_j a_j^2$ converges, then what can you say about $\sum_j a_j^4$? How about $\sum_j a_j^3$?

10. Let $b_j > 0$. If $\sum b_j$ converges, then what can you say about convergence of $\sum \sqrt{b_j}$?

11. If $a_j > 0$, $b_j > 0$ and $\sum a_j$ converges and $\sum b_j$ converges, then what can you say about $\sum a_j b_j$?

* **12.** Discuss convergence and divergence for the series $\sum_j (\sin j)/j$ and $\sum_j (\sin j)^2/j$.

7.2   Elementary Convergence Tests

Preliminary Remarks

One of the elegant features of the theory of numerical series is that there are several convergence tests that are easy to apply to get specific, concrete information about convergence. While these tests are not exhaustive nor comprehensive, they do give a good deal of useful information.

The tests that we study in this section are for series of positive terms. More advanced tests will be treated in later sections.

As previously noted, a series may converge because its terms diminish in size fairly rapidly (thus causing its partial sums to grow slowly) or it may converge because of cancellation among the terms. The tests which measure the first type of convergence are the most obvious and these are the "elementary" ones that we discuss in the present section.

Proposition 7.15 (The Comparison Test) *Suppose that $\sum_{j=1}^{\infty} a_j$ is a convergent series of nonnegative terms. If $\{b_j\}$ are real numbers and if $|b_j| \leq a_j$ for every j then the series $\sum_{j=1}^{\infty} b_j$ converges.*

Proof: Because the first series converges, it satisfies the Cauchy criterion for series. Hence, given $\epsilon > 0$, there is an N so large that if $n \geq m > N$ then

$$\sum_{j=m}^{n} a_j = \left| \sum_{j=m}^{n} a_j \right| < \epsilon.$$

But then

$$\left| \sum_{j=m}^{n} b_j \right| \leq \sum_{j=m}^{n} |b_j| \leq \sum_{j=m}^{n} a_j < \epsilon.$$

It follows that the series $\sum b_j$ satisfies the Cauchy criterion for series. Therefore it converges.   □

Corollary 7.16 *If $\sum_{j=1}^{\infty} a_j$ is as in the proposition and if $0 \leq b_j \leq a_j$ for every j then the series $\sum_{j=1}^{\infty} b_j$ converges.*

Proof: Obvious.   □

EXAMPLE 7.17 The series $\sum_{j=1}^{\infty} 2^{-j} \sin j$ is seen to converge by comparing it with the series $\sum_{j=1}^{\infty} 2^{-j}$.   ∎

Theorem 7.18 (The Cauchy Condensation Test) *Assume that* $a_1 \geq a_2 \geq \cdots \geq a_j \geq \ldots 0$. *The series*

$$\sum_{j=1}^{\infty} a_j$$

converges if and only if the series

$$\sum_{k=1}^{\infty} 2^k \cdot a_{2^k}$$

converges.

Proof: First assume that the series $\sum_{j=1}^{\infty} a_j$ converges. Notice that, for each $k \geq 1$,

$$
\begin{aligned}
2^{k-1} \cdot a_{2^k} &= \underbrace{a_{2^k} + a_{2^k} + \cdots + a_{2^k}}_{2^{k-1} \text{ times}} \\
&\leq a_{2^{k-1}+1} + a_{2^{k-1}+2} + \cdots + a_{2^k}. \\
&= \sum_{m=2^{k-1}+1}^{2^k} a_m
\end{aligned}
$$

Therefore

$$\sum_{k=1}^{N} 2^{k-1} \cdot a_{2^k} \leq \sum_{k=1}^{N} \sum_{m=2^{k-1}+1}^{2^k} a_m = \sum_{m=2}^{2^N} a_m .$$

Since the partial sums on the right are bounded (because the series $\sum_j a_j$ converges), so are the partial sums on the left. It follows that the series

$$\sum_{k=1}^{\infty} 2^k \cdot a_{2^k} = 2 \sum_{k=1}^{\infty} 2^{k-1} \cdot a_{2^k}$$

converges.

For the converse, assume that the series

$$\sum_{k=1}^{\infty} 2^k \cdot a_{2^k} \tag{7.18.1}$$

converges. Observe that, for $k \geq 1$,

$$
\begin{aligned}
\sum_{m=2^{k-1}+1}^{2^k} a_j &= a_{2^{k-1}+1} + a_{2^{k-1}+2} + \cdots + a_{2^k} \\[2ex]
&\leq \underbrace{a_{2^{k-1}} + a_{2^{k-1}} + \cdots + a_{2^{k-1}}}_{2^{k-1} \text{ times}} \\[2ex]
&= 2^{k-1} \cdot a_{2^{k-1}} .
\end{aligned}
$$

It follows that

$$
\begin{aligned}
\sum_{m=2}^{2^N} a_j &= \sum_{k=1}^{N} \sum_{m=2^{k-1}+1}^{2^k} a_m \\[2ex]
&\leq \sum_{k=1}^{N} 2^{k-1} \cdot a_{2^{k-1}} .
\end{aligned}
$$

By the hypothesis that the series (7.18.1) converges, the partial sums on the right must be bounded. But then the partial sums on the left are bounded as well. Since the summands a_j are nonnegative, the series on the left converges. $\square$

POINT OF CONFUSION 7.19 In order to apply the Cauchy condensation test correctly, and effectively, it must be that the terms of the series decrease to zero. The proof shows why this needs to be true.

EXAMPLE 7.20 We apply the Cauchy condensation test to the harmonic series

$$
\sum_{j=1}^{\infty} \frac{1}{j} .
$$

It leads us to examine the series

$$
\sum_{k=1}^{\infty} 2^k \cdot \frac{1}{2^k} = \sum_{k=1}^{\infty} 1 .
$$

Since the latter series diverges, the harmonic series diverges as well.   ∎

Proposition 7.21 (Geometric Series) *Let α be a real number. The series*

$$
\sum_{j=0}^{\infty} \alpha^j
$$

is called a geometric series. It converges if and only if $|\alpha| < 1$. In this circumstance, the sum of the series (that is, the limit of the partial sums) is $1/(1-\alpha)$.

Proof: Let S_N denote the Nth partial sum of the geometric series. Then

$$\alpha \cdot S_N = \alpha(1 + \alpha + \alpha^2 + \dots \alpha^N)$$
$$= \alpha + \alpha^2 + \dots \alpha^{N+1} .$$

It follows that $\alpha \cdot S_N$ and S_N are nearly the same: in fact

$$\alpha \cdot S_N + 1 - \alpha^{N+1} = S_N .$$

Solving this equation for the quantity S_N yields

$$S_N = \frac{1 - \alpha^{N+1}}{1 - \alpha}$$

when $\alpha \neq 1$.

If $|\alpha| < 1$ then $\alpha^{N+1} \to 0$, hence the sequence of partial sums tends to the limit $1/(1 - \alpha)$. If $|\alpha| > 1$ then α^{N+1} diverges, hence the sequence of partial sums diverges. This completes the proof for $|\alpha| \neq 1$. But the divergence in case $|\alpha| = 1$ follows because the summands will not tend to zero. □

EXAMPLE 7.22 Consider the series

$$\sum_{j=0}^{\infty} \frac{2^j}{3^j} .$$

Writing the series as

$$\sum_{j=0}^{\infty} \left(\frac{2}{3}\right)^j ,$$

we see that it is a geometric series. Since $|2/3| < 1$, the series converges. Its sum is $1/(1 - 2/3) = 3$. ∎

Corollary 7.23 *The series*

$$\sum_{j=1}^{\infty} \frac{1}{j^r}$$

converges if r is a real number that exceeds 1 and diverges otherwise.

Proof: When $r > 0$ we can apply the Cauchy Condensation Test. This leads us to examine the series

$$\sum_{k=1}^{\infty} 2^k \cdot 2^{-kr} = \sum_{k=1}^{\infty} \left(2^{1-r}\right)^k .$$

This last is a geometric series, with the role of α played by the quantity $\alpha = 2^{1-r}$. When $r > 1$ then $|\alpha| < 1$ so the series converges. Otherwise, by the Cauchy test, it diverges.

Later on, in Proposition 7.33, we learn the Integral Test. This gives another nice way to think about this example. □

EXAMPLE 7.24 Let us apply the Cauchy Condensation Test to the series

$$\sum_{j=1}^{\infty} \frac{1}{j(\log_2 j)^2}.$$

This leads us to examine the series

$$\sum_{k=1}^{\infty} 2^k \cdot \frac{1}{2^k \cdot k^2} = \sum_{k=1}^{\infty} \frac{1}{k^2}.$$

By the preceding corollary, this is a convergent series. So the original series converges.

Theorem 7.25 (The Root Test for Convergence) *Consider the series*

$$\sum_{j=1}^{\infty} a_j.$$

If

$$\limsup_{j \to \infty} |a_j|^{1/j} < 1$$

then the series converges.

Proof: Refer again to the discussion of the concept of limit superior in Chapter 6. By our hypothesis, there is a number $0 < \beta < 1$ and an integer $N > 0$ such that, for all $j > N$, it holds that

$$|a_j|^{1/j} < \beta.$$

In other words,

$$|a_j| < \beta^j.$$

Since $0 < \beta < 1$ the sum of the terms on the right constitutes a convergent geometric series. By the Comparison Test, the sum of the terms on the left converges.   □

Theorem 7.26 (The Ratio Test for Convergence) *Consider a series*

$$\sum_{j=1}^{\infty} a_j.$$

If

$$\limsup_{j \to \infty} \left| \frac{a_{j+1}}{a_j} \right| < 1$$

then the series converges.

Proof: It is possible to supply a proof similar to that of the Root Test. We leave such a proof for the exercises, and instead supply an argument which relates the two tests in an interesting fashion.

Let

$$\lambda = \limsup_{j \to \infty} \left| \frac{a_{j+1}}{a_j} \right| < 1 \,.$$

Select a real number μ such that $\lambda < \mu < 1$. By the definition of lim sup, there is an N so large that, if $j > N$, then

$$\left| \frac{a_{j+1}}{a_j} \right| < \mu \,.$$

This may be rewritten as

$$|a_{j+1}| < \mu \cdot |a_j| \quad , \quad j \geq N \,.$$

Thus (much as in the proof of the Root Test) we have for $k \geq 0$ that

$$|a_{N+k}| \leq \mu \cdot |a_{N+k-1}| \leq \mu \cdot \mu \cdot |a_{N+k-2}| \leq \cdots \leq \mu^k \cdot |a_N| \,.$$

It is convenient to denote $N + k$ by $n, n \geq N$. Thus the last inequality reads

$$|a_n| < \mu^{n-N} \cdot |a_N|$$

or

$$|a_n|^{1/n} < \mu^{(n-N)/n} \cdot |a_N|^{1/n} \,.$$

Remembering that N has been fixed once and for all, we pass to the lim sup as $n \to \infty$. The result is

$$\limsup_{n \to \infty} |a_n|^{1/n} \leq \mu \,.$$

Since $\mu < 1$, we find that our series satisfies the hypotheses of the Root Test. Hence it converges. □

POINT OF CONFUSION 7.27 The proof of the Ratio Test shows that *if* a series passes the Ratio Test then it passes the Root Test (the converse is not true, as you will learn in Exercise 2). Put another way, the Root Test is a better test than the Ratio Test because it will give information whenever the Ratio Test does and also in some circumstances when the Ratio Test does not.

Why do we therefore learn the Ratio Test? The answer is that there are circumstances when the Ratio Test is easier to apply than the Root Test.

EXAMPLE 7.28 The series

$$\sum_{j=1}^{\infty} \frac{2^j}{j!}$$

is easily studied using the Ratio Test (recall that $j! \equiv j \cdot (j-1) \cdot \ldots \cdot 2 \cdot 1$). Indeed $a_j = 2^j/j!$ and

$$\left| \frac{a_{j+1}}{a_j} \right| = \frac{2^{j+1}/(j+1)!}{2^j/j!}.$$

We can perform the division to see that

$$\left| \frac{a_{j+1}}{a_j} \right| = \frac{2}{j+1}.$$

The lim sup of the last expression is 0. By the Ratio Test, the series converges.

Notice that in this example, while the Root Test applies in principle, it would be difficult to use in practice. ∎

POINT OF CONFUSION 7.29 How do we know when to apply the Root Test and when to apply the Ratio Test? The most definitive answer to this question is that you learn from experience.

But we can give these guidelines. If the summands involve products, such as factorials, then the Ratio Test is probably most appropriate. If instead the summands involve powers, then the Root Test is probably most appropriate.

EXAMPLE 7.30 We apply the Root Test to the series

$$\sum_{j=1}^{\infty} \frac{j^2}{2^j}.$$

Observe that

$$a_j = \frac{j^2}{2^j}$$

hence that

$$|a_j|^{1/j} = \frac{\left(j^{1/j}\right)^2}{2}.$$

As $j \to \infty$, we see that

$$\limsup_{j \to \infty} |a_j|^{1/j} = \frac{1}{2}.$$

By the Root Test (refer to Theorem 7.25), the series converges. ∎

It is natural to ask whether the Ratio and Root Tests can detect divergence. Neither test is necessary and sufficient: there are series which elude the analysis of both tests. However, the arguments that we used to establish Theorems 7.25 and 7.26 can also be used to establish the following (the proofs are left as exercises):

Theorem 7.31 (The Root Test for Divergence) *Consider the series*

$$\sum_{j=1}^{\infty} a_j$$

of nonzero terms. If

$$\limsup_{j \to \infty} |a_j|^{1/j} > 1$$

then the series diverges.

Theorem 7.32 (The Ratio Test for Divergence) *Consider the series*

$$\sum_{j=1}^{\infty} a_j$$

of nonzero terms. If

$$\liminf_{j \to \infty} \left| \frac{a_{j+1}}{a_j} \right| > 1,$$

then the series diverges.

In the Root Test, if the lim sup is equal to 1, then no conclusion is possible. In the Ratio Test, if the lim inf is equal to 1, then no conclusion is possible. The exercises give examples of series, some of which converge and some of which do not, in which these tests give lim sup or lim inf equal to 1.

We conclude this section by saying a word about the integral test.

Proposition 7.33 (The Integral Test) *Let f be a continuous, nonnegative function on $[0, \infty)$ that is monotonically decreasing. The series*

$$\sum_{j=1}^{\infty} f(j)$$

converges if and only if the integral

$$\int_{1}^{\infty} f(x)\,dx$$

converges.

We have not treated the integral yet in this book, so we shall not prove the result here. We note that it is easy to apply the integral test to the function $f(x) = 1/x$ to see that the harmonic series diverges.

A Look Back

1. Explain the Cauchy Condensation Test in the language of comparison of series.

2. Why would the Comparison Test not be relevant to a series with both positive and negative terms?

3. What is the slowest converging series that you know?

4. What is the fastest converging series that you know?

Exercises

1. Let p be a polynomial with no constant term. If $b_j > 0$ for every j and if $\sum_{j=1}^{\infty} b_j$ converges then prove that the series $\sum_{j=1}^{\infty} p(b_j)$ converges.

2. Examine the series

$$\frac{1}{3} + \frac{1}{5} + \frac{1}{3^2} + \frac{1}{5^2} + \frac{1}{3^3} + \frac{1}{5^3} + \frac{1}{3^4} + \frac{1}{5^4} + \cdots$$

   Prove that the Root Test shows that the series converges while the Ratio Test gives no information.

3. Check that both the Root Test and the Ratio Test give no information for the series $\sum_{j=1}^{\infty} \frac{1}{j}$, $\sum_{j=1}^{\infty} \frac{1}{j^2}$. However, one of these series is divergent and the other is convergent.

4. Prove Theorem 7.31.

5. Prove Theorem 7.32.

6. Let a_j be a sequence of real numbers. Define

$$m_j = \frac{a_1 + a_2 + \ldots a_j}{j} .$$

   Prove that, if $\lim_{j \to \infty} a_j = \ell$, then $\lim_{j \to \infty} m_j = \ell$. Give an example to show that the converse is not true.

7. Imitate the proof of the Root Test to give a direct proof of the Ratio Test.

8. Let $\sum_j a_j$ and $\sum_j b_j$ be series of positive terms. Prove that, if there is a constant $C > 0$ such that

$$\frac{1}{C} \le \frac{a_j}{b_j} \le C$$

   for all j large, then either both series diverge or both series converge.

9. TRUE or FALSE: If the a_j are positive and $\sum a_j$ converges then $\sum a_j/j$ converges.

10. Use the integral test to determine whether each of these series converges:

   (a) $\sum 1/(j \log j)$
   (b) $\sum 1/j^2$
   (c) $\sum e^{-j}$
   (d) $\sum j/2^j$

11. The Ratio Test tells us nothing about the series

$$\sum_j \frac{1}{j^\alpha |\log j|^\beta}$$

   for $\alpha > 0$, $\beta > 0$. Can you use some other reasoning to comment on the convergence of this series?

12. Is there any value of $\alpha > 0$ for which

$$\sum_j \frac{1}{j |\log j|^\alpha}$$

   converges?

7.3 Advanced Convergence Tests

Preliminary Remarks

Whereas the previous section treated convergence tests for positive series, we now study convergence tests for series that contain both positive and negative terms. These are both more mysterious and more interesting.

The types of series studied here arise in the theory of Fourier series and in the study of solutions of partial differential equations.

In this section we consider convergence tests for series which depend on cancellation among the terms of the series. One of the most profound of these depends on a technique called *Summation by Parts*. You may wonder whether this process is at all related to the "integration by parts" procedure that you learned in calculus—it certainly has a similar form. Indeed it will turn out (and we shall see the details of this assertion as the book develops) that summing a series and performing an integration are two aspects of the same limiting process. The Summation by Parts method is merely our first glimpse of this relationship.

Proposition 7.34 (Summation by Parts) *Let $\{a_j\}_{j=0}^{\infty}$ and $\{b_j\}_{j=0}^{\infty}$ be two sequences of real numbers. For $N = 0, 1, 2, \ldots$ set*

$$A_N = \sum_{j=0}^{N} a_j$$

(we adopt the convention that $A_{-1} = 0$). Then, for any $0 \leq m \leq n < \infty$, it holds that

$$\sum_{j=m}^{n} a_j \cdot b_j = [A_n \cdot b_n - A_{m-1} \cdot b_m]$$
$$+ \sum_{j=m}^{n-1} A_j \cdot (b_j - b_{j+1}).$$

Proof: We write

$$\sum_{j=m}^{n} a_j \cdot b_j = \sum_{j=m}^{n} (A_j - A_{j-1}) \cdot b_j$$

$$= \sum_{j=m}^{n} A_j \cdot b_j - \sum_{j=m}^{n} A_{j-1} \cdot b_j$$

$$= \sum_{j=m}^{n} A_j \cdot b_j - \sum_{j=m-1}^{n-1} A_j \cdot b_{j+1}$$

$$= \sum_{j=m}^{n-1} A_j \cdot (b_j - b_{j+1}) + A_n \cdot b_n - A_{m-1} \cdot b_m .$$

This is what we wished to prove.   □

Now we apply Summation by Parts to prove a convergence test due to Niels Henrik Abel (1802–1829).

Theorem 7.35 (Abel's Convergence Test) *Consider the series*

$$\sum_{j=0}^{\infty} a_j \cdot b_j .$$

Suppose that

1. *The partial sums $A_N = \sum_{j=0}^{N} a_j$ form a bounded sequence;*

2. *$b_0 \geq b_1 \geq b_2 \geq \ldots$;*

3. *$\lim_{j \to \infty} b_j = 0.$*

Then the original series

$$\sum_{j=0}^{\infty} a_j \cdot b_j$$

converges.

Proof: Suppose that the partial sums A_N are bounded in absolute value by a number M. Pick $\epsilon > 0$ and choose an integer N so large that $b_N < \epsilon/(2M)$. For $N < m \leq n < \infty$ we use the partial summation formula to write

$$\left| \sum_{j=m}^{n} a_j \cdot b_j \right| = \left| A_n \cdot b_n - A_{m-1} \cdot b_m + \sum_{j=m}^{n-1} A_j \cdot (b_j - b_{j+1}) \right|$$

$$\leq M \cdot |b_n| + M \cdot |b_m| + M \cdot \sum_{j=m}^{n-1} |b_j - b_{j+1}| .$$

Now we take advantage of the facts that $b_j \geq 0$ for all j and that $b_j \geq b_{j+1}$ for all j to estimate the last expression by

$$M \cdot \left[b_n + b_m + \sum_{j=m}^{n-1} (b_j - b_{j+1}) \right] .$$

[Notice that the expressions $b_j - b_{j+1}, b_m$, and b_n are all nonnegative.] Now the sum collapses and the last line is estimated by

$$M \cdot [b_n + b_m - b_n + b_m] = 2 \cdot M \cdot b_m .$$

By our choice of N, the right side is smaller than ϵ. Thus our series satisfies the Cauchy criterion and therefore converges. $\square$

EXAMPLE 7.36 (THE ALTERNATING SERIES TEST) As a first application of Abel's convergence test, we examine alternating series. Consider a series of the form

$$\sum_{j=1}^{\infty} (-1)^j \cdot b_j , \qquad (7.36.1)$$

with $b_1 \geq b_2 \geq b_3 \geq \cdots \geq 0$ and $b_j \to 0$ as $j \to \infty$. We set $a_j = (-1)^j$ and apply Abel's test. We see immediately that all partial sums A_N are either -1 or 0. In particular, this sequence of partial sums is bounded. And the b_j terms are decreasing and tending to zero. By Abel's convergence test, the alternating series (7.36.1) converges.   ∎

Proposition 7.37 Let $b_1 \geq b_2 \geq \ldots$ and assume that $b_j \to 0$. Consider the alternating series $\sum_{j=1}^{\infty} (-1)^j b_j$ as in the last example. It is convergent: let S be its sum. Then the partial sums S_N satisfy $|S - S_N| \leq b_{N+1}$.

Proof: Observe that

$$|S - S_N| = |b_{N+1} - b_{N+2} + b_{N+3} - + \ldots| .$$

But

$$\begin{aligned} b_{N+2} - b_{N+3} + - \ldots &\leq b_{N+2} + (-b_{N+3} + b_{N+3}) \\ &\qquad + (-b_{N+5} + b_{N+5}) + \ldots \\ &= b_{N+2} \end{aligned}$$

and

$$\begin{aligned} b_{N+2} - b_{N+3} + - \ldots &\geq (b_{N+2} - b_{N+2}) + (b_{N+4} - b_{N+4}) + \ldots \\ &= 0 . \end{aligned}$$

It follows that

$$|S - S_N| \leq |b_{N+1}|$$

as claimed. $\square$

EXAMPLE 7.38 Consider the series

$$\sum_{j=1}^{\infty}(-1)^j\frac{1}{j}\,.$$

Then the partial sum $S_{100} = -.688172$ is within 0.01 (in fact within 1/101) of the full sum S and the partial sum $S_{10000} = -.6930501$ is within 0.0001 (in fact within 1/10001) of the sum S. ∎

EXAMPLE 7.39 Next we examine a series which is important in the study of Fourier analysis. Consider the series

$$\sum_{j=1}^{\infty}\frac{\sin j}{j}\,. \tag{7.39.1}$$

We already know that the series $\sum\frac{1}{j}$ diverges. However, the expression $\sin j$ changes sign in a rather sporadic fashion. We might hope that the series (7.39.1) converges because of cancellation of the summands. We take $a_j = \sin j$ and $b_j = 1/j$. Abel's test will apply if we can verify that the partial sums A_N of the a_j terms are bounded. To see this we use a trick:

Observe that

$$\cos(j + 1/2) = \cos j \cdot \cos 1/2 - \sin j \cdot \sin 1/2$$

and

$$\cos(j - 1/2) = \cos j \cdot \cos 1/2 + \sin j \cdot \sin 1/2\,.$$

Subtracting these equations and solving for $\sin j$ yields that

$$\sin j = \frac{\cos(j - 1/2) - \cos(j + 1/2)}{2 \cdot \sin 1/2}\,.$$

We conclude that

$$A_N = \sum_{j=1}^{N} a_j = \sum_{j=1}^{N} \frac{\cos(j - 1/2) - \cos(j + 1/2)}{2 \cdot \sin 1/2}\,.$$

Of course this sum collapses and we see that

$$A_N = \frac{-\cos(N + 1/2) + \cos 1/2}{2 \cdot \sin 1/2}\,.$$

Thus

$$|A_N| \le \frac{2}{2 \cdot \sin 1/2} = \frac{1}{\sin 1/2}\,,$$

independent of N.

Thus the hypotheses of Abel's test are verified and the series

$$\sum_{j=1}^{\infty}\frac{\sin j}{j}$$

converges. ∎

Remark 7.40 It is interesting to notice that both the series

$$\sum_{j=1}^{\infty} \frac{|\sin j|}{j} \quad \text{and} \quad \sum_{j=1}^{\infty} \frac{\sin^2 j}{j}$$

diverge. The proofs of these assertions are left as exercises for you.

We turn next to the topic of absolute and conditional convergence. A series of real constants

$$\sum_{j=1}^{\infty} a_j$$

is said to be *absolutely convergent* if

$$\sum_{j=1}^{\infty} |a_j|$$

converges. We have:

Proposition 7.41 *If the series $\sum_{j=1}^{\infty} a_j$ is absolutely convergent, then it is convergent.*

Proof: This is an immediate corollary of the Comparison Test.   □

Definition 7.42 A series $\sum_{j=1}^{\infty} a_j$ is said to be *conditionally convergent* if $\sum_{j=1}^{\infty} a_j$ converges, but it does not converge absolutely.

We see that absolutely convergent series are convergent but the next example shows that the converse is not true.

EXAMPLE 7.43 The series

$$\sum_{j=1}^{\infty} \frac{(-1)^j}{j}$$

converges by the Alternating Series Test. However, it is not absolutely convergent because the harmonic series

$$\sum_{j=1}^{\infty} \frac{1}{j}$$

diverges.   ∎

There is a remarkable robustness result for absolutely convergent series that fails dramatically for conditionally convergent series. This result is enunciated in the next theorem. We first need a definition.

Definition 7.44 Let $\sum_{j=1}^{\infty} a_j$ be a given series. Let $\{p_j\}_{j=1}^{\infty}$ be a sequence in which every positive integer occurs once and only once (but not necessarily in the usual order). We call $\{p_j\}$ a *permutation* of the natural numbers.

Then the series

$$\sum_{j=1}^{\infty} a_{p_j}$$

is said to be a *rearrangement* of the given series.

Theorem 7.45 (Riemann, Weierstrass) *If the series $\sum_{j=1}^{\infty} a_j$ of real numbers is absolutely convergent to a (limiting) sum ℓ, then every rearrangement of the series converges also to ℓ.*

If the real series $\sum_{j=1}^{\infty} b_j$ is conditionally convergent and if β is any real number or $\pm\infty$ then there is a rearrangement of the series that converges to β.

Proof: We prove the first assertion here and explore the second in the exercises.

Let us choose a rearrangement of the given series and denote it by $\sum_{j=1}^{\infty} a_{p_j}$, where p_j is a permutation of the positive integers. Pick $\epsilon > 0$. By the hypothesis that the original series converges absolutely we may choose an integer $N > 0$ such that $N < m \leq n < \infty$ implies that

$$\sum_{j=m}^{n} |a_j| < \epsilon. \qquad\qquad (7.45.1)$$

[The presence of the absolute values in the left side of this inequality will prove crucial in a moment.] Choose a positive integer M such that $M \geq N$ and the integers $1, \ldots, N$ are all contained in the list $p_1, p_2, \ldots, p_M$. If $K > M$ then the partial sum $\sum_{j=1}^{K} a_j$ will trivially contain the summands $a_1, a_2, \ldots a_N$. Also the partial sum $\sum_{j=1}^{K} a_{p_j}$ will contain the summands $a_1, a_2, \ldots a_N$. It follows that

$$\sum_{j=1}^{K} a_j - \sum_{j=1}^{K} a_{p_j}$$

will contain only summands *after* the Nth one in the original series. By inequality (7.45.1) we may conclude that

$$\left| \sum_{j=1}^{K} a_j - \sum_{j=1}^{K} a_{p_j} \right| \leq \sum_{j=N+1}^{\infty} |a_j| \leq \epsilon.$$

We conclude that the rearranged series converges; and it converges to the same sum as the original series. $\qquad\qquad \square$

A Look Back

1. What does Summation by Parts say? How does it work?

2. How is Summation by Parts similar to integration by parts?

3. What is Abel's convergence test?

4. Why does an alternating series converge?

Exercises

1. If $1/2 > b_j > 0$ for every j and if $\sum_{j=1}^{\infty} b_j$ converges then prove that $\sum_{j=1}^{\infty} \frac{b_j}{1-b_j}$ converges.

2. Follow these steps to give another proof of the Alternating Series Test: a) prove that the odd partial sums form an increasing sequence; b) prove that the even partial sums form a decreasing sequence; c) prove that every even partial sum majorizes all subsequent odd partial sums; d) use a pinching principle.

3. If $b_j > 0$ and $\sum_{j=1}^{\infty} b_j$ converges then prove that

$$\sum_{j=1}^{\infty} (b_j)^{1/2} \cdot \frac{1}{j^{\alpha}}$$

converges for any $\alpha > 1/2$. Give an example to show that the assertion is false if $\alpha = 1/2$.

4. Let p be a polynomial with integer coefficients and degree at least 1. Let $b_1 \geq b_2 \geq \cdots \geq 0$ and assume that $b_j \to 0$. Prove that if $(-1)^{p(j)}$ is not always positive and not always negative then in fact it will alternate in sign so that $\sum_{j=1}^{\infty} (-1)^{p(j)} \cdot b_j$ will converge.

5. Use Abel's test to see that a series of the form

$$\sum_{j=1}^{\infty} (-1)^{3j} a_j,$$

with the a_j positive numbers tending monotonically to zero, converges.

6. Apply Summation by Parts to the series

$$\sum_{j=1}^{\infty} j 2^{-j}.$$

What can you say about the sum of this series?

* 7. Assume that $\sum_{j=1}^{\infty} b_j$ is a convergent series of positive real numbers. Let $s_j = \sum_{\ell=1}^{j} b_\ell$. Discuss convergence or divergence for the series $\sum_{j=1}^{\infty} s_j \cdot b_j$. Discuss convergence or divergence for the series $\sum_{j=1}^{\infty} \frac{b_j}{1+s_j}$.

* 8. If $b_j > 0$ for every j and if $\sum_{j=1}^{\infty} b_j$ diverges then define $s_j = \sum_{\ell=1}^{j} b_\ell$. Discuss convergence or divergence for the series $\sum_{j=1}^{\infty} \frac{b_j}{s_j}$.

* **9.** Let $\sum_{j=1}^{\infty} b_j$ be a conditionally convergent series of real numbers. Let β be a real number. Prove that there is a rearrangement of the series that converges to β. (**Hint:** First observe that the positive terms of the given series must form a divergent series. Also, the negative terms form a divergent series. Now build the rearrangement by choosing finitely many positive terms whose sum "just exceeds" β. Then add on enough negative terms so that the sum is "just less than" β. Repeat this oscillatory procedure.)

* **10.** What can you say about the convergence or divergence of

$$\sum_{j=1}^{\infty} \frac{(2j+3)^{1/2} - (2j)^{1/2}}{j^{3/4}} \,?$$

* **11.** Find a rearrangement of the series $\sum_j (-1)^j / j$ that converges to 10. Find a rearrangement that converges to π.

7.4   Some Special Series

Preliminary Remarks

When we studied sequences we collected a library of basic sequences to which we could compare other sequences. Just so, in the study of series we want some basic series to which we can compare new series. That is what we shall address in the present section.

We begin with a series that defines a special constant of mathematical analysis.

Definition 7.46 The series

$$\sum_{j=0}^{\infty} \frac{1}{j!} ,$$

where $j! \equiv j \cdot (j-1) \cdot (j-2) \cdots 1$ for $j \geq 1$ and $0! \equiv 1$, is convergent (by the Ratio Test, for instance). Its sum is denoted by the symbol e in honor of the Swiss mathematician Leonhard Euler, who first studied it (see also Example 6.43, where the number e is studied by way of a sequence). We shall see in Proposition 7.47 that these two approaches to the number e are equivalent.

Like the number π, to be considered later in this book, the number e is one which arises repeatedly in a number of contexts in mathematics. It has many special properties. We first relate the series definition of e to the sequence definition:

Proposition 7.47 The limit

$$\lim_{n \to \infty} \left(1 + \frac{1}{n}\right)^n$$

exists and equals e.

Proof: We need to compare the quantities

$$A_N \equiv \sum_{j=0}^{N} \frac{1}{j!} \quad \text{and} \quad B_N \equiv \left(1 + \frac{1}{N}\right)^N.$$

We use the Binomial theorem to expand B_N:

$$
\begin{aligned}
B_N &= 1 + \frac{N}{1} \cdot \frac{1}{N} + \frac{N \cdot (N-1)}{2 \cdot 1} \cdot \frac{1}{N^2} + \frac{N \cdot (N-1) \cdot (N-2)}{3 \cdot 2 \cdot 1} \cdot \frac{1}{N^3} + \cdots \\
&\quad \frac{N}{1} \cdot \frac{1}{N^{N-1}} + 1 \cdot \frac{1}{N^N} \\
&= 1 + 1 + \frac{1}{2!} \cdot \frac{N-1}{N} + \frac{1}{3!} \cdot \frac{N-1}{N} \cdot \frac{N-2}{N} + \cdots \\
&\quad + \frac{1}{(N-1)!} \cdot \frac{N-1}{N} \cdot \frac{N-2}{N} \cdots \frac{2}{N} \\
&\quad + \frac{1}{N!} \cdot \frac{N-1}{N} \cdot \frac{N-2}{N} \cdots \frac{1}{N} \\
&= 1 + 1 + \frac{1}{2!} \cdot \left(1 - \frac{1}{N}\right) + \frac{1}{3!} \cdot \left(1 - \frac{1}{N}\right) \cdot \left(1 - \frac{2}{N}\right) + \cdots \\
&\quad + \frac{1}{(N-1)!} \cdot \left(1 - \frac{1}{N}\right) \cdot \left(1 - \frac{2}{N}\right) \cdots \left(1 - \frac{N-2}{N}\right) \\
&\quad + \frac{1}{N!} \cdot \left(1 - \frac{1}{N}\right) \cdot \left(1 - \frac{2}{N}\right) \cdots \left(1 - \frac{N-1}{N}\right).
\end{aligned}
$$

Notice that every summand that appears in this last equation is positive. Thus, for $0 \leq M \leq N$,

$$
\begin{aligned}
B_N &\geq 1 + 1 + \frac{1}{2!} \cdot \left(1 - \frac{1}{N}\right) + \frac{1}{3!} \cdot \left(1 - \frac{1}{N}\right) \cdot \left(1 - \frac{2}{N}\right) \\
&\quad + \cdots + \frac{1}{M!} \cdot \left(1 - \frac{1}{N}\right) \left(1 - \frac{2}{N}\right) \cdots \left(1 - \frac{M-1}{N}\right).
\end{aligned}
$$

In this last inequality we hold M fixed and let N tend to infinity. The result is that

$$\liminf_{N \to \infty} B_N > 1 + 1 + \frac{1}{2!} + \frac{1}{3!} + \cdots + \frac{1}{M!} = A_M.$$

Now, as $M \to \infty$, the quantity A_M converges to e (by the *definition* of e). So we obtain

$$\liminf_{N \to \infty} B_N \geq e. \tag{7.47.1}$$

On the other hand, our expansion for B_N allows us to observe that $B_N \leq A_N$. Thus

$$\limsup_{N \to \infty} B_N \leq e. \tag{7.47.2}$$

Combining (7.47.1) and (7.47.2), we find that

$$e \leq \liminf_{N \to \infty} B_N \leq \limsup_{N \to \infty} B_N \leq e$$

hence that $\lim_{N \to \infty} B_N$ exists and equals e. This is the desired result.   □

Remark 7.48 The last proof illustrates the value of the concepts of lim inf and lim sup. For we do not know in advance that the limit of the expressions B_N exists, much less that the limit equals e. However, the lim inf and the lim sup always exist. So we estimate those instead, and find that they are equal and that they equal e.

A Look Back

1. What is the series representation for e^x?

2. What is the series representation for $\sin x$?

3. Why is the power series representation for e^x unique?

4. Give another series representation for e^x besides the power series representation.

Exercises

1. Use induction to prove a formula for the sum of the first N perfect squares.

2. A real number s is called *algebraic* if it satisfies a polynomial equation of the form

$$a_0 + a_1 x + a_2 x^2 + \cdots + a_m x^m = 0$$

with the coefficients a_j being integers and $a_m \neq 0$. Prove that, if we replace the word "integers" in this definition with "rational numbers," then the set of algebraic numbers remains the same. Prove that $n^{p/q}$ is algebraic for any positive integers p, q, n. A number which is not algebraic is called *transcendental.*

3. Refer to Example 7.39 and Remark 7.40. What can you say about the convergence of $\sum_j [\sin j]^k / j$ for k a positive integer?

4. Discuss convergence of $\sum_j 1/[\ln j]^k$ for k a positive integer.

5. Discuss convergence of $\sum_j 1/p(j)$ for p a polynomial.

6. Discuss convergence of $\sum_j \exp(p(j))$ for p a polynomial.

7. Give a series expansion for $\ln n$.

* 8. At least one of the numbers $e + \pi$ and $e - \pi$ is transcendental. Explain why.

9. Give a power series expansion for $\tan x$.

10. Discuss convergence of the series

$$\sum_j \frac{2^j}{j!} .$$

11. Give sufficient conditions on $b_j > 0$ to guarantee that

$$\sum_j \frac{b_j}{j}$$

converges.

* **12.** Use the power series expansion of $\log x$ to prove that $\log(a \cdot b) = \log a + \log b$.

7.5 Operations on Series

Preliminary Remarks

Since series are algebraic objects, it is natural that we would want to perform arithmetic operations on them. In the present section we examine these operations, and see how series behave in this context.

Some operations on series, such as addition, subtraction, and scalar multiplication, are straightforward. Others, such as multiplication, entail subtleties. This section treats all these matters.

Proposition 7.49 *Let*

$$\sum_{j=1}^{\infty} a_j \quad and \quad \sum_{j=1}^{\infty} b_j$$

be convergent series of real numbers; assume that the series sum to limits α and β respectively. Then

(a) *The series $\sum_{j=1}^{\infty} (a_j + b_j)$ converges to the limit $\alpha + \beta$.*

(b) *If c is a constant then the series $\sum_{j=1}^{\infty} c \cdot a_j$ converges to $c \cdot \alpha$.*

Proof: We shall prove assertion **(a)** and leave the easier assertion **(b)** as an exercise.

Pick $\epsilon > 0$. Choose an integer N_1 so large that $n > N_1$ implies that the partial sum $S_n \equiv \sum_{j=1}^{n} a_j$ satisfies $|S_n - \alpha| < \epsilon/2$. Choose N_2 so large that $n > N_2$ implies that the partial sum $T_n \equiv \sum_{j=1}^{n} b_j$ satisfies $|T_n - \beta| < \epsilon/2$. If U_n is the nth partial sum of the series $\sum_{j=1}^{\infty} (a_j + b_j)$ and if $n > N_0 \equiv \max(N_1, N_2)$ then

$$|U_n - (\alpha + \beta)| \leq |S_n - \alpha| + |T_n - \beta| < \frac{\epsilon}{2} + \frac{\epsilon}{2} = \epsilon.$$

Thus the sequence $\{U_n\}$ converges to $\alpha + \beta$. This proves part **(a)**. The proof of **(b)** is similar. $\square$

Of course subtraction of series is covered by part **(a)** of the last proposition.

In order to keep our discussion of multiplication of series as straightforward as possible, we deal at first with absolutely convergent series. It is convenient in this discussion to begin our sums at $j = 0$ instead of $j = 1$. If we wish to multiply

$$\sum_{j=0}^{\infty} a_j \quad \text{and} \quad \sum_{j=0}^{\infty} b_j,$$

then we need to specify what the partial sums of the product series should be. An obvious necessary condition that we wish to impose is that, if the first series converges to α and the second converges to β, then the product series, whatever we define it to be, should converge to $\alpha \cdot \beta$.

The naive method for defining the summands of the product series $\sum_j c_j$ is to let $c_j = a_j \cdot b_j$. However, a glance at the product of two partial sums of the given series shows that such a definition would be ignoring the distributivity of multiplication over addition.

Cauchy's idea was that the summands for the product series should be

$$c_n \equiv \sum_{j=0}^{n} a_j \cdot b_{n-j}.$$

This particular form for the summands can be easily motivated using power series considerations. For now we concentrate on verifying that this "Cauchy product" of two series really works.

Theorem 7.50 *Let $\sum_{j=0}^{\infty} a_j$ and $\sum_{j=0}^{\infty} b_j$ be two absolutely convergent series which converge to limits α and β respectively. Define the series $\sum_{m=0}^{\infty} c_m$ with summands $c_m = \sum_{j=0}^{m} a_j \cdot b_{m-j}$. Then the series $\sum_{m=0}^{\infty} c_m$ converges to $\alpha \cdot \beta$.*

Proof: Let $A_n, B_n,$ and C_n be the partial sums of the three series in question. We calculate that

$$
\begin{aligned}
C_n &= (a_0 b_0) + (a_0 b_1 + a_1 b_0) + (a_0 b_2 + a_1 b_1 + a_2 b_0) \\
&\quad + \cdots + (a_0 b_n + a_1 b_{n-1} + \cdots + a_n b_0) \\
&= a_0 \cdot B_n + a_1 \cdot B_{n-1} + a_2 \cdot B_{n-2} + \cdots + a_n \cdot B_0.
\end{aligned}
$$

We set $\lambda_n = B_n - \beta$, each n, and rewrite the last line as

$$
\begin{aligned}
C_n &= a_0(\beta + \lambda_n) + a_1(\beta + \lambda_{n-1}) + \cdots a_n(\beta + \lambda_0) \\
&= A_n \cdot \beta + [a_0 \lambda_n + a_1 \cdot \lambda_{n-1} + \cdots + a_n \cdot \lambda_0].
\end{aligned}
$$

Denote the expression in square brackets by the symbol ρ_n. Suppose that we could show that $\lim_{n \to \infty} \rho_n = 0$. Then we would have

$$
\begin{aligned}
\lim_{n \to \infty} C_n &= \lim_{n \to \infty} (A_n \cdot \beta + \rho_n) \\
&= (\lim_{n \to \infty} A_n) \cdot \beta + (\lim_{n \to \infty} \rho_n) \\
&= \alpha \cdot \beta + 0 \\
&= \alpha \cdot \beta.
\end{aligned}
$$

Thus it is enough to examine the limit of the expressions ρ_n.

Since $\sum_{j=1}^{\infty} a_j$ is absolutely convergent, we know that $A = \sum_{j=1}^{\infty} |a_j|$ is a finite number. Choose $\epsilon > 0$. Since $\sum_{j=1}^{\infty} b_j$ converges to β it follows that $\lambda_n \to 0$. Thus we may choose an integer $N > 0$ such that $n > N$ implies that $|\lambda_n| < \epsilon$. Thus, for $n = N + k$, $k > 0$, we may estimate

$$
\begin{aligned}
|\rho_{N+k}| &\leq |\lambda_0 a_{N+k} + \lambda_1 a_{N+k-1} + + \cdots + \lambda_N a_k| \\
&\quad + |\lambda_{N+1} a_{k-1} + \lambda_{N+2} a_{k-2} + \cdots + \lambda_{N+k} a_0| \\
&\leq |\lambda_0 a_{N+k} + \lambda_1 a_{N+k-1} + + \cdots + \lambda_N a_k| \\
&\quad + \max_{p \geq 1}\{|\lambda_{N+p}|\} \cdot (|a_{k-1}| + |a_{k-2}| + \cdots + |a_0|) \\
&\leq (N+1) \cdot \max_{\ell \geq k}|a_\ell| \cdot \max_{0 \leq j \leq N}|\lambda_j| + \epsilon \cdot A.
\end{aligned}
$$

With N fixed, we let $k \to \infty$ in the last inequality. Since $\max_{\ell \geq k}|a_\ell| \to 0$, we find that

$$
\limsup_{n \to \infty} |\rho_n| \leq \epsilon \cdot A.
$$

Since $\epsilon > 0$ was arbitrary, we conclude that

$$
\lim_{n \to \infty} |\rho_n| \to 0.
$$

This completes the proof. $\qquad\qquad\qquad\qquad\qquad\qquad\qquad\qquad\square$

POINT OF CONFUSION 7.51 The idea of the product of series is sophisticated. But, if we keep in mind how we multiply polynomials, then the ideas will fall into place. Namely,

$$
\begin{aligned}
&(a_0 + a_1 x + a_2 x^2 + a_3 x^3 + \cdots + a_k x^k) \cdot \\
&(b_0 + b_1 x + b_2 x^2 + b_3 x^3 + \cdots + a_k x_k) \\
= \;&a_0 b_0 + (a_0 b_1 + a_1 b_0)x + (a_0 b_2 + a_1 b_1 + a_2 b_0)x^2 \\
&+ (a_0 b_3 + a_1 b_2 + a_2 b_1 + a_3 b_0)x^3 \\
&+ \cdots + (a_0 b_k + a_1 b_{k-1} + a_2 b_{k-2} \\
&+ \cdots + a_k b_0)x^k + \cdots.
\end{aligned}
$$

And we recognize the coefficients c_n as the coefficients of this product.

The main point to remember is that we do *not* multiply series termwise.

Notice that, in the proof of the theorem, we really only used the fact that one of the given series was absolutely convergent, not that both were absolutely convergent. Some hypothesis of this nature is necessary, as the following example shows.

EXAMPLE 7.52 Consider the Cauchy product of the two conditionally convergent series

$$\sum_{j=0}^{\infty} \frac{(-1)^j}{\sqrt{j+1}} \quad \text{and} \quad \sum_{j=0}^{\infty} \frac{(-1)^j}{\sqrt{j+1}}.$$

Observe that

$$
\begin{aligned}
c_m &= \frac{(-1)^0(-1)^m}{\sqrt{1}\sqrt{m+1}} + \frac{(-1)^1(-1)^{m-1}}{\sqrt{2}\sqrt{m}} + \cdots \\
&\quad + \frac{(-1)^m(-1)^0}{\sqrt{m+1}\sqrt{1}} \\
&= \sum_{j=0}^{m} (-1)^m \frac{1}{\sqrt{(j+1)\cdot(m+1-j)}}.
\end{aligned}
$$

However, for $0 \le j \le m$,

$$(j+1)\cdot(m+1-j) \le (m+1)\cdot(m+1) = (m+1)^2.$$

Thus

$$|c_m| \ge \sum_{j=0}^{m} \frac{1}{m+1} = 1.$$

We thus see that the terms of the series $\sum_{m=0}^{\infty} c_m$ do not tend to zero, so the series cannot converge. ∎

A Look Back

1. Suppose that $\sum_j a_j$ and $\sum_j b_j$ are absolutely convergent series. Can you make sense of the Cauchy product of $\sum_j a_j^2$ and $\sum_j b_j^2$?

2. Where does the hypothesis of absolute convergence come into play in our argument for the validity of the Cauchy product?

3. Does the Cauchy product distribute over addition?

4. Is the Cauchy product commutative?

Exercises

1. Let $\sum_{j=1}^{\infty} a_j$ and $\sum_{j=1}^{\infty} b_j$ be convergent series of positive real numbers. Discuss convergence of $\sum_{j=1}^{\infty} a_j b_j$.

2. Prove Proposition 7.49(b).

3. Calculate the Cauchy product of the series $\sum_j 1/j^2$ and the series $\sum_j 1/j^4$.

4. Look up the definition of "ring" on Google. Prove that the set of all absolutely convergent series forms a ring.

5. How many different divergent series are there? How many different absolutely convergent series are there?

6. Calculate the Cauchy product of $\sum_j x^j$ and $\sum_j x^{2j}$.

7. Give an example of terms $a_j > 0$ so that $\sum_j a_j$ converges and also $\sum_j \log a_j$ converges.

8. Show that it is never the case that if $a_j > 0$ and $\sum a_j$ converges then $\sum_j e^{a_j}$ converges.

9. TRUE or FALSE: If $a_j > 0$ and $\sum a_j$ converges then $\sum \sin a_j$ converges.

*   10. Let $\sum_{j=1}^{\infty} a_j$ and $\sum_{j=1}^{\infty} b_j$ be convergent series of positive real numbers. Discuss division of these two series. Use the idea of the Cauchy product.

*   11. Discuss the concept of the exponential of a power series.

Chapter 8

Basic Topology

8.1 Open and Closed Sets

Preliminary Remarks

Topology is, in a sense, a generalization of classical Euclidean geometry. But, whereas classical geometry studies rigid equivalences of triangles and rectangles, topology studies more flexible equivalences of a great variety of shapes. The subject of topology as we know it today was founded largely by Henri Poincaré. Over the course of the twentieth century it has developed into a lively and intensely studied discipline. This section is your introduction to the subject.

To specify a topology on a set is to describe certain subsets that will play the role of neighborhoods. These are called *open sets*.

In what follows, we will use "interval notation": If $a \leq b$ are real numbers then we define

$$
\begin{aligned}
(a,b) &= \{x \in \mathbb{R} : a < x < b\}, \\
[a,b] &= \{x \in \mathbb{R} : a \leq x \leq b\}, \\
[a,b) &= \{x \in \mathbb{R} : a \leq x < b\}, \\
(a,b] &= \{x \in \mathbb{R} : a < x \leq b\}.
\end{aligned}
$$

Intervals of the form (a,b) are called *open*. Those of the form $[a,b]$ are called *closed*. The other two are termed *half-open* or *half-closed*. See Figure 8.1.

Now we extend the terms "open" and "closed" to more general sets.

Definition 8.1 A set $U \subset \mathbb{R}$ is called *open* if, for each $x \in U$, there is an $\epsilon > 0$ such that the interval $(x - \epsilon, x + \epsilon)$ is contained in U. See Figure 8.2.

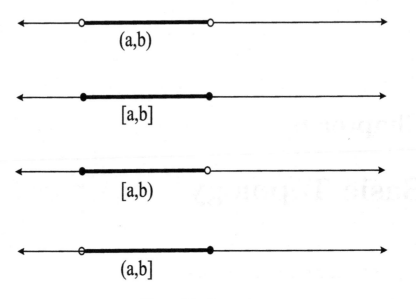

Figure 8.1: Intervals.

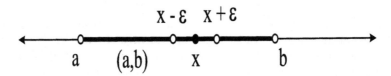

Figure 8.2: An open set.

Remark 8.2 The interval $(x - \epsilon, x + \epsilon)$ is frequently termed a *neighborhood* of x.

EXAMPLE 8.3 The set $U = \{x \in \mathbb{R} : |x - 3| < 2\}$ is open. To see this, choose a point $x \in U$. Let $\epsilon = 2 - |x - 3| > 0$. Then we claim that the interval $I = (x - \epsilon, x + \epsilon) \subset U$.

For, if $t \in I$, then

$$
\begin{aligned}
|t - 3| &= |(t - x) + (x - 3)| \\
&\leq |t - x| + |x - 3| \\
&< \epsilon + |x - 3| \\
&= (2 - |x - 3|) + |x - 3| \\
&= 2.
\end{aligned}
$$

But this means that $t \in U$.

We have shown that $t \in I$ implies $t \in U$. Therefore $I \subset U$. It follows from the definition that U is open. ∎

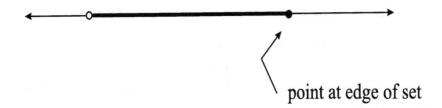

point at edge of set

Figure 8.3: A set that is not open.

POINT OF CONFUSION 8.4 The way to think about the definition of open set is that a set is open when none of its elements is at the "edge" of the set—each element is surrounded by other elements of the set, indeed a whole interval of them. See Figure 8.3. The remainder of this section will make these comments precise.

Proposition 8.5 *If U_α are open sets, for α in some (possibly uncountable) index set A, then*

$$U = \bigcup_{\alpha \in A} U_\alpha$$

is open.

Proof: Let $x \in U$. By definition of union, the point x must lie in some U_α. But U_α is open. Therefore there is an interval $I = (x - \epsilon, x + \epsilon)$ such that $I \subset U_\alpha$. Therefore certainly $I \subset U$. This proves that U is open. $\square$

Proposition 8.6 *If $U_1, U_2, \ldots, U_k$ are open sets then the set*

$$V = \bigcap_{j=1}^{k} U_j$$

is also open.

Proof: Let $x \in V$. Then $x \in U_j$ for each j. Since each U_j is open there is for each j a positive number ϵ_j such that $I_j = (x - \epsilon_j, x + \epsilon_j)$ lies in U_j. Set $\epsilon = \min\{\epsilon_1, \ldots, \epsilon_k\}$. Then $\epsilon > 0$ and $(x - \epsilon, x + \epsilon) \subset I_j \subset U_j$ for every j. But that just means that $(x - \epsilon, x + \epsilon) \subset V$. Therefore V is open. $\square$

Notice the difference between these two propositions: arbitrary unions of open sets are open. But, in order to guarantee that an intersection of open sets is still open, we had to assume that we were only intersecting finitely many such sets. To understand this matter, bear in mind the example of the open sets

$$U_j = \left(-\frac{1}{j}, \frac{1}{j} \right) , \quad j = 1, 2, \ldots .$$

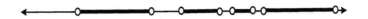

Figure 8.4: Structure of an open set.

These are infinitely many open sets. The intersection of the open sets U_j is the singleton $\{0\}$, which is not open.

POINT OF CONFUSION 8.7 It is natural to think of an open set as an interval without endpoints—what we usually call an open interval. And that is not far from the whole truth. We shall prove below that absolutely any open set is the disjoint union of at most countably many open intervals.

The same analysis as in the first example shows that, if $a < b$, then the interval (a, b) is an open set. On the other hand, intervals of the form $(a, b]$ or $[a, b)$ or $[a, b]$ are *not* open. In the first instance, the point b is the center of no interval $(b - \epsilon, b + \epsilon)$ contained in $(a, b]$. Think about the other two intervals to understand why they are not open. We call intervals of the form (a, b) *open intervals*.

We are now in a position to give a complete description of all open sets.

Proposition 8.8 *Let $U \subset \mathbb{R}$ be a nonempty open set. Then there are either finitely many or countably many pairwise disjoint open intervals I_j such that*

$$U = \bigcup_{j=1}^{\infty} I_j .$$

See Figure 8.4.

Proof: Assume that U is an open subset of the real line. We define an equivalence relation on the set U. The resulting equivalence classes (see Section 4.1) will be the open intervals I_j.

Let a and b be elements of U. We say that a is related to b if all real numbers between a and b are also elements of U. It is obvious that this relation is both reflexive and symmetric. For transitivity notice that if a is related to b and b is related to c then (assuming that a, b, c are distinct) one of the numbers a, b, c must lie between the other two. Assume for simplicity that $a < b < c$. Then all numbers between a and c lie in U, for all such numbers are either between a and b or between b and c or are b itself. Thus a is related to c. (The other possible orderings of a, b, c are left for you to consider.)

Thus we have an equivalence relation on the set U. Call the equivalence classes $\{U_\alpha\}_{\alpha \in A}$. We claim that each U_α is an open interval. In fact if a, b are elements of some U_α then all points between a and b are in U. But then a moment's thought shows that each of those "in between" points is related to both a and b. Therefore all points between a and b are elements of U_α. We conclude that U_α is an interval. Is it an *open* interval?

Figure 8.5: A closed set.

Let $x \in U_\alpha$. Then $x \in U$ so that there is an open interval $I = (x - \epsilon, x + \epsilon)$ contained in U. But x is related to all the elements of I; it follows that $I \subset U_\alpha$. Therefore U_α is open.

We have exhibited the set U as a union of open intervals. These intervals are pairwise disjoint because they arise as the equivalence classes of an equivalence relation. Finally, each of these open intervals contains a (different) rational number (why?). Therefore there can be at most countably many of the intervals U_α.   □

Definition 8.9 A subset $F \subset \mathbb{R}$ is called *closed* if the complement $\mathbb{R} \setminus F$ is open. See Figure 8.5.

EXAMPLE 8.10 The set $[0, 1]$ is closed. For its complement is

$$(-\infty, 0) \cup (1, \infty),$$

which is certainly open.   ∎

EXAMPLE 8.11 An interval of the form $[a, b] = \{x : a \leq x \leq b\}$ is closed. For its complement is $(-\infty, a) \cup (b, \infty)$, which is the union of two open intervals.

The finite set $A = \{-4, -2, 5, 13\}$ is closed because its complement is

$$(-\infty, -4) \cup (-4, -2) \cup (-2, 5) \cup (5, 13) \cup (13, \infty),$$

which is open.

The set $B = \{1, 1/2, 1/3, 1/4, \ldots\} \cup \{0\}$ is closed, for its complement is the set

$$(-\infty, 0) \cup \left\{ \bigcup_{j=1}^{\infty} (1/(j+1), 1/j) \right\} \cup (1, \infty),$$

which is open.

Verify for yourself that if the point 0 is omitted from the set B, then the set is no longer closed.   ∎

POINT OF CONFUSION 8.12 A very common point of confusion among beginners is to think that *any* set is either open or closed. This is simply not true. For example, the set $S = [0, 1)$ is not open (because it contains its left endpoint) and not closed (because it does not contain its right endpoint). The fact is that most sets of reals are *neither* open nor closed.

Proposition 8.13 *If E_α are closed sets, for α in some (possibly uncountable) index set A, then*

$$E = \bigcap_{\alpha \in A} E_\alpha$$

is closed.

Proof: This is just the contrapositive of Proposition 8.5 above: if U_α is the complement of E_α, each α, then U_α is open. Then $U = \cup U_\alpha$ is also open. But then

$$E = \bigcap E_\alpha = \bigcap {}^c(U_\alpha) = {}^c\left(\bigcup U_\alpha\right) = {}^cU$$

is closed. Here ${}^cS$ denotes the complement of a set S. □

The fact that the set B in the last example is closed, but that $B \setminus \{0\}$ is not, is placed in perspective by the next proposition.

Proposition 8.14 *Let S be a set of real numbers. Then S is closed if and only if every Cauchy sequence $\{s_j\}$ of elements of S has a limit which is also an element of S.*

Proof: First suppose that S is closed and let $\{s_j\}$ be a Cauchy sequence in S. We know, since the reals are complete, that there is an element $s \in \mathbb{R}$ such that $s_j \to s$. The point of this half of the proof is to see that $s \in S$. If this statement were false then $s \in U = \mathbb{R} \setminus S$. But U must be open since it is the complement of a closed set. Thus there is an $\epsilon > 0$ such that the interval $I = (s - \epsilon, s + \epsilon) \subset U$. This means that no element of S lies in I. In particular, $|s - s_j| \geq \epsilon$ for every j. This contradicts the statement that $s_j \to s$. We conclude that $s \in S$.

Conversely, assume that every Cauchy sequence in S has its limit in S. If S were not closed then its complement would not be open. Hence there would be a point $t \in \mathbb{R} \setminus S$ with the property that no interval $(t - \epsilon, t + \epsilon)$ lies in $\mathbb{R} \setminus S$. In other words, $(t - \epsilon, t + \epsilon) \cap S \neq \emptyset$ for every $\epsilon > 0$. Thus for $j = 1, 2, 3, \ldots$ we may choose a point $s_j \in (t - 1/j, t + 1/j) \cap S$. It follows that $\{s_j\}$ is a sequence of elements of S that converge to $t \in \mathbb{R} \setminus S$. That contradicts our hypothesis. We conclude that S must be closed. □

Let S be a subset of $\mathbb{R}$. A point x is called an *accumulation point* of S if every neighborhood of x contains infinitely many distinct elements of S. See Figure 8.6. In particular, x is an accumulation point of S if it is the limit of a sequence of distinct elements in S. The last proposition tells us that closed sets are characterized by the property that they contain all of their accumulation points.

A Look Back

1. Give a verbal description of an open set.

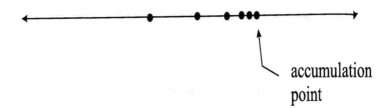

Figure 8.6: The idea of an accumulation point.

2. Give a verbal description of a closed set.

3. Give an example of a set that is neither open nor closed.

4. What is an accumulation point?

Exercises

1. Let S be any set and $\epsilon > 0$. Define $T = \{t \in \mathbb{R} : |t - s| < \epsilon \text{ for some } s \in S\}$. Prove that T is open.

2. Give an example of nonempty *closed* sets $X_1 \supseteq X_2 \supseteq \ldots$ such that $\cap_j X_j = \emptyset$.

3. Give an example of nonempty *closed* sets $X_1 \subset X_2 \ldots$ such that $\cup_j X_j$ is open.

4. Give an example of open sets $U_1 \supseteq U_2 \ldots$ such that $\cap_j U_j$ is closed and nonempty.

5. Exhibit a countable collection of open sets U_j such that each open set $\mathcal{O} \subset \mathbb{R}$ can be written as a union of some of the sets U_j.

6. Let S be any closed set and define, for $x \in \mathbb{R}$,
$$\text{dis}(x, S) = \inf\{|x - s| : s \in S\}.$$
Prove that, if $x \notin S$, then $\text{dis}(x, S) > 0$. If $x, y \in \mathbb{R}$ then prove that
$$|\text{dis}(x, S) - \text{dis}(y, S)| \leq |x - y|.$$

7. Let S be a set of real numbers. If S is not open then must it be closed? If S is not closed then must it be open?

8. Let E be a closed set and F a closed and bounded set. Assume that $E \cap F = \emptyset$. Show that there is an $\epsilon > 0$ so that
$$|e - f| > \epsilon$$
for all $e \in E$ and $f \in F$.

9. Show that the conclusion of Exercise 8 is false if E and F are both closed but not bounded.

* 10. The *closure* of a set S is the intersection of all closed sets that contain S. We denote the closure of S by $\overline{S}$. Call a set S *robust* if it is the closure of its interior (where the *interior* of S is the set of all $x \in S$ so that there is an $\epsilon > 0$ with $(x - \epsilon, x + \epsilon) \subset S$). Which sets of reals are robust?

* 11. Let S be an uncountable subset of $\mathbb{R}$. Prove that S must have infinitely many accumulation points. Must it have uncountably many?

* 12. Let S be any set and define $V = \{t \in \mathbb{R} : |t - s| \leq 1 \text{ for some } s \in S\}$. Is V necessarily closed?

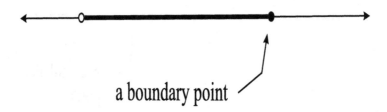

a boundary point

Figure 8.7: The idea of a boundary point.

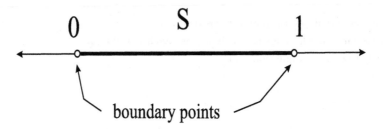

boundary points

Figure 8.8: Boundary of the open unit interval.

8.2   Further Properties of Open and Closed Sets

Preliminary Remarks

Open and closed sets are the basic elements of topology. Understanding the topology of a space consists in understanding its open and closed sets. In this section we move beyond the basics and explore some of the deeper properties of open and closed sets.

Let $S \subset \mathbb{R}$ be a set. We call $b \in \mathbb{R}$ a *boundary point* of S if every nonempty neighborhood $(b - \epsilon, b + \epsilon)$ contains both points of S and points of $\mathbb{R} \setminus S$. See Figure 8.7. We denote the set of boundary points of S by ∂S.

A boundary point b might lie in S and might lie in the complement of S. The next example serves to illustrate the concept:

EXAMPLE 8.15 Let S be the interval $(0, 1)$. Then no point of $(0, 1)$ is in the boundary of S since every point of $(0, 1)$ has a neighborhood that lies entirely inside $(0, 1)$ (in other words, every point is an interior point—see Exercise 10 of the last section). Also, no point of the complement of $T = [0, 1]$ lies in the boundary of T for a similar reason.

Indeed, the only candidates for elements of the boundary of S are 0 and 1. See Figure 8.8. The point 0 *is* an element of the boundary since every neighborhood $(0 - \epsilon, 0 + \epsilon)$ contains the point $\epsilon/2 \in S$ and the point $-\epsilon/2 \in \mathbb{R} \setminus S$. A similar calculation shows that 1 lies in the boundary of S.

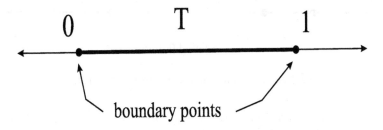

Figure 8.9: Boundary of the closed unit interval.

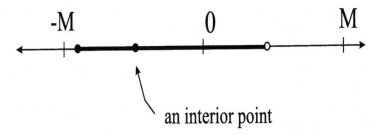

Figure 8.10: The idea of an interior point.

Now consider the set $T = [0, 1]$. Certainly there are no boundary points in $(0, 1)$, for the same reason as in the first paragraph. And there are no boundary points in $\mathbb{R} \setminus [0, 1]$, since that set is open. Thus the only candidates for elements of the boundary are 0 and 1. As in the first paragraph, these are both indeed boundary points for T. See Figure 8.9.

Notice that neither of the boundary points of S lie in S while both of the boundary points of T lie in T. ∎

EXAMPLE 8.16 The boundary of the set $\mathbb{Q} \subset \mathbb{R}$ is the entire real line. For if x is any element of $\mathbb{R}$ then every interval $(x - \epsilon, x + \epsilon)$ contains both rational numbers and irrational numbers. ∎

The union of a set S with its boundary is the *closure* of S, denoted $\overline{S}$ (Exercise 10 at the end of the last section discusses this idea from a different point of view). The next example illustrates the concept.

EXAMPLE 8.17 Let S be the set of rational numbers in the interval $[0, 1]$. Then the closure $\overline{S}$ of S is the entire interval $[0, 1]$.

Let T be the open interval $(0, 1)$. Then the closure $\overline{T}$ of T is the closed interval $[0, 1]$. ∎

Definition 8.18 Let $S \subset \mathbb{R}$. A point $s \in S$ is called an *interior point* of S if there is an $\epsilon > 0$ such that the interval $(s - \epsilon, s + \epsilon)$ lies in S. See Figure 8.10. We call the set of all interior points the *interior* of S, and we denote this set by $\overset{\circ}{S}$.

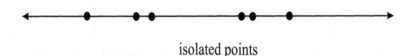

isolated points

Figure 8.11: The idea of an isolated point.

A point $t \in S$ is called an *isolated point* of S if there is an $\epsilon > 0$ such that the intersection of the interval $(t - \epsilon, t + \epsilon)$ with S is just the singleton $\{t\}$. See Figure 8.11.

By the definitions given here, an isolated point t of a set $S \subset \mathbb{R}$ is a boundary point. For any interval $(t - \epsilon, t + \epsilon)$ contains a point of S (namely, t itself) and points of $\mathbb{R} \setminus S$ (since t is isolated).

Proposition 8.19 *Let $S \subset \mathbb{R}$. Then each point of S is either an interior point or a boundary point of S.*

Proof: Fix $s \in S$. If s is not an interior point then no open interval centered at s contains only elements of s. Thus any interval centered at s contains an element of S (namely, s itself) and also contains points of $\mathbb{R} \setminus S$. Thus s is a boundary point of S. □

POINT OF CONFUSION 8.20 Let $S = \mathbb{Q}$. Then the interior of S is empty and the boundary of S is all of $\mathbb{R}$.

By contrast, let $S = \mathbb{R}$. Then the interior of S is all of $\mathbb{R}$ and the boundary of S is empty.

EXAMPLE 8.21 Let $S = [0, 1]$. Then the interior points of S are the elements of $(0, 1)$. The boundary points of S are the points 0 and 1. The set S has no isolated points.

Let $T = \{1, 1/2, 1/3, \dots\} \cup \{0\}$. Then the points $1, 1/2, 1/3, \dots$ are isolated points of T. The point 0 is an accumulation point of T. Every element of T is a boundary point, and there are no others. ∎

Remark 8.22 Observe that the interior points of a set S are *elements* of S—by their very definition. Also isolated points of S are elements of S. However, a boundary point of S may or may not be an element of S.

If x is an accumulation point of S then every open neighborhood of x contains infinitely many elements of S. Hence x is either a boundary point of S or an interior point of S; it *cannot* be an isolated point of S.

Proposition 8.23 *Let S be a subset of the real numbers. Then the boundary of S equals the boundary of $\mathbb{R} \setminus S$.*

Proof: Exercise. □

The next theorem allows us to use the concept of boundary to distinguish open sets from closed sets.

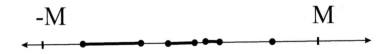

Figure 8.12: A bounded set.

Theorem 8.24 *A closed set contains all of its boundary points. An open set contains none of its boundary points.*

Proof: Let S be closed and let x be an element of its boundary. If every neighborhood of x contains points of S *other than x itself* then x is an accumulation point of S hence $x \in S$. If not every neighborhood of x contains points of S other than x itself, then there is an $\epsilon > 0$ such that $\{(x - \epsilon, x) \cup (x, x + \epsilon)\} \cap S = \emptyset$. The only way that x can be an element of ∂S in this circumstance is if $x \in S$. That is what we wished to prove.

For the other half of the theorem notice that if U is open then ^cU is closed. But then ^cU will contain all its boundary points, which are the same as the boundary points of U itself (why is this true?). Thus U can contain none of its boundary points. □

Proposition 8.25 *Every nonisolated boundary point of a set S is an accumulation point of the set S.*

Proof: This proof is treated in the exercises. □

Definition 8.26 A subset S of the real numbers is called *bounded* if there is a positive number M such that $|s| \leq M$ for every element s of S. See Figure 8.12.

The next result is one of the great theorems of nineteenth century analysis. It is essentially a restatement of the Bolzano-Weierstrass theorem (Theorem 6.24).

Theorem 8.27 (Bolzano-Weierstrass) *Every bounded, infinite subset of $\mathbb{R}$ has an accumulation point.*

Proof: Let S be a bounded, infinite set of real numbers. Let $\{a_j\}$ be a sequence of distinct elements of S. By Theorem 6.24, there is a subsequence $\{a_{j_k}\}$ that converges to a limit α. Then α is an accumulation point of S. □

Corollary 8.28 *Let $S \subset \mathbb{R}$ be a nonempty, closed, and bounded set. If $\{a_j\}$ is any sequence in S, then there is a Cauchy subsequence $\{a_{j_k}\}$ that converges to an element of S.*

Proof: Merely combine the Bolzano-Weierstrass theorem with Proposition 8.14 of the last section. □

A Look Back

1. Say in words what an interior point is.
2. Say in words what a boundary point is.
3. Say in words what an isolated point is.
4. Explain why an isolated point is always a boundary point.

Exercises

1. Let S be any set of real numbers. Prove that $S \subset \overline{S}$. Prove that $\overline{S}$ is a closed set. Prove that $\overline{S} \setminus \overset{\circ}{S}$ is the boundary of S.

3. Prove Proposition 8.23.

4. The union of infinitely many closed sets need not be closed. It need not be open either. Give examples to illustrate the possibilities.

5. The intersection of infinitely many open sets need not be open. It need not be closed either. Give examples to illustrate the possibilities.

6. Give an example of a one-to-one, onto, continuous function f with a continuous inverse from the halfline $(0, \infty)$ to the full line $(-\infty, \infty)$.

7. Prove Proposition 8.25.

8. Let S be any set of real numbers. Prove that $\overset{\circ}{S}$ is open. Prove that S is open if and only if S equals its interior.

* 9. Let U be an open set in the plane. Show that its projection on the x-axis is open.

10. Let $E \subset \mathbb{R}$ be closed. Let U be the complement of E. Prove that U is the countable union of open intervals.

11. Show that every nonisolated boundary point of a set S is an accumulation point of the set S.

* 12. Give an example of a closed set in the plane whose projection on the x-axis is not closed.

* 13. Show that the projection of a closed, bounded set in the plane into the x-axis will be closed. Contrast this problem with Exercise 12 above.

8.3   Compact Sets

Preliminary Remarks

Compact sets are a relatively recent development in mathematics. A compact set is an infinite set that, in certain key ways, behaves like a finite set. While this may sound like a conundrum, the proof is in the pudding. This section will introduce you to a remarkable and fascinating collection of sets of real numbers.

Compact sets are sets (usually infinite) which share many of the most important properties of finite sets. They play an important role in real analysis.

Definition 8.29 A set $S \subset \mathbb{R}$ is called *compact* if every sequence in S has a subsequence that converges *to an element of S*.

Theorem 8.30 (Heine-Borel) *A set $S \subset \mathbb{R}$ is compact if and only if it is closed and bounded.*

Proof: That a closed, bounded set has the property of compactness is the content of Corollary 8.28.

Now let S be a set that is compact. If S is not bounded, then there is an element s_1 of S that has absolute value larger than 1. Also there must be an element s_2 of S that has absolute value larger than $2 + |s_1|$. Continuing, we find elements $s_j \in S$ satisfying

$$|s_j| > j + |s_{j-1}|$$

for each j. But then no subsequence of the sequence $\{s_j\}$ can be Cauchy. This contradiction shows that S must be bounded.

If S is compact but S is not closed, then there is a point x which is the limit of a sequence $\{s_j\} \subset S$ but which is not itself in S. But every sequence in S is, by definition of "compact," supposed to have a subsequence converging *to an element of S*. For the sequence $\{s_j\}$ that we are considering, x is the only possibility for the limit of a subsequence. Thus it must be that $x \in S$. That contradiction establishes that S is closed. $\square$

EXAMPLE 8.31 The last theorem makes it particularly easy to identify compact sets. The set $[0,1]$ is closed and bounded, hence compact. The set $\{0, 2, 4, 8\}$ is closed and bounded, hence compact.

By contrast, the set $[0,1)$ is bounded but *not* closed. So it is not compact.

EXAMPLE 8.32 If $A \subset B$ and both sets are nonempty then $A \cap B = A \neq \emptyset$. A similar assertion holds when intersecting *finitely many* nonempty sets $A_1 \supseteq A_2 \supseteq \cdots \supseteq A_k$; it holds in this circumstance that $\cap_{j=1}^{k} A_j = A_k$.

However, it is possible to have infinitely many nonempty nested sets with null intersection. An example is the sets $I_j = (0, 1/j)$. Certainly $I_j \supseteq I_{j+1}$ for all j yet

$$\bigcap_{j=1}^{\infty} I_j = \emptyset \,.$$

By contrast, if we take $K_j = [0, 1/j]$ then

$$\bigcap_{j=1}^{\infty} K_j = \{0\}.$$

The next proposition shows that compact sets have the intuitively appealing property of the collection of sets K_j rather than the unsettling property of the collection of sets I_j.   ∎

Proposition 8.33 *Let*

$$K_1 \supseteq K_2 \supseteq \cdots \supseteq K_j \supseteq \cdots$$

be nonempty compact sets of real numbers. Set

$$\mathcal{K} = \bigcap_{j=1}^{\infty} K_j.$$

Then $\mathcal{K}$ is compact and $\mathcal{K} \neq \emptyset$.

Proof: Each K_j is closed and bounded hence $\mathcal{K}$ is closed and bounded. Thus $\mathcal{K}$ is compact. Let $x_j \in K_j$, each j. Then $\{x_j\} \subset K_1$. By compactness, there is a convergent subsequence $\{x_{j_k}\}$ with limit $x_0 \in K_1$. However, $\{x_{j_k}\}_{k=2}^{\infty} \subset K_2$. Thus $x_0 \in K_2$. Similar reasoning shows that $x_0 \in K_m$ for all $m = 1, 2, \ldots$. In conclusion, $x_0 \in \cap_j K_j = \mathcal{K}$.   □

A Look Back

1. Describe in words what a compact set is.

2. What does the complement of a compact set look like? Is it open? Is it bounded?

3. Is a finite set compact?

4. Is the intersection of two compact sets compact? Is the union of two compact sets compact?

Exercises

1. Let K be a compact set and let U be an open set that contains K. Prove that there is an $\epsilon > 0$ such that, if $k \in K$, then the interval $(k - \epsilon, k + \epsilon)$ is contained in U.

Figure 8.13: Construction of the Cantor set.

2. Let K be a compact set. Let $\mathcal{U} = \{U_j\}_{j=1}^k$ be a finite covering of K by open sets. Show that there is a $\delta > 0$ so that, if x is any point of K, then the disc or interval of center x and radius δ lies entirely in one of the U_j.

3. Prove that the intersection of a compact set and a closed set is compact.

4. Assume that we have intervals $[a_1, b_1] \supseteq [a_2, b_2] \supseteq \cdots$ and that $\lim_{j \to \infty} |a_j - b_j| = 0$. Prove that there is a point x such that $x \in [a_j, b_j]$ for every j.

5. If K in $\mathbb{R}$ is compact then show that ^cK is not compact.

6. Prove that the intersection of any number of compact sets is compact. The analogous statement for unions is false.

7. Let $U \subset \mathbb{R}$ be any open set. Show that there exist compact sets $K_1 \subset K_2 \subset \cdots$ so that $\cup_j K_j = U$.

8. Produce an open set U in the real line so that U may *not* be written as the decreasing intersection of compact sets.

9. Prove that the union of finitely many compact sets is compact.

10. Prove that the union of countably many compact sets is not necessarily compact.

* **11.** Let K be compact and L closed, and assume that the two sets are disjoint. Show that there is a positive distance between the two sets.

* **12.** Let K be a compact set. Let $\delta > 0$. Prove that there is a finite collection of intervals of radius δ that covers K.

8.4 The Cantor Set

Preliminary Remarks

Certainly one of the most amazing and mysterious sets ever constructed is the Cantor set. While elementary to define, the Cantor set has a fractal-like character and offers many mysteries. It is one hundred years old, but is still intensely studied today.

In this section we describe the construction of a remarkable subset of $\mathbb{R}$ with many pathological properties. It only begins to suggest the richness of the structure of the real number system.

We begin with the unit interval $S_0 = [0, 1]$. We extract from S_0 its open middle third; thus $S_1 = S_0 \setminus (1/3, 2/3)$. Observe that S_1 consists of two closed intervals of equal length $1/3$: $S_1 = [0, 1/3] \cup [2/3, 1]$. See Figure 8.13.

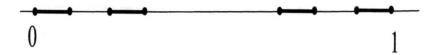

0 1

Figure 8.14: Second step in the construction of the Cantor set.

Now we construct S_2 from S_1 by extracting from each of its two intervals the middle third: $S_2 = [0, 1/9] \cup [2/9, 3/9] \cup [6/9, 7/9] \cup [8/9, 1]$. Figure 8.14 shows S_2.

Continuing in this fashion, we construct S_{j+1} from S_j by extracting the middle third from each of its component subintervals. We define the Cantor set C to be

$$C = \bigcap_{j=1}^{\infty} S_j \,.$$

Notice that each of the sets S_j is closed and bounded, hence compact. By Proposition 8.33 of the last section, C is therefore not empty. The set C is closed and bounded, hence compact.

Proposition 8.34 *The Cantor set C has zero length, in the sense that the complementary set $[0, 1] \setminus C$ has length 1.*

Proof: In the construction of S_1, we removed from the unit interval one interval of length 3^{-1}. In constructing S_2, we further removed two intervals of length 3^{-2}. In constructing S_j, we removed 2^{j-1} intervals of length 3^{-j}. Thus the total length of the intervals removed from the unit interval is

$$\sum_{j=1}^{\infty} 2^{j-1} \cdot 3^{-j} \,.$$

This last equals

$$\frac{1}{3} \sum_{j=0}^{\infty} \left(\frac{2}{3}\right)^{j} \,.$$

The geometric series sums easily and we find that the total length of the intervals removed is

$$\frac{1}{3} \left(\frac{1}{1 - 2/3}\right) = 1 \,.$$

Thus the Cantor set has length zero because its complement in the unit interval has length one.   □

POINT OF CONFUSION 8.35 The Cantor set is uncountable, as we shall see below. It contains no intervals. In fact it is disconnected in a very strong sense to be discussed below. Only countably many of the elements of the Cantor set are endpoints of the component intervals. Uncountably many of the points are non-endpoints.

Proposition 8.36 *The Cantor set is uncountable.*

Proof: We assign to each element of the Cantor set a "label" consisting of a sequence of 0s and 1s that identifies its location in the set.

Fix an element x in the Cantor set. Then certainly x is in S_1. If x is in the left half of S_1, then the first digit in the "label" of x is 0; otherwise it is 1. Likewise $x \in S_2$.

By the first part of this argument, it is either in the left half S_1^0 of S_1 (when the first digit in the label is 0) or the right half S_1^1 of S_1 (when the first digit of the label is 1). Whichever of these is correct, that half will consist of two intervals of length 3^{-2}. If x is in the leftmost of these two intervals then the second digit of the "label" of x is 0. Otherwise the second digit is 1. Continuing in this fashion, we may assign to x an infinite sequence of 0s and 1s.

Conversely, if $a, b, c, \ldots$ is a sequence of 0s and 1s, then we may locate a unique corresponding element y of the Cantor set. If the first digit is a zero then y is in the left half of S_1; otherwise y is in the right half of S_1. Likewise the second digit locates y within S_2, and so forth.

Thus we have a one-to-one correspondence between the Cantor set and the collection of all infinite sequences of 0s and 1s. [Notice that we are in effect thinking of the point assigned to a sequence $c_1 c_2 c_3 \ldots$ of 0s and 1s as the limit of the points assigned to $c_1, c_1 c_2, c_1 c_2 c_3, \ldots$ Thus we are using the fact that C is closed.] However, as we learned in Section 4.5, the set of all infinite sequences of 0s and 1s is uncountable. Thus we see that the Cantor set is uncountable. $\square$

Remark 8.37 A useful way to think about the Cantor set is in terms of series. Namely, C is the set of all numbers between 0 and 1 inclusive which can be written in the form

$$\sum_{j=1}^{\infty} \frac{a_j}{3^j}, \tag{8.37.1}$$

where each a_j is either 0 or 2. This representation is simply an interpretation of the labeling that we used in the last proof. We invite the reader to write out some expressions like (8.37.1) (only finite sums, of course), just to see what elements of the Cantor set arise. As you read the proof of the next theorem, you should think about it in terms of this series representation.

The Cantor set is quite thin (it has zero length) but it is large in the sense that it has uncountably many elements. Also it is compact. The next result reveals a surprising, and not generally well known, property of this "thin" set.

Theorem 8.38 *Let C be the Cantor set and define*

$$S = \{x + y : x \in C, y \in C\}.$$

Then $S = [0, 2]$.

Proof: We sketch the proof here and treat the details in the exercises.

Since $C \subset [0, 1]$ it is clear that $S \subset [0, 2]$. For the reverse inclusion, fix an element $t \in [0, 2]$. Our job is to find two elements c and d in C such that $c + d = t$.

First observe that $\{x + y : x \in S_1, y \in S_1\} = [0, 2]$. Therefore there exist $x_1 \in S_1$ and $y_1 \in S_1$ such that $x_1 + y_1 = t$.

Similarly, $\{x + y : x \in S_2, y \in S_2\} = [0, 2]$. Therefore there exist $x_2 \in S_2$ and $y_2 \in S_2$ such that $x_2 + y_2 = t$.

Continuing in this fashion we may find for each j numbers x_j and y_j such that $x_j, y_j \in S_j$ and $x_j + y_j = t$. Of course $\{x_j\} \subset C$ and $\{y_j\} \subset C$ hence there are subsequences $\{x_{j_k}\}$ and $\{y_{j_k}\}$ which converge to real numbers c and d respectively. Since C is compact, we can be sure that $c \in C$ and $d \in C$. But the operation of addition respects limits, thus we may pass to the limit as $k \to \infty$ in the equation

$$x_{j_k} + y_{j_k} = t$$

to obtain

$$c + d = t.$$

Therefore $[0, 2] \subset \{x + y : x \in C\}$. This completes the proof. $\square$

In the exercises at the end of the section we shall explore constructions of other Cantor sets, some of which have zero length and some of which have positive length. The Cantor set that we have discussed in detail in the present section is sometimes distinguished with the name "the Cantor ternary set." We shall also consider in the exercises other ways to construct the Cantor ternary set.

Observe that, whereas any open set is the countable or finite disjoint union of open intervals, the existence of the Cantor set shows us that there is no such structure theorem for closed sets. That is to say, we cannot hope to write an arbitrary closed set as the disjoint union of closed intervals. In fact closed intervals are atypically simple when considered as examples of closed sets.

A Look Back

1. Why is the Cantor set uncountable?

2. Why is the Cantor set compact?

3. How many connected components does the complement of the Cantor set have? [Here a "connected component" is a maximal connected piece of the complement.]

4. What is the boundary of the Cantor set?

Exercises

1. What is the interior of the Cantor set?

2. Fix the sequence $a_j = 3^{-j}, j = 1, 2, \ldots$. Consider the set S of all sums

$$\sum_{j=1}^{\infty} \mu_j a_j,$$

where each μ_j is one of the numbers 0 or 2. Show that S is the Cantor set. If s is an element of $S, s = \sum \mu_j a_j$, and if $\mu_j = 0$ for all j sufficiently large, then show that s is an endpoint of one of the intervals in one of the sets S_j that were used to construct the Cantor set in the text.

3. Construct a Cantor-like set by removing the middle *fifth* from the unit interval, removing the middle fifth of each of the remaining intervals, and so on. What is the length of the set that you construct in this fashion? Is it uncountable? Is it perfect (see Section 8.6)? Is it different from the Cantor set constructed in the text?

4. Refer to Exercise 3. Construct a Cantor set by removing, at the jth step, a middle subinterval of length 3^{-2j+1} from each existing interval. The Cantor-like set that results should have positive length. What is that length? Does this Cantor set have the other properties of the Cantor set constructed in the text?

*   5. Describe how to produce a two-dimensional Cantor-like set in the plane.

6. How many endpoints of intervals are there in the Cantor set? How many non-endpoints?

7. How many points in the Cantor set have finite ternary expansions? How many have infinite ternary expansions?

*   8. Let $0 < \lambda < 1$. Imitate the construction of the Cantor set to produce a perfect subset (see Section 8.6) of the unit interval whose complement has length λ.

*   9. Discuss which sequences a_j of positive numbers could be used as in Exercise 2 to construct sets which are like the Cantor set.

*   10. Let us examine the proof that $\{x + y : x \in C, y \in C\}$ equals $[0, 2]$ more carefully:

   a) Prove for each j that $\{x + y : x \in S_j, y \in S_j\}$ equals the interval $[0, 2]$.

   b) For $t \in C$, explain how the subsequences $\{x_{j_k}\}$ and $\{y_{j_k}\}$ can be chosen to satisfy $x_{j_k} + y_{j_k} = t$. Observe that it is important for the proof that the index j_k be the same for both subsequences.

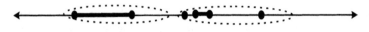

a disconnected set

Figure 8.15: The idea of disconnected.

    c) Formulate a suitable statement concerning the assertion that the binary operation of addition "respects limits" as required in the argument in the text. Prove this statement and explain how it allows us to pass to the limit in the equation $x_{j_k} + y_{j_k} = t$.

*   **11.** Use the characterization of the Cantor set from Exercise 2 to give a new proof of the fact that $\{x + y : x \in C, y \in C\}$ equals the interval $[0, 2]$.

8.5 Connected and Disconnected Sets

> ### Preliminary Remarks
>
> We assign various attributes to sets in order to help us understand their shape and form. One of those attributes is connectedness. Connected sets are very natural objects for us to study, and we want to see how they behave under the action of continuous functions. That is the purpose of our study in the present section.

    Let S be a set of real numbers. We say that S is *disconnected* if it is possible to find a pair of disjoint open sets U and V such that

$$U \cap S \neq \emptyset, V \cap S \neq \emptyset,$$

$$(U \cap S) \cap (V \cap S) = \emptyset,$$

and

$$S = (U \cap S) \cup (V \cap S).$$

See Figure 8.15. If no such U and V exist then we call S *connected.*

EXAMPLE 8.39 The set $T = \{x \in \mathbb{R} : |x| < 1, x \neq 0\}$ is disconnected. For take $U = \{x : x < 0\}$ and $V = \{x : x > 0\}$. Then U and V are disjoint and

$$U \cap T = \{x : -1 < x < 0\} \neq \emptyset$$

and

$$V \cap T = \{x : 0 < x < 1\} \neq \emptyset.$$

Also $(U \cap T) \cap (V \cap T) = \emptyset$. Clearly $T = (U \cap T) \cup (V \cap T)$, hence T is disconnected.

∎

Figure 8.16: A closed interval is connected.

EXAMPLE 8.40 The set $X = [-1, 1]$ is connected. To see this, suppose to the contrary that there exist open sets U and V such that $U \cap X \neq \emptyset, V \cap X \neq \emptyset, (U \cap X) \cap (V \cap X) = \emptyset$, and

$$X = (U \cap X) \cup (V \cap X).$$

Choose $a \in U \cap X$ and $b \in V \cap X$. We may assume that $a < b$. Set

$$\alpha = \sup (U \cap [a, b]).$$

Now $[a, b] \subset X$ hence $U \cap [a, b]$ is disjoint from V. Thus $\alpha \leq b$. But ^cV is closed hence $\alpha \notin V$. It follows that $\alpha < b$.

If $\alpha \in U$ then, because U is open, there exists an $\widetilde{\alpha} \in U$ such that $\alpha < \widetilde{\alpha} < b$. The existence of $\widetilde{\alpha}$ contradicts the definition of α as the supremum of $U \cap [a, b]$. So $\alpha \notin U$. But $\alpha \notin U$ and $\alpha \notin V$ means $\alpha \notin X$. On the other hand, α is the supremum of a subset of X (since $a \in X, b \in X$, and X is an interval). Since X is a closed interval, we conclude that $\alpha \in X$. This contradiction shows that X must be connected. ∎

POINT OF CONFUSION 8.41 As we shall see in detail below, the connected subsets of the real numbers are the intervals. There are no other connected subsets. The disconnected sets are much more profuse, and much more varied.

With small modifications, the discussion in the last example demonstrates that any closed interval is connected (Exercise 1). See Figure 8.16. Also (see Exercise 2), we may similarly see that any open interval or half-open interval is connected. In fact the converse is true as well:

Theorem 8.42 *A subset S of $\mathbb{R}$ is connected if and only if S is an interval.*

Proof: We have already noted that an interval is connected.

For the converse, note that if S is not an interval, then there exist $a \in S, b \in S$ and a point t between a and b such that $t \notin S$. Define $U = \{x \in \mathbb{R} : x < t\}$ and $V = \{x \in \mathbb{R} : t < x\}$. Then U and V are open and disjoint, $U \cap S \neq \emptyset$, $V \cap S \neq \emptyset$, and

$$S = (U \cap S) \cup (V \cap S).$$

Thus S is disconnected.

We have proved the contrapositive of the statement for this direction of the theorem, hence we are finished. □

The Cantor set is not connected; indeed it is disconnected in a special sense. Call a set S *totally disconnected* if, for each distinct $x \in S$, $y \in S$, there exist disjoint open sets U and V such that $x \in U, y \in V$, and $S = (U \cap S) \cup (V \cap S)$.

Proposition 8.43 *The Cantor set is totally disconnected.*

Proof: Let $x, y \in C$ be distinct and assume that $x < y$. Set $\delta = |x - y|$. Choose j so large that $3^{-j} < \delta$. Then $x, y \in S_j$, but x and y cannot both be in the same interval of S_j (since the intervals will have length equal to 3^{-j}). It follows that there is a point t between x and y that is not an element of S_j, hence certainly not an element of C. Set $U = \{s : s < t\}$ and $V = \{s : s > t\}$. Then $x \in U \cap C$ hence $U \cap C \neq \emptyset$; likewise $V \cap C \neq \emptyset$. Also $(U \cap C) \cap (V \cap C) = \emptyset$. Finally $C = (U \cap C) \cup (V \cap C)$. Thus C is totally disconnected. $\qquad\square$

A Look Back

1. Describe verbally what a connected set is.

2. Describe verbally what a disconnected set is.

3. What is the simplest thing you can do to a connected set to make it disconnected?

4. The entire real line is connected, but the real line with one point removed is disconnected. Explain.

Exercises

1. Imitate the example in the text to prove that any closed interval is connected.

2. Imitate the example in the text to prove that any open interval or half-open interval is connected.

3. Give an example of a totally disconnected set $S \subset [0, 1]$ such that $\overline{S} = [0, 1]$.

4. If A is connected and B is connected then will $A \cap B$ be connected?

5. If A is connected and B is connected then will $A \cup B$ be connected?

6. If A is connected and B is disconnected then what can you say about $A \cap B$?

7. If sets U_j form the basis of a topology on a space X (that is to say, each open set in X can be written as a union of some of the U_j) and if each U_j is connected, then what can you say about X?

8. Is the set-theoretic difference of connected sets connected?

9. Prove that the union of two connected sets is connected provided that the two sets have at least one point in common.

*   10. Let $S \subset \mathbb{R}$ be a set. Let $s, t \in S$. We say that s and t are in the same *connected component* of S if the entire interval $[s, t]$ lies in S. What are the connected components of the Cantor set? Is it possible to have a set S with countably many distinct connected components? With uncountably many distinct connected components?

*   11. Write the real line as the union of two totally disconnected sets.

*   12. If A is connected and B is connected then does it follow that $A \times B$ is connected?

8.6 Perfect Sets

Preliminary Remarks

A perfect set is a very special type of set that makes it better than closed, better than compact, and remarkable in a number of ways. The Cantor set is perfect. There are many examples of perfect sets. We learn their lore in this section.

A set $S \subset \mathbb{R}$ is called *perfect* if it is closed and if every point of S is an accumulation point of S. The property of being perfect is a rather special one: it implies that the set has no isolated points.

Obviously a closed interval $[a, b]$ is perfect. After all, a point x in the interior of the interval is surrounded by an entire open interval $(x - \epsilon, x + \epsilon)$ of elements of the interval; moreover a is the limit of elements from the right and b is the limit of elements from the left.

Perhaps more surprising is that the Cantor set, *a totally disconnected set*, is perfect. It is certainly closed. Now fix $x \in C$. Then certainly $x \in S_1$. Thus x is in one of the two intervals composing S_1. One (or perhaps both) of the endpoints of that interval does not equal x. Call that endpoint a_1. Likewise $x \in S_2$. Therefore x lies in one of the intervals of S_2. Choose an endpoint a_2 of that interval which does not equal x. Continuing in this fashion, we construct a sequence $\{a_j\}$. Notice that *each of the elements of this sequence lies in the Cantor set* (why?). Finally, $|x - a_j| \leq 3^{-j}$ for each j. Therefore x is the limit of the sequence. We have thus proved that the Cantor set is perfect.

The fundamental theorem about perfect sets tells us that such a set must be rather large. We have

Theorem 8.44 *A nonempty perfect set must be uncountable.*

Proof: Let S be a nonempty perfect set. Since S has accumulation points, it cannot be finite. Therefore it is either countable or uncountable.

Seeking a contradiction, we suppose that S is countable. Write $S = \{s_1, s_2, \ldots\}$. Set $U_1 = (s_1 - 1, s_1 + 1)$. Then U_1 is a neighborhood of s_1. Now s_1 is a limit point of S so there must be infinitely many elements of S lying in U_1. We select a bounded open interval U_2 such that $\overline{U}_2 \subset U_1$, $\overline{U}_2$ does not contain s_1, and U_2 *does* contain some element of S.

Continuing in this fashion, assume that $s_1, \ldots, s_j$ have been selected and choose a bounded interval U_{j+1} such that (i) $\overline{U}_{j+1} \subset U_j$, (ii) $s_j \notin \overline{U}_{j+1}$, and (iii) U_{j+1} contains some element of S.

Observe that each set $V_j = \overline{U}_j \cap S$ is closed and bounded, hence compact. Also each V_j is nonempty by construction but V_j does not contain s_{j-1}. It follows that $V \equiv \cap_j V_j$ cannot contain s_1 (since V_2 does not), cannot contain s_2 (since V_3 does not), indeed cannot contain any element of S. Hence V, being a subset of S, is empty. But V is the decreasing intersection of nonempty compact sets, hence cannot be empty!

This contradiction shows that S cannot be countable. So it must be uncountable.   □

Corollary 8.45 *If $a < b$ then the closed interval $[a, b]$ is uncountable.*

Proof: The interval $[a, b]$ is perfect.   □

We also have a new way of seeing that the Cantor set is uncountable, since it is perfect:

Corollary 8.46 *The Cantor set is uncountable.*

Proof: The Cantor set is nonempty and perfect.   □

A Look Back

1. Describe verbally what a perfect set is.

2. What characteristics does the complement of a perfect set have (see Exercise 9 below)?

3. Explain why the rational numbers $\mathbb{Q}$ do not form a perfect set.

4. Explain why the real numbers $\mathbb{R}$ *do* form a perfect set.

Exercises

1. Let $U_1 \subset U_2 \ldots$ be open sets and assume that each of these sets has bounded, nonempty complement. Can it be that $\cup_j U_j = \mathbb{R}$?

2. Let $X_1, X_2, \ldots$ each be perfect sets and suppose that $X_1 \supseteq X_2 \supseteq \ldots$. Set $X = \cap_j X_j$. Is X perfect?

* **3.** Is the product of perfect sets perfect?

4. If $A \cap B$ is perfect, then what may we conclude about A and B?

5. If $A \cup B$ is perfect, then what may we conclude about A and B?

6. Call a set imperfect if its complement is perfect. Which sets are imperfect? Can you specify a connected imperfect set?

7. A Cantor set is formed by removing not middle thirds but rather middle ninths. Which properties of the ternary Cantor set will this new set have? What will be the length of this new set?

* **8.** Let $S_1, S_2, \ldots$ be closed sets and assume that $\cup_j S_j = \mathbb{R}$. Prove that at least one of the sets S_j has nonempty interior. (**Hint:** Use an idea from the proof that perfect sets are uncountable.)

*   **9.** Let S be a nonempty set of real numbers. A point x is called a *condensation point* of S if every neighborhood of x contains uncountably many points of S. Prove that the set of condensation points of S is closed. Is it necessarily nonempty? Is it nonempty when S is uncountable?

   If T is an uncountable set then show that the set of its condensation points is perfect.

*   **10.** Prove that any closed set can be written as the union of a perfect set and a countable set. (**Hint:** Refer to Exercise 9.)

Chapter 9

Limits and Continuity of Functions

9.1 Definition and Basic Properties of the Limit of a Function

Preliminary Remarks

Questions about limits go back to the ancient Greeks. The Greeks really did not understand limits (witness Zeno's paradoxes). The question of limits arose even more intensely in the development of calculus. Isaac Newton did not understand limits, and neither did Leibniz. It took the combined efforts of a number of nineteenth-century mathematical geniuses—including Cauchy, Riemann, Dirichlet, Weierstrass, and others—to finally nail down the concept of limit. Here we present the fruits of their efforts.

In this chapter we are going to treat some topics that you have seen before in your calculus class. However, we shall use the deep properties of the real numbers that we have developed in this text to obtain important new insights. Therefore you should *not* think of this chapter as review. Look at the concepts introduced here with the power of your new understanding of analysis.

Definition 9.1 Let f be a real-valued function whose domain E contains adjoining intervals (a, c) and (c, b). Let ℓ be a real number. We say that

$$\lim_{x \to c} f(x) = \ell$$

if, for each $\epsilon > 0$, there is a $\delta > 0$ such that, when $0 < |x - c| < \delta$, then

$$|f(x) - \ell| < \epsilon.$$

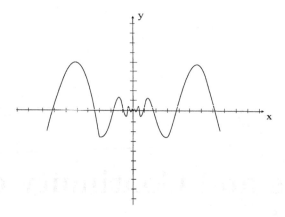

Figure 9.1: The limit of an oscillatory function.

POINT OF CONFUSION 9.2 The definition makes precise the notion that we can force $f(x)$ to be just as close as we please to ℓ by making x sufficiently close to c. Notice that the definition puts the condition $0 < |x - c|$ on x, so that x is *not* allowed to take the value c. In other words we do not look at $x = c$, but rather at x *near* to c.

Also observe that we only consider the limit of f at a point c that is not isolated. In the exercises you will be asked to discuss why it would be nonsensical to use the above definition to study the limit at an isolated point.

EXAMPLE 9.3 Let $E = \mathbb{R} \setminus \{0\}$ and

$$f(x) = x \cdot \sin(1/x) \ \ \text{if} \ \ x \in E .$$

See Figure 9.1. Then $\lim_{x \to 0} f(x) = 0$. To see this, let $\epsilon > 0$. Choose $\delta = \epsilon$. If $0 < |x - 0| < \delta$ then

$$|f(x) - 0| = |x \cdot \sin(1/x)| \leq |x| < \delta = \epsilon ,$$

as desired. Thus the limit exists and equals 0.   ∎

EXAMPLE 9.4 Let $E = \mathbb{R}$ and

$$g(x) = \begin{cases} 1 & \text{if} \ \ x \text{ is rational} \\ 0 & \text{if} \ \ x \text{ is irrational.} \end{cases}$$

Then $\lim_{x \to c} g(x)$ does not exist for any point c of E.

To see this, fix $c \in \mathbb{R}$. Seeking a contradiction, assume that there is a limiting value ℓ for g at c. If this is so then we take $\epsilon = 1/2$ and we can find a $\delta > 0$ such that $0 < |x - c| < \delta$ implies

$$|g(x) - \ell| < \epsilon = \frac{1}{2} . \tag{9.4.1}$$

If we take x to be rational then (9.4.1) says that

$$|1 - \ell| < \frac{1}{2}, \tag{9.4.2}$$

while if we take x irrational then (9.4.1) says that

$$|0 - \ell| < \frac{1}{2}. \tag{9.4.3}$$

But then the triangle inequality gives that

$$
\begin{aligned}
1 &= |1 - 0| \\
&= |(1 - \ell) + (\ell - 0)| \\
&\leq |1 - \ell| + |\ell - 0|,
\end{aligned}
$$

which by (9.4.2) and (9.4.3) is

$$< 1.$$

This contradiction, that $1 < 1$, allows us to conclude that the limit does not exist at c. ∎

Proposition 9.5 Let f be a function whose domain contains adjoining intervals (a, c) and (c, b). If $\lim_{x \to c} f(x) = \ell$ and $\lim_{x \to c} f(x) = m$, then $\ell = m$.

Proof: Let $\epsilon > 0$. Let E be the domain of f. Choose $\delta_1 > 0$ such that, if $x \in E$ and $0 < |x - c| < \delta_1$, then $|f(x) - \ell| < \epsilon/2$. Similarly choose $\delta_2 > 0$ such that, if $x \in E$ and $0 < |x - c| < \delta_2$, then $|f(x) - m| < \epsilon/2$. Define δ to be the minimum of δ_1 and δ_2. If $x \in E$ and $0 < |x - c| < \delta$, then the triangle inequality tells us that

$$
\begin{aligned}
|\ell - m| &= |(\ell - f(x)) + (f(x) - m)| \\
&\leq |(\ell - f(x)| + |f(x) - m)| \\
&< \frac{\epsilon}{2} + \frac{\epsilon}{2} \\
&= \epsilon.
\end{aligned}
$$

Since $|\ell - m| < \epsilon$ for every positive ϵ we conclude that $\ell = m$. That is the desired result. □

POINT OF CONFUSION 9.6 The point of the last proposition is that, if a limit is calculated by two different methods, then the same answer will result. While of primarily philosophical interest now, this will be important information later when we establish the existence of certain limits.

This is a good time to observe that the limits

$$\lim_{x \to c} f(x)$$

and

$$\lim_{h \to 0} f(c + h)$$

are equal in the sense that, if one limit exists then so does the other, and they both have the same value.

In order to facilitate checking that certain limits exist, we now record some elementary properties of the limit. This requires that we first recall how functions are combined.

Suppose that f and g are each functions which have domain E. We define the *sum* or *difference* of f and g to be the function

$$(f \pm g)(x) = f(x) \pm g(x),$$

the *product* of f and g to be the function

$$(f \cdot g)(x) = f(x) \cdot g(x),$$

and the quotient of f and g to be

$$\left(\frac{f}{g}\right)(x) = \frac{f(x)}{g(x)}.$$

Notice that the quotient is only defined at points x for which $g(x) \neq 0$. Now we have:

Theorem 9.7 (Elementary Properties of Limits of Functions) *Let f and g be functions whose domains contain adjoining intervals (a, c) and (c, b). Assume that*

(i) $\lim_{x \to c} f(x) = \ell$

(ii) $\lim_{x \to c} g(x) = m$.

Then

(a) $\lim_{x \to c} (f \pm g)(x) = \ell \pm m$

(b) $\lim_{x \to c} (f \cdot g)(x) = \ell \cdot m$

(c) $\lim_{x \to c} (f/g)(x) = \ell/m$ *provided* $m \neq 0$.

Proof: We prove part **(b)**. Parts **(a)** and **(c)** are treated in the exercises.

Let E be the common domain of f and g. Let $\epsilon > 0$. We may also assume that $\epsilon < 1$. Choose $\delta_1 > 0$ such that, if $x \in E$ and $0 < |x - c| < \delta_1$, then

$$|f(x) - \ell| < \frac{\epsilon}{2(|m| + 1)}.$$

Choose $\delta_2 > 0$ such that, if $x \in E$ and $0 < |x - c| < \delta_2$ then

$$|g(x) - m| < \frac{\epsilon}{2(|\ell| + 1)}.$$

(Notice that this last inequality implies that $|g(x)| < |m| + |\epsilon|$.) Let δ be the minimum of δ_1 and δ_2. If $x \in E$ and $0 < |x - c| < \delta$, then

$$
\begin{aligned}
|f(x) \cdot g(x) - \ell \cdot m| &= |(f(x) - \ell) \cdot g(x) + (g(x) - m) \cdot \ell| \\
&\leq |(f(x) - \ell) \cdot g(x)| + |(g(x) - m) \cdot \ell| \\
&< \left(\frac{\epsilon}{2(|m| + 1)} \right) \cdot |g(x)| + \left(\frac{\epsilon}{2(|\ell| + 1)} \right) \cdot |\ell| \\
&\leq \left(\frac{\epsilon}{2(|m| + 1)} \right) \cdot (|m| + |\epsilon|) + \frac{\epsilon}{2} \\
&< \frac{\epsilon}{2} + \frac{\epsilon}{2} \\
&= \epsilon.
\end{aligned}
$$

This shows that the limit of $f \cdot g$ at c equals $\ell \cdot m$. $\qquad\square$

EXAMPLE 9.8 It is a simple matter to check that, if $f(x) = x$, then

$$\lim_{x \to c} f(x) = c$$

for every real c. (Indeed, for $\epsilon > 0$ we may take $\delta = \epsilon$.) Also, if $g(x) \equiv \alpha$ is the constant function taking value α, then

$$\lim_{x \to c} g(x) = \alpha.$$

It then follows from parts **(a)** and **(b)** of the theorem that, if $p(x)$ is any polynomial function, then

$$\lim_{x \to c} p(x) = p(c).$$

Moreover, if $r(x)$ is any *rational function* (quotient of polynomials) then we may also use part **(c)** of the theorem to conclude that

$$\lim_{x \to c} r(x) = r(c)$$

for all points c at which the rational function $r(x)$ is defined. $\qquad\blacksquare$

EXAMPLE 9.9 If x is a small, positive real number then $0 < \sin x < x$. This is true because $\sin x$ is the nearest distance from the point $(\cos x, \sin x)$ to the x-axis while x is the distance from that point to the x-axis along an arc. If $\epsilon > 0$, then we set $\delta = \epsilon$. We conclude that, if $0 < |x - 0| < \delta$, then

$$|\sin x - 0| < |x| < \delta = \epsilon.$$

Since $\sin(-x) = -\sin x$, the same result holds when x is a negative number with small absolute value. Therefore

$$\lim_{x \to 0} \sin x = 0.$$

Notice that

$$1 - \cos x = 2 \sin^2(x/2).$$

Certainly $\lim_{x \to 0} \sin(x/2) = 0$ (exercise). So we may apply Theorem 9.7 to conclude that $\lim_{x \to 0} \cos x = 1$.

Now fix any real number c. We have

$$
\begin{aligned}
\lim_{x \to c} \sin x &= \lim_{h \to 0} \sin(c + h) \\
&= \lim_{h \to 0} \left(\sin c \cos h + \cos c \sin h \right) \\
&= \sin c \cdot 1 + \cos c \cdot 0 \\
&= \sin c.
\end{aligned}
$$

We of course have used parts **(a)** and **(b)** of the theorem to commute the limit process with addition and multiplication. A similar argument shows that

$$\lim_{x \to c} \cos x = \cos c.$$

We conclude that sine and cosine are continuous functions (see the next section for the precise definition). ∎

Remark 9.10 In the last example, we have used the definition of the sine function and the cosine function that you learned in calculus. In Chapters 12 and 13, when we learn about series of functions, we will learn a more rigorous method for treating the trigonometric functions.

We conclude by giving a characterization of the limit of a function using sequences.

Proposition 9.11 *Let f be a function whose domain E contains adjoining intervals (a, c) and (c, b). Then*

$$\lim_{x \to c} f(x) = \ell \tag{9.11.1}$$

if and only if, for any sequence $\{a_j\}$ satisfying $\lim_{j \to \infty} a_j = c$, it holds that

$$\lim_{j \to \infty} f(a_j) = \ell. \tag{9.11.2}$$

Proof: Assume that condition (9.11.1) fails. Then there is an $\epsilon > 0$ such that for no $\delta > 0$ is it the case that when $0 < |x - c| < \delta$ then $|f(x) - \ell| < \epsilon$. Thus, for each $\delta = 1/j$, we may choose a number $a_j \in E \setminus \{c\}$ with $0 < |a_j - c| < 1/j$ and $|f(a_j) - \ell| \geq \epsilon$. But then condition (9.11.2) fails for this sequence $\{a_j\}$.

If condition (9.11.2) fails then there is a sequence $\{a_j\}$ such that $\lim_{j\to\infty} a_j = c$ but $\lim_{j\to\infty} f(a_j) \neq \ell$. This means that there is an $\epsilon > 0$ such that, for infinitely many a_j, it holds that $|f(a_j) - \ell| \geq \epsilon$. But then, no matter how small $\delta > 0$, there will be an a_j satisfying $0 < |a_j - c| < \delta$ (since $a_j \to c$) and $|f(a_j) - \ell| \geq \epsilon$. Thus (9.11.1) fails. □

A Look Back

1. Give a verbal description of what a limit is.

2. Give an example of a function on $\mathbb{R}$ that does not have a limit at any point.

3. Give an example of a function on $\mathbb{R}$ that has a limit at every point except the origin.

4. Explain how the rigorous definition of limit using ϵ and δ relates to the intuitive definition of limit that uses the phrase "can draw the graph without lifting your pencil from the paper."

Exercises

1. Let f and g be functions on a set $A = (a, c) \cup (c, b)$ and assume that $f(x) \leq g(x)$ for all $x \in A$. Assuming that both limits exist, show that

$$\lim_{x\to c} f(x) \leq \lim_{x\to c} g(x).$$

   Does the conclusion improve if we assume that $f(x) < g(x)$ for all $x \in A$?

2. Prove parts **(a)** and **(c)** of Theorem 9.7.

3. Give a definition of limit using the concept of distance.

4. If $\lim_{x\to c} f(x) = \ell > 0$ then prove that there is a $\delta > 0$ so small that $|x - c| < \delta$ guarantees that $f(x) > \ell/2$.

5. Show that, if f is a monotone function, then f has a limit at "most" points. What does the word "most" mean in this context?

6. Give an example of a function f such that $\lim_{x\to c} f(x)$ exists at every point but f is discontinuous at every point.

7. Prove that $\lim_{x\to c} f(x) = \lim_{h\to 0} f(c+h)$ whenever both expressions make sense.

8. Give an example of a function $f : (-1, 0) \cup (0, 1) \to \mathbb{R}$ with the property that, if $-1 < \alpha < 1$, then there is a sequence $x_j \to 0$ so that $f(x_j) \to \alpha$.

9. Suppose that f and g are functions and each has a discontinuity at 0. Then what can you say about the continuity properties of $f + g$ at 0?

* 10. Give a definition of limit using the concept of open set.

* 11. Give an example of a function $f : \mathbb{R} \to \mathbb{R}$ so that $\lim_{x\to c} f(x)$ exists when c is irrational but does not exist when c is rational.

9.2   Continuous Functions

Preliminary Remarks

The concept of continuous function is intuitively appealing. But what we need is a rigorous definition. Having the idea of limit under control enables us to give a precise and accurate definition of continuity. Then we can prove some results about continuous functions and begin to develop a cogent theory.

Definition 9.12 Let $E \subset \mathbb{R}$ be a set and let f be a real-valued function with domain E. Fix a point c which is in E. We say that f is *continuous* at c if

$$\lim_{x \to c} f(x) = f(c).$$

We learned from the penultimate example of Section 9.1 that polynomial functions are continuous at every real x. So are the transcendental functions $\sin x$ and $\cos x$ (see Example 9.9). A rational function is continuous at every point of its domain.

EXAMPLE 9.13 The function

$$h(x) = \begin{cases} \sin(1/x) & \text{if} \quad x \neq 0 \\ 1 & \text{if} \quad x = 0 \end{cases}$$

is discontinuous at 0. See Figure 9.2. The reason is that

$$\lim_{x \to 0} h(x)$$

does not exist. (Details of this assertion are left for you: notice that $h(1/(j\pi)) = 0$ while $h(2/[(4j + 1)\pi]) = 1$ for $j = 1, 2, \ldots$.)

The function

$$k(x) = \begin{cases} x \cdot \sin(1/x) & \text{if} \quad x \neq 0 \\ 1 & \text{if} \quad x = 0 \end{cases}$$

is also discontinuous at $x = 0$. This time the limit $\lim_{x \to 0} k(x)$ exists (see Example 9.3), but the limit does not agree with $k(0)$.

However, the function

$$m(x) = \begin{cases} x \cdot \sin(1/x) & \text{if} \quad x \neq 0 \\ 0 & \text{if} \quad x = 0 \end{cases}$$

is continuous at $x = 0$ because the limit at 0 exists and agrees with the value of the function there. See Figure 9.3.   ∎

POINT OF CONFUSION 9.14 As we shall see in detail below, a function can be discontinuous because it oscillates, or it can be discontinuous because the limit at c disagrees with the value at c. We must learn to distinguish these two cases.

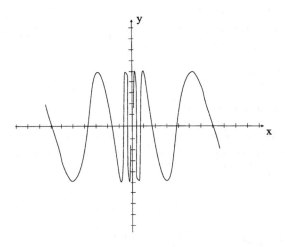

Figure 9.2: A function discontinuous at 0.

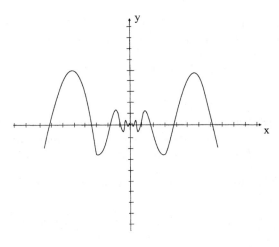

Figure 9.3: A function continuous at 0.

The arithmetic operations $+, -, \times$, and $\div$ preserve continuity (so long as we avoid division by zero). We now formulate this assertion as a theorem.

Theorem 9.15 *Let f and g be functions with domain E and let c be a point of E. If f and g are continuous at c then so are $f \pm g, f \cdot g$, and (provided $g(c) \neq 0$) f/g.*

Proof: Apply Theorem 9.7 of Section 9.1.   □

Continuous functions may also be characterized using sequences:

Proposition 9.16 *Let f be a function with domain E and fix $c \in E$. The function f is continuous at c if and only if, for every sequence $\{a_j\} \subset E$ satisfying $\lim_{j \to \infty} a_j = c$, it holds that*

$$\lim_{j \to \infty} f(a_j) = f(c).$$

Proof: Apply Proposition 9.11 of Section 9.1.   □

POINT OF CONFUSION 9.17 A continuous function is one with the property that what it does at a point c is what we *anticipate* that it will do at that point c. What does this mean?

What we anticipate that the function will do is the *limit* as $x \to c$. What it actually does at the point is take the *value* $f(c)$. We demand for continuity that the two agree.

Recall that, if g is a function with domain D and range E, and if f is a function with domain E and range F, then the *composition* of f and g is

$$f \circ g(x) = f(g(x)).$$

See Figure 9.4.

Proposition 9.18 *Let g have domain D and range E and let f have domain E and range F. Let $c \in D$. Assume that g is continuous at c and that f is continuous at $g(c)$. Then $f \circ g$ is continuous at c.*

Proof: Let $\{a_j\}$ be any sequence in D such that $\lim_{j \to \infty} a_j = c$. Then

$$\lim_{j \to \infty} f \circ g(a_j) = \lim_{j \to \infty} f(g(a_j)) = f\left(\lim_{j \to \infty} g(a_j)\right)$$

$$= f\left(g\left(\lim_{j \to \infty} a_j\right)\right) = f(g(c)) = f \circ g(c).$$

Now apply Proposition 9.11.   □

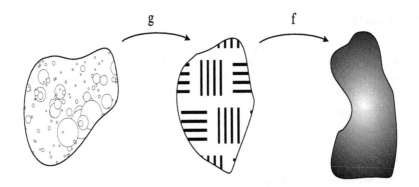

Figure 9.4: Composition of functions.

POINT OF CONFUSION 9.19 Continuity is a robust attribute for a function. It is preserved under most of our standard operations. Limits are a bit more delicate, and must be handled more carefully. See the next remark.

Remark 9.20 It is not the case that if

$$\lim_{x \to c} g(x) = \ell$$

and

$$\lim_{t \to \ell} f(t) = m$$

then

$$\lim_{x \to c} f \circ g(x) = m \,.$$

A counterexample is given by the functions

$$g(x) = 0$$

$$f(x) = \begin{cases} 2 & \text{if } x \neq 0 \\ 5 & \text{if } x = 0. \end{cases}$$

Notice that $\lim_{x \to 0} g(x) = 0$, $\lim_{t \to 0} f(t) = 2$, yet $\lim_{x \to 0} f \circ g(x) = 5$.

The additional hypothesis that f be continuous at ℓ is necessary in order to guarantee that the limit of the composition will behave as expected.

A Look Back

1. Describe in words what a continuous function is.

2. Describe what the role of limits is in the definition of continuity.

3. Why do we need to use the concept of limit to define continuity?

4. Why is it the case that a continuous function is not necessarily differentiable?

Exercises

1. Let $0 < \alpha \leq 1$. A function f with domain E is said to satisfy a *Lipschitz condition* of order α if there is a constant $C > 0$ such that, for any $s, t \in E$, it holds that $|f(s) - f(t)| \leq C \cdot |s - t|^{\alpha}$. Prove that such a function must be uniformly continuous (see the next section for the definition of this concept).

2. Define the function
$$g(x) = \begin{cases} 0 & \text{if} \quad x \text{ is irrational} \\ x & \text{if} \quad x \text{ is rational} \end{cases}$$

   At which points x is g continuous? At which points is it discontinuous?

3. Explain why it would be foolish to define the concept of limit at an isolated point.

4. Define an onto, continuous function from $\mathbb{R}$ to $[0, 2]$.

5. Explain why a continuous function defined on a compact set must be bounded.

6. Explain why a continuous function defined on a compact set actually assumes its maximum and minimum values.

7. Give an example of a closed set E and a continuous function f so that $f(E)$ is open.

8. Give an example of an open set U and a continuous function f so that $f(U)$ is closed.

* 9. Let f be a continuous function whose domain contains an open interval (a, b). What form can $f((a, b))$ have? (**Hint:** There are just four possibilities.)

* 10. Let f be a continuous function on the open interval (a, b). Under what circumstances can f be extended to a continuous function on $[a, b]$?

* 11. Define continuity using the notion of closed set.

* 12. The image of a compact set under a continuous function is compact (see the next section). But the image of a closed set need not be closed. Explain. The *inverse image* of a compact set under a continuous function need not be compact. Explain.

9.3   Topological Properties and Continuity

Preliminary Remarks

Continuous functions are, from the point of view of topology, very natural objects to study. Continuous functions interact very naturally with open sets, with compact sets, and with other standard artifacts of this theory. We explore these connections in the present section.

Definition 9.21 Let f be a function with domain E and let L be a subset of E. We define
$$f(L) = \{f(x) : x \in L\}.$$
The set $f(L)$ is called the *image* of L under f. See Figure 9.5.

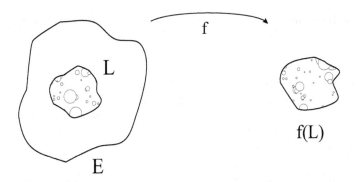

Figure 9.5: The image of the set L under the function f.

Theorem 9.22 *The image of a compact set under a continuous function is also compact.*

Proof: Let K be a compact set and f a continuous function. Consider the set $K' \equiv f(K)$. Let $\{x_j\}$ be a sequence in K'. Then each x_j has a pre-image t_j (that is to say, $f(t_j) = x_j$) in K. Since K is compact, there is a subsequence t_{j_k} that converges to a point $t_0 \in K$. But then, by continuity, the points $f(t_{j_k})$ converge to $f(t_0) \in K'$. So we see that $\{x_j\}$ has a convergent subsequence $x_{j_k} = f(t_{j_k})$. Hence K' is compact. $\qquad\square$

POINT OF CONFUSION 9.23 It is not the case that the continuous image of a closed set is closed. For instance, take $f(x) = 1/(1 + x^2)$ and $E = \mathbb{R}$: the set E is closed and f is continuous but $f(E) = (0, 1]$ is not closed.

It is also not the case that the continuous image of a bounded set is bounded. As an example, take $f(x) = 1/x$ and $E = (0, 1)$. Then E is bounded and f continuous but $f(E) = (1, \infty)$ is unbounded.

However, the combined properties of closedness *and* boundedness (that is, compactness) are preserved. That is the content of the preceding theorem.

Corollary 9.24 *Let f be a continuous, real-valued function with compact domain $\subset \mathbb{R}$. Then there is a number L such that*

$$|f(x)| \le L$$

for all $x \in K$.

Proof: We know from the theorem that $f(K)$ is compact. By Theorem 8.30, we conclude that $f(K)$ is bounded. Thus there is a number L such that $|t| \le L$ for all $t \in f(K)$. But that is just the assertion that we wish to prove. $\qquad\square$

In fact we can prove an important strengthening of the corollary. Since $f(K)$ is compact, it contains its supremum M and its infimum m. Therefore there must be a number $C \in K$ such that $f(C) = M$ and a number $c \in K$ such that $f(c) = m$. In other words, $f(c) \leq f(x) \leq f(C)$ for all $x \in K$. We summarize:

Theorem 9.25 *Let f be a continuous function on a compact set $K \subset \mathbb{R}$. Then there exist numbers c and C in K such that $f(c) \leq f(x) \leq f(C)$ for all $x \in K$. We call c an* absolute minimum *for f on K and C an* absolute maximum *for f on K. We call $f(c)$ the* absolute minimum value *for f on K and $f(C)$ the* absolute maximum value *for f on K.*

Notice that, in the last theorem, the location of the absolute maximum and absolute minimum need not be unique. For instance, the function $\sin x$ on the compact interval $[0, 4\pi]$ has an absolute minimum at $3\pi/2$ and $7\pi/2$. It has an absolute maximum at $\pi/2$ and at $5\pi/2$.

Now we define a refined type of continuity called "uniform continuity." We shall learn that this new notion of continuous function arises naturally for a continuous function on a compact set. It will also play an important role in our later studies, especially in the context of the integral.

Definition 9.26 Let f be a function with domain $E \subset \mathbb{R}$. We say that f is *uniformly continuous* on E if, for each $\epsilon > 0$, there is a $\delta > 0$ such that, whenever $s, t \in E$ and $|s - t| < \delta$, then $|f(s) - f(t)| < \epsilon$.

POINT OF CONFUSION 9.27 Observe that the terminology "uniform continuity" differs from "continuity" in that it treats all points of the domain simultaneously: the $\delta > 0$ that is chosen is independent of the points $s, t \in E$. This difference is highlighted by the next two examples.

EXAMPLE 9.28 Suppose that a function $f : \mathbb{R} \to \mathbb{R}$ satisfies the condition

$$|f(s) - f(t)| \leq C \cdot |s - t|, \tag{9.28.1}$$

where C is some positive constant. This is called a *Lipschitz condition*, and it arises frequently in analysis. We refer to the collection of functions satisfying such a condition as the space Lip_1. Let $\epsilon > 0$ and set $\delta = \epsilon/C$. If $|x - y| < \delta$ then, by (9.28.1),

$$|f(x) - f(y)| \leq C \cdot |x - y| < C \cdot \delta = C \cdot \frac{\epsilon}{C} = \epsilon.$$

It follows that f is uniformly continuous.   ∎

EXAMPLE 9.29 Consider the function $f(x) = x^2$. Fix a point $c \in \mathbb{R}, c > 0$, and let $\epsilon > 0$. In order to guarantee that $|f(x) - f(c)| < \epsilon$ we must have (for $x > 0$)

$$|x^2 - c^2| < \epsilon$$

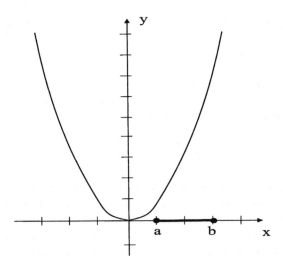

Figure 9.6: Uniform continuity on the interval $[a, b]$.

or

$$|x - c| < \frac{\epsilon}{x + c}.$$

Since x will range over a neighborhood of c, we see that the required δ in the definition of continuity cannot be larger than $\epsilon/(2c)$. In fact the choice $|x - c| < \delta = \epsilon/(2c + 1)$ will do the job. We see that the choice of δ depends decisively on c.

Put in slightly different words, let $\epsilon = 1$. Then

$$|f(j + 1/j) - f(j)| \geq |(j + 1/j)^2 - j^2| > 1 = \epsilon$$

for any j. Thus, for this ϵ, we may not take δ to be $1/j$ for any j. So no uniform δ exists.

Thus the choice of δ depends not only on ϵ (which we have come to expect) but also on c. In particular, f is not uniformly continuous on $\mathbb{R}$. This a quantitative reflection of the fact that the graph of f becomes ever steeper as the variable is positive and moves to the right.

Notice that the same calculation shows that the function f with restricted domain $[a, b], 0 < a < b < \infty$, *is* uniformly continuous. That is because, when the function is restricted to $[a, b]$, its rate of increase (sometimes called the "slope") does not become arbitrarily large. See Figure 9.6. ∎

Now the main result about uniform continuity is the following:

Theorem 9.30 *Let f be a continuous function with compact domain K. Then f is uniformly continuous on K.*

Proof: Suppose not. Then there is an $\epsilon > 0$ and points x_j, t_j such that $|x_j - t_j| < 1/j$ yet $|f(x_j) - f(t_j)| > \epsilon$. Since K is compact, there are subsequences x_{j_k} and t_{j_k} that converge respectively to points x_0 and t_0 in K. But in fact x_0 and t_0 must be the same point, yet $f(x_0) \neq f(t_0)$. That is a contradiction.   □

EXAMPLE 9.31 The function $f(x) = \sin(1/x)$ is continuous on the domain $E = (0, \infty)$ since it is the composition of continuous functions (refer again to Figure 9.2). However, it is not uniformly continuous since

$$\left| f\left(\frac{1}{2j\pi}\right) - f\left(\frac{1}{\frac{(4j+1)\pi}{2}}\right) \right| = 1$$

for $j = 1, 2, \ldots$. Thus, even though the arguments are becoming arbitrarily close together, the images of these arguments remain bounded apart. We conclude that f cannot be uniformly continuous. See Figure 9.2.

However, if f is considered as a function on any interval of the form $[a, b], 0 < a < b < \infty$, then the preceding theorem tells us that the function f is uniformly continuous.   ∎

As an exercise, you should check that

$$g(x) = \begin{cases} x \sin(1/x) & \text{if } x \neq 0 \\ 0 & \text{if } x = 0 \end{cases}$$

is uniformly continuous on any interval of the form $[-N, N]$. See Figure 9.3.

Let us note that a function f is said to be continuous on a closed interval $[a, b]$ if f is continuous at each point of $[a, b]$. Refer back to our original definition of continuity.

Corollary 9.32 (The Intermediate Value Theorem) *Let f be a continuous function whose domain contains the interval $[a, b]$. Let γ be a number that lies between $f(a)$ and $f(b)$. Then there is a number c between a and b such that $f(c) = \gamma$. Refer to Figure 9.7.*

Proof: We merely sketch the proof.

Assume without loss of generality that $f(a) < 0$ and $f(b) > 0$ and $\gamma = 0$. Let

$$S = \{x \in [a, b] : f(x) < 0\}.$$

Then S is bounded and nonempty (because $a \in S$), so S has a least upper bound c. We claim that $f(c) = 0$.

Clearly $f(c) \leq 0$ by the continuity of f. If in fact $f(c) < 0$, then points x to the right of c will satisfy $f(x) < 0$, contradicting the fact that c was chosen to be the least upper bound of S. So it must be that $f(c) = 0$ as desired.   □

Remark 9.33 Another way to think about the Intermediate Value Theorem is in terms of connectedness. The interval $[a, b]$ is of course connected. So its image $f([a, b])$ is connected. Therefore $f([a, b])$ must be an interval. Since that interval contains the points $f(a)$ and $f(b)$, it must therefore also contain the point γ which lies between $f(a)$ and $f(b)$.

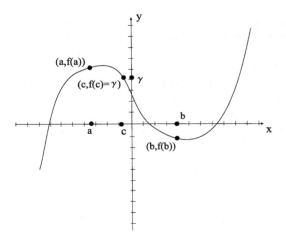

Figure 9.7: The Intermediate Value Theorem.

A Look Back

1. Say in words why the continuous image of a compact set is compact.

2. Say in words why the continuous image of a connected set is connected.

3. Say in words why the continuous image of a bounded set is not necessarily bounded.

4. Say in words why the continuous image of a closed set is not necessarily closed.

Exercises

1. If f is continuous on $[0,1]$ and if $f(x)$ is positive for each rational x, then does it follow that f is positive at all x?

2. Let S be any subset of $\mathbb{R}$. Define the function
$$f(x) = \inf\{|x - s| : s \in S\}.$$
[We think of $f(x)$ as the distance of x to S.] Prove that f is uniformly continuous.

3. Define the function $g(x)$ to take the value 0 at irrational values of x and to take the value $1/q$ when $x = p/q$ is a rational number in lowest terms, $q > 0$. At which points is g continuous? At which points is the function discontinuous?

4. Let f be any function whose domain and range is the entire real line. If A and B are disjoint sets, does it follow that $f(A)$ and $f(B)$ are disjoint sets? If C and D are disjoint sets, does it follow that $f^{-1}(C)$ and $f^{-1}(D)$ are disjoint?

5. Let f be any function whose domain is the entire real line. If A and B are sets then is $f(A \cup B) = f(A) \cup f(B)$? If C and D are sets then is $f^{-1}(C \cup D) = f^{-1}(C) \cup f^{-1}(D)$? What is the answer to these questions if we replace $\cup$ by $\cap$?

6. Prove that the function $f(x) = \sin x$ can be written, on the interval $(0, 2\pi)$, as the difference of two increasing functions.

7. Let f be a continuous function with domain $[0, 1]$ and range $[0, 1]$. Prove that there exists a point $c \in [0, 1]$ such that $f(c) = c$. (**Hint:** Apply the Intermediate Value Theorem to the function $g(x) = f(x) - x$.) Prove that this result is false if the domain and range of the function are both $(0, 1)$.

8. Let f be a continuous function and let $\{a_j\}$ be a Cauchy sequence in the domain of f. Does it follow that $\{f(a_j)\}$ is a Cauchy sequence? What if we assume instead that f is uniformly continuous?

9. Let E and F be disjoint closed sets of real numbers. Prove that there is a continuous function f with domain the real numbers such that $\{x : f(x) = 0\} = E$ and $\{x : f(x) = 1\} = F$.

10. If K and L are sets then define

$$K + L = \{k + \ell : k \in K \text{ and } \ell \in L\}.$$

If K and L are compact then prove that $K + L$ is compact. If K and L are merely closed, does it follow that $K + L$ is closed?

* 11. Prove that the composition of continuous functions is continuous.

* 12. Give an example of a continuous function f and a connected set E such that $f^{-1}(E)$ is not connected. Is there a condition you can add that will force $f^{-1}(E)$ to be connected?

* 13. Give an example of a continuous function f and an open set U so that $f(U)$ is not open.

* 14. A function f from an interval (a, b) to an interval (c, d) is called *proper* if, for any compact set $K \subset (c, d)$, it holds that $f^{-1}(K)$ is compact in (a, b). Prove that, if f is proper, then either

$$\lim_{x \to a^+} f(x) = c \quad \text{or} \quad \lim_{x \to a^+} f(x) = d.$$

Likewise prove that either

$$\lim_{x \to b^-} f(x) = c \quad \text{or} \quad \lim_{x \to b^-} f(x) = d.$$

* 15. A function f with domain A and range B is called a *homeomorphism* if it is one-to-one, onto, continuous, and has a continuous inverse. If such an f exists then we say that A and B are *homeomorphic*. Which sets of reals are homeomorphic to the open unit interval $(0, 1)$? Which sets of reals are homeomorphic to the closed unit interval $[0, 1]$?

9.4 Classifying Discontinuities and Monotonicity

Preliminary Remarks

Just by using a little logic, we can classify the discontinuities of a real function. There are jump discontinuities and oscillatory discontinuities. This gives an elegant way to understand discontinuities.

We begin by refining our notion of limit:

Definition 9.34 Fix $c \in \mathbb{R}$. Let f be a function whose domain contains an interval (a, c). We say that f has *left limit* ℓ at c, and write

$$\lim_{x \to c^-} f(x) = \ell,$$

if, for every $\epsilon > 0$, there is a $\delta > 0$ such that, whenever $c - \delta < x < c$, then it holds that

$$|f(x) - \ell| < \epsilon.$$

Now suppose instead that the domain of f contains an interval (c, b). We say that f has *right limit* m at c, and write

$$\lim_{x \to c^+} f(x) = m$$

if, for every $\epsilon > 0$, there is a $\delta > 0$ such that, whenever $c < x < c + \delta$, then it holds that

$$|f(x) - m| < \epsilon.$$

This definition simply formalizes the notion of either letting x tend to c from the left only or from the right only.

Definition 9.35 Fix $c \in \mathbb{R}$. Let f be a function with domain E. Suppose that c is a limit point of $E \cap [c - \delta, c)$ for some $\delta > 0$ and that c is an element of E. We say that f is *left continuous* at c if

$$\lim_{x \to c^-} f(x) = f(c).$$

Likewise, in case c is a limit point of $E \cap (c, c + \delta]$ for some $\delta > 0$ and is also an element of E, we say that f is *right continuous* at c if

$$\lim_{x \to c^+} f(x) = f(c).$$

Let f be a function with domain E. Let c in E and assume that f is discontinuous at c. There are two ways in which this discontinuity can occur:

I. If $\lim_{x \to c^-} f(x)$ and $\lim_{x \to c^+} f(x)$ both exist but either do not equal each other or do not equal $f(c)$ then we say that f has a *discontinuity of the first kind* (or sometimes a *simple discontinuity*) at c.

II. If either $\lim_{x \to c^-}$ does not exist or $\lim_{x \to c^+}$ does not exist then we say that f has a *discontinuity of the second kind* at c.

Refer to Figure 9.8.

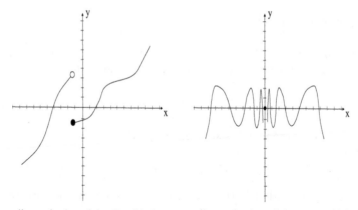

discontinuity of the first kind   discontinuity of the second kind

Figure 9.8: Discontinuities of the first and second kind.

EXAMPLE 9.36 Define

$$f(x) = \begin{cases} \sin(1/x) & \text{if} \quad x \neq 0 \\ 0 & \text{if} \quad x = 0 \end{cases}$$

$$g(x) = \begin{cases} 1 & \text{if} \quad x > 0 \\ 0 & \text{if} \quad x = 0 \\ -1 & \text{if} \quad x < 0 \end{cases}$$

$$h(x) = \begin{cases} 1 & \text{if } x \text{ is irrational} \\ 0 & \text{if } x \text{ is rational.} \end{cases}$$

Then f has a discontinuity of the second kind at 0 while g has a discontinuity of the first kind at 0. The function h has a discontinuity of the second kind at every point. ∎

POINT OF CONFUSION 9.37 For a bounded function, a discontinuity of the second kind can be thought of as an oscillatory discontinuity. A discontinuity of the first kind can be thought of as a case of the limiting value not equaling the actual value (sometimes we think of this as a jump discontinuity).

Definition 9.38 Let f be a function whose domain contains an open interval (a, b). We say that f is *increasing* on (a, b) if, whenever $a < s < t < b$, it holds that $f(s) \leq f(t)$. We say that f is *decreasing* on (a, b) if, whenever $a < s < t < b$, it holds that $f(s) \geq f(t)$.

We say that f is *strictly increasing* on (a, b) if, whenever $a < s < t < b$, it holds that $f(s) < f(t)$. We say that f is *strictly decreasing* on (a, b) if, whenever $a < s < t < b$, it holds that $f(s) > f(t)$.

If a function is either increasing or decreasing then we call it *monotone* or *monotonic*. If the function is strictly increasing or strictly decreasing then we

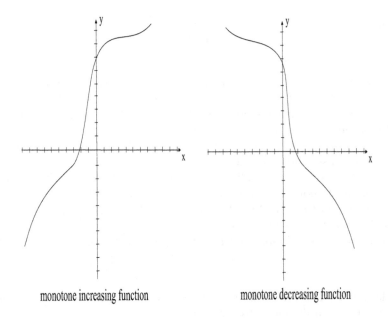

monotone increasing function   monotone decreasing function

Figure 9.9: Increasing and decreasing functions.

call the function *strictly monotonic*. Compare with the definition of monotonic sequences in Section 6.1.

As with sequences, the word "monotonic" is superfluous in many contexts. But its use is traditional and occasionally convenient.

Proposition 9.39 *Let f be a monotonic function on an open interval (a, b). Then all of the discontinuities of f are of the first kind.*

Proof: It is enough to show that, for each $c \in (a, b)$, the limits

$$\lim_{x \to c^-} f(x)$$

and

$$\lim_{x \to c^+} f(x)$$

exist.

Let us first assume that f is monotonically increasing. Fix $c \in (a, b)$. If $a < s < c$ then $f(s) \leq f(c)$. Therefore $S = \{f(s) : a < s < c\}$ is bounded above. Let M be the least upper bound of S. Pick $\epsilon > 0$. By definition of least upper bound there must be an $f(s) \in S$ such that $|f(s) - M| < \epsilon$. Let $\delta = |c - s|$. If $c - \delta < t < c$ then $s < t < c$ and $f(s) \leq f(t) \leq M$ or $|f(t) - M| < \epsilon$. Thus $\lim_{x \to c^-} f(x)$ exists and equals M.

If we set m equal to the infimum of the set $T = \{f(t) : c < t < b\}$ then a similar argument shows that $\lim_{x \to c^+} f(x)$ exists and equals m.

So we see that the function f has both a left and a right limit at c. So either f is continuous at c or f has a discontinuity of the first kind.

The argument for f monotonically decreasing is the same, and we omit the details. □

Corollary 9.40 *Let f be a monotonic function on an interval (a, b). Then f has at most countably many discontinuities.*

Proof: Assume for simplicity that f is monotonically increasing. If c is a discontinuity then the proposition tells us that

$$\lim_{x \to c^-} f(x) < \lim_{x \to c^+} f(x).$$

Therefore there is a rational number q_c between $\lim_{x \to c^-} f(x)$ and $\lim_{x \to c^+} f(x)$. Notice that different discontinuities will have different rational numbers associated to them because if $\widehat{c}$ is another discontinuity and, say, $\widehat{c} < c$ then

$$\lim_{x \to \widehat{c}^-} f(x) < q_{\widehat{c}} < \lim_{x \to \widehat{c}^+} f(x) \leq \lim_{x \to c^-} f(x) < q_c < \lim_{x \to c^+} f(x).$$

Thus we have exhibited a one-to-one function of the set of discontinuities of f into the set of rational numbers. It follows (see Section 4.5) that the set of discontinuities is countable. □

POINT OF CONFUSION 9.41 A function can have arbitrarily many (even uncountably many) discontinuities. But a monotone function has at most countably many.

Theorem 9.42 *Let f be a strictly monotone, continuous function with domain $[a, b]$. Then f^{-1} exists and is continuous.*

Proof: Assume without loss of generality that f is strictly monotone *increasing*. Let us extend f to the entire real line by defining

$$f(x) = \begin{cases} (x - a) + f(a) & \text{if} \quad x < a \\ \text{as given} & \text{if} \quad a \leq x \leq b \\ (x - b) + f(b) & \text{if} \quad x > b. \end{cases}$$

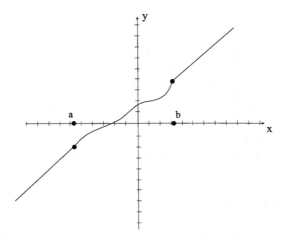

Figure 9.10: A strictly monotonically increasing function.

See Figure 9.10. Then it is easy to see that this extended version of f is still continuous and is strictly monotone increasing on all of $\mathbb{R}$.

The definition of strict monotonicity implies certainly that f is one-to-one. That it is onto follows from the continuity and from the way that we extended the function to have domain all of $\mathbb{R}$. Thus f^{-1} exists.

The extended function f takes any open interval (c, d) to the open interval $(f(c), f(d))$. Since any open set is a union of open intervals, we see that f takes any open set to an open set. In other words, $\left[f^{-1}\right]^{-1}$ takes open sets to open sets. But this just says that f^{-1} is continuous.

Since the inverse of the extended function f is continuous, then so is the inverse of the original function f. That completes the proof. $\qquad\square$

A Look Back

1. What is a discontinuity of the first kind?

2. What is a discontinuity of the second kind?

3. Why can a discontinuity not be both of the first kind and the second kind?

4. Why is there no other kind of discontinuity besides first and second?

Exercises

1. Give an example of two functions, discontinuous at $x = 0$, whose sum *is* continuous at $x = 0$. Give an example of two such functions whose product is

continuous at $x = 0$. How does the problem change if we replace "product" by "quotient"?

2. Let f be a function with domain $\mathbb{R}$. If $f^2(x) = f(x) \cdot f(x)$ is continuous, then does it follow that f is continuous? If $f^3(x) = f(x) \cdot f(x) \cdot f(x)$ is continuous, then does it follow that f is continuous? What about if both f^2 and f^3 are continuous?

3. Fix an interval (a, b). Is the collection of increasing functions on (a, b) closed under $+, -, \times,$ or $\div$?

4. Let f be a continuous function whose domain contains a closed, bounded interval $[a, b]$. What topological properties does $f([a, b])$ possess? Is this set necessarily an interval?

5. Refer to Exercise 15 of Section 9.3 for terminology. Show that there is no homeomorphism from the real line to the interval $[0, 1)$.

6. Let f be a differentiable function on the interval (a, b). Show that f' has the Intermediate Value Property. Thus, in particular, all the discontinuities of f' are of the second kind. This result is known as *Darboux's theorem*.

7. Let f be a function with domain $(-1, 1)$. Can the set of discontinuities of f of the first kind be countable? Uncountable? What about the set of discontinuities of the second kind?

8. Let f and g be functions and assume that each has a discontinuity of the first kind at the origin. What can you say about the behavior of $f + g$ at the origin?

* 9. Let A be any left-to-right ordered, countable subset of the reals. Construct an increasing function whose set of points of discontinuity is precisely the set A. Explain why this is, in general, impossible for an uncountable set A.

* 10. TRUE or FALSE: If f is a function with domain and range the real numbers and which is both one-to-one and onto, then f must be either increasing or decreasing. Does your answer change if we assume that f is continuously differentiable?

* 11. Let $I \subset \mathbb{R}$ be an open interval and $f : I \to \mathbb{R}$ a function. We say that f is *convex* if whenever $\alpha, \beta \in I$ and $0 \leq t \leq 1$ then

$$f((1 - t)\alpha + t\beta) \leq (1 - t)f(\alpha) + tf(\beta).$$

Prove that a convex function must be continuous. What does this definition of convex function have to do with the notion of "concave up" that you learned in calculus?

* 12. Refer to Exercise 11 for terminology. What can you say about differentiability of a convex function?

Chapter 10

Differentiation of Functions

10.1 The Concept of Derivative

Preliminary Remarks

Of course the derivative is a significant idea from calculus, and it is important that we establish a rigorous and precise understanding of the concept. Our aim is to develop the derivative as a precise analytic tool, and also to get a rigorous proof of the Fundamental Theorem of Calculus.

Let f be a function with domain an open interval I. If $x \in I$ then the quantity

$$\frac{f(t) - f(x)}{t - x}$$

measures the slope of the chord of the graph of f that connects the points $(x, f(x))$ and $(t, f(t))$. See Figure 10.1. If we let $t \to x$ then the limit of the quantity represented by this "Newton quotient" should represent the slope of the graph *at the point* x. These considerations motivate the definition of the derivative:

Definition 10.1 If f is a function with domain an open interval I and if $x \in I$, then the limit

$$\lim_{t \to x} \frac{f(t) - f(x)}{t - x},$$

when it exists, is called the *derivative* of f at x. See Figure 10.2. If the derivative of f at x exists then we say that f is *differentiable* at x. If f is differentiable at every $x \in I$ then we say that f is *differentiable on I*.

We write the derivative of f at x either as

$$f'(x) \quad \text{or} \quad \frac{d}{dx}f \quad \text{or} \quad \frac{df}{dx} \quad \text{or} \quad \dot{f}.$$

217

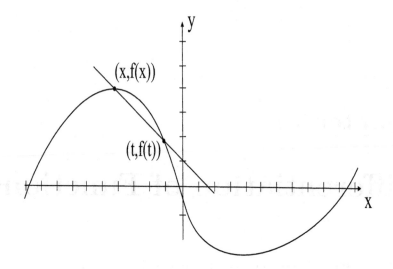

Figure 10.1: The Newton quotient.

The last of these is particularly popular among physicists (as an homage to Isaac Newton).

We begin our discussion of the derivative by establishing some basic properties and relating the notion of derivative to continuity.

Lemma 10.2 *If f is differentiable at a point x then f is continuous at x. In particular, $\lim_{t \to x} f(t) = f(x)$.*

Proof: We use Theorem 9.7(b) about limits to see that

$$
\begin{aligned}
\lim_{t \to x} (f(t) - f(x)) &= \lim_{t \to x} \left((t - x) \cdot \frac{f(t) - f(x)}{t - x} \right) \\
&= \lim_{t \to x} (t - x) \cdot \lim_{t \to x} \frac{f(t) - f(x)}{t - x} \\
&= 0 \cdot f'(x) \\
&= 0.
\end{aligned}
$$

Therefore $\lim_{t \to x} f(t) = f(x)$ and f is continuous at x.   □

POINT OF CONFUSION 10.3 All differentiable functions are continuous: differentiability is a stronger property than continuity. Observe that the function $f(x) = |x|$ is continuous at every x but is not differentiable at 0. So continuity does not imply differentiability. Details appear in Example 10.6 below.

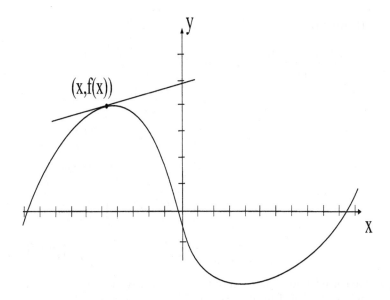

Figure 10.2: The derivative.

POINT OF CONFUSION 10.4 You might think, from your experience in other math courses, that most functions are differentiable at most points. As we shall learn below, this expectation is woefully incorrect. It turns out that "most" continuous functions are not differentiable at any point.

Theorem 10.5 *Assume that f and g are functions with domain an open interval I and that f and g are differentiable at $x \in I$. Then $f \pm g, f \cdot g$, and f/g are differentiable at x (for f/g we assume that $g(x) \neq 0$). Moreover*

(a) $(f \pm g)'(x) = f'(x) \pm g'(x);$

(b) $(f \cdot g)'(x) = f'(x) \cdot g(x) + f(x) \cdot g'(x);$

(c) $\left(\dfrac{f}{g}\right)'(x) = \dfrac{g(x) \cdot f'(x) - f(x) \cdot g'(x)}{g^2(x)}.$

Proof: Assertion **(a)** is easy and we leave it as an exercise for you.

For **(b)**, we write

$$\lim_{t \to x} \frac{(f \cdot g)(t) - (f \cdot g)(x)}{t - x} = \lim_{t \to x} \left(\frac{(f(t) - f(x)) \cdot g(t)}{t - x} \right.$$
$$\left. + \frac{(g(t) - g(x)) \cdot f(x)}{t - x} \right)$$
$$= \lim_{t \to x} \left(\frac{(f(t) - f(x)) \cdot g(t)}{t - x} \right)$$
$$+ \lim_{t \to x} \left(\frac{(g(t) - g(x)) \cdot f(x)}{t - x} \right)$$
$$= \lim_{t \to x} \left(\frac{(f(t) - f(x))}{t - x} \right) \cdot \left(\lim_{t \to x} g(t) \right)$$
$$+ \lim_{t \to x} \left(\frac{(g(t) - g(x))}{t - x} \right) \cdot \left(\lim_{t \to x} f(x) \right),$$

where we have used Theorem 9.7 about limits. Now the first limit is the derivative of f at x, while the third limit is the derivative of g at x. Also notice that the limit of $g(t)$ equals $g(x)$ by the lemma. The result is that the last line equals

$$f'(x) \cdot g(x) + g'(x) \cdot f(x),$$

as desired.

To prove **(c)**, write

$$\lim_{t \to x} \frac{(f/g)(t) - (f/g)(x)}{t - x} = \lim_{t \to x} \frac{1}{g(t) \cdot g(x)} \left(\frac{f(t) - f(x)}{t - x} \cdot g(x) \right.$$
$$\left. - \frac{g(t) - g(x)}{t - x} \cdot f(x) \right).$$

The proof is now completed by using Theorem 9.7 about limits to evaluate the individual limits in this expression.   □

EXAMPLE 10.6 That $f(x) = x$ is differentiable follows from

$$\lim_{t \to x} \frac{t - x}{t - x} = 1.$$

Any constant function is differentiable (with derivative identically zero) by a similar argument. It follows from the theorem that any polynomial function is differentiable.

On the other hand, the continuous function $f(x) = |x|$ is *not* differentiable at the point $x = 0$. This is so because

$$\lim_{t \to 0^-} \frac{|t| - |0|}{t - x} = \lim_{t \to 0^-} \frac{-t - 0}{t - 0} = -1$$

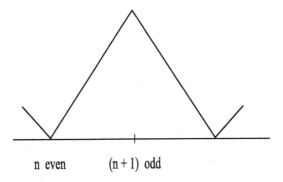

n even   (n + 1) odd

Figure 10.3: The van der Waerden example.

while

$$\lim_{t \to 0^+} \frac{|t| - |0|}{t - x} = \lim_{t \to 0^+} \frac{t - 0}{t - 0} = 1.$$

So the required limit does not exist. ∎

Since the subject of differential calculus is concerned with learning uses of the derivative, it concentrates on functions which *are* differentiable. One comes away from the subject with the impression that most functions are differentiable except at a few isolated points—as is the case with the function $f(x) = |x|$. Indeed this was what the mathematicians of the nineteenth century thought. Therefore it came as a shock when Karl Weierstrass produced a continuous function that is not differentiable at *any point*. In fact *most* continuous functions are of this nature: their graphs "wiggle" so much that they cannot have a tangent line at any point. Now we turn to an elegant variant of the example of Weierstrass that is due to B. L. van der Waerden (1903–1996).

Theorem 10.7 *Define a function ψ with domain $\mathbb{R}$ by the rule*

$$\psi(x) = \begin{cases} x - n & \text{if } n \le x < n + 1 \text{ and } n \text{ is even} \\ n + 1 - x & \text{if } n \le x < n + 1 \text{ and } n \text{ is odd} \end{cases}$$

for every integer n. The graph of this function is exhibited in Figure 10.3. Then the function

$$f(x) = \sum_{j=1}^{\infty} \left(\frac{3}{4}\right)^j \psi\left(4^j x\right)$$

is continuous at every real x and differentiable at no real x.

The proof of this theorem is intricate, and we put it in an Appendix to this chapter. What is important for you to understand right now is that this remarkable nowhere differentiable function exists, and can be constructed explicitly.

POINT OF CONFUSION 10.8 The proof of Weierstrass's theorem is long, but the idea is simple: the function f is built by piling oscillations on top of oscillations. When the ℓth oscillation is added, it is made very small in size so that it does not cancel the previous oscillations. But it is made very steep so that it will cause the derivative to become large. The practical meaning of the examples of Weierstrass and van der Waerden is that we should realize that differentiability is a very strong and special property of functions. Most continuous functions are not differentiable at any point. Just as an instance, if φ is *any* differentiable function and f is the function from Theorem 10.7, then $h = \varphi + \epsilon f$ will be nowhere differentiable for any $\epsilon > 0$.

When we are proving theorems about continuous functions, we should not think of them in terms of properties of differentiable functions.

Next we turn to the Chain Rule.

Theorem 10.9 *Let g be a differentiable function on an open interval I and let f be a differentiable function on an open interval that contains the range of g. Then $f \circ g$ is differentiable on the interval I and*

$$(f \circ g)'(x) = f'(g(x)) \cdot g'(x)$$

for each $x \in I$.

Proof: We use the notation Δt to stand for an increment in the variable t. Let us use the symbol $\mathcal{V}(r)$ to stand for any expression which tends to 0 as $\Delta r \to 0$. Fix $x \in I$. Set $r = g(x)$. By hypothesis,

$$\lim_{\Delta r \to 0} \frac{f(r + \Delta r) - f(r)}{\Delta r} = f'(r)$$

or

$$\frac{f(r + \Delta r) - f(r)}{\Delta r} - f'(r) = \mathcal{V}(r)$$

or

$$f(r + \Delta r) = f(r) + \Delta r \cdot f'(r) + \Delta r \cdot \mathcal{V}(r). \tag{10.9.1}$$

Notice that equation (10.9.1) is valid even when $\Delta r = 0$. Since Δr in equation (10.9.1) can be any small quantity, we set

$$\Delta r = \Delta x \cdot [g'(x) + \mathcal{V}(x)].$$

Substituting this expression into (10.9.1) and using the fact that $r = g(x)$ yields

$$\begin{aligned}
f(g(x) + &\Delta x[g'(x) + \mathcal{V}(x)]) \\
&= f(r) + (\Delta x \cdot [g'(x) + \mathcal{V}(x)]) \cdot f'(r) \\
&\quad + (\Delta x \cdot [g'(x) + \mathcal{V}(x)]) \cdot \mathcal{V}(r) \\
&= f(g(x)) + \Delta x \cdot f'(g(x)) \cdot g'(x) + \Delta x \cdot \mathcal{V}(x). \tag{10.9.2}
\end{aligned}$$

Notice that, in this last line, we have combined terms to create a new term $\mathcal{V}(x)$ that vanishes as Δx tends to 0.

Just as we derived (10.9.1), we may also obtain

$$
\begin{aligned}
g(x + \Delta x) &= g(x) + \Delta x \cdot g'(x) + \Delta x \cdot \mathcal{V}(x) \\
&= g(x) + \Delta x [g'(x) + \mathcal{V}(x)] \,.
\end{aligned}
$$

We may substitute this equality into the left side of (10.9.2) to obtain

$$
f(g(x + \Delta x)) = f(g(x)) + \Delta x \cdot f'(g(x)) \cdot g'(x) + \Delta x \cdot \mathcal{V}(x) \,.
$$

With some algebra this can be rewritten as

$$
\frac{f(g(x + \Delta x)) - f(g(x))}{\Delta x} - f'(g(x)) \cdot g'(x) = \mathcal{V}(x) \,.
$$

But this just says that

$$
\lim_{\Delta x \to 0} \frac{(f \circ g)(x + \Delta x) - (f \circ g)(x)}{\Delta x} = f'(g(x)) \cdot g'(x) \,.
$$

That is, $(f \circ g)'(x)$ exists and equals $f'(g(x)) \cdot g'(x)$, as desired. $\qquad\square$

POINT OF CONFUSION 10.10 It is tempting, in trying to prove the Chain Rule, to reason that

$$
(f \circ g)' = \lim_{\Delta x \to 0} \frac{\triangle(f \circ g)}{\triangle x} = \lim_{\Delta x \to 0} \frac{\triangle(f \circ g)}{\triangle g} \cdot \lim_{\Delta x \to 0} \frac{\triangle g}{\triangle x} = f'(g) \cdot g' \,.
$$

The trouble with this intuitively appealing way of thinking is that $\triangle g$ could vanish. As a result, we must work harder. That is why the proof of the proof of the Chain Rule is a bit complicated.

A Look Back

1. Say in words what the derivative signifies.

2. Give a verbal explanation of the Chain Rule.

3. What is the significance of the Weierstrass Nowhere Differentiable Function?

4. What properties does a function with positive derivative have?

Exercises

1. For which positive integers k is it true that if $f^k = f \cdot f \cdots f$ is differentiable at x then f is differentiable at x?

2. Let $f(x)$ equal 0 if x is irrational; let $f(x)$ equal $1/q$ if x is a rational number that can be expressed in lowest terms as p/q. Is f differentiable at any x?

3. Formulate notions of "left differentiable" and "right differentiable" for functions defined on suitable half-open intervals. Also formulate definitions of "left continuous" and "right continuous." If you have done things correctly, then you should be able to prove that a left differentiable (right differentiable) function is left continuous (right continuous).

4. Give an example of a function $f : \mathbb{R} \to \mathbb{R}$ which is differentiable at every point but so that the derivative function is not continuous (at least at one point).

5. Prove part **(a)** of Theorem 10.5.

6. Prove that, if f is differentiable on an open interval I and if $f'(x) > 0$, then it is not necessarily the case that f' is positive at points near x.

7. Give an example of a function f on $\mathbb{R}$ so that f' takes on all possible real values.

8. Prove that $f(x) = x^2 \sin(1/x)$ is differentiable at the origin, but the derivative is not continuous.

* **9.** Assume that f is a continuous function on $(-1, 1)$ and that f is differentiable on $(-1, 0) \cup (0, 1)$. If the limit $\lim_{x \to 0} f'(x)$ exists then is f differentiable at $x = 0$?

* **10.** Prove that the Weierstrass Nowhere Differentiable Function f satisfies

$$\frac{|f(x + h) + f(x - h) - 2f(x)|}{|h|} \leq C|h|$$

for all nonzero h but f is *not* in Lip_1.

* **11.** Prove that the Nowhere Differentiable Function constructed in Theorem 10.7 is in Lip_α for all $\alpha < 1$. [Here $f \in \mathrm{Lip}_\alpha$ if $|f(x) - f(t)| \leq C|x - t|^\alpha$ for all x, t.]

* **12.** Let $E \subset \mathbb{R}$ be a closed set. Fix a nonnegative integer k. Show that there is a function f in $C^k(\mathbb{R})$ (that is, a k-times continuously differentiable function) such that $E = \{x : f(x) = 0\}$.

10.2   The Mean Value Theorem and Applications

Preliminary Remarks

The Mean Value Theorem is a powerful analytic tool. It has intuitive appeal as well. You learned about the Mean Value Theorem in calculus class. But this result is best perceived as a technique of real analysis. In particular, it is Cauchy's version of the Mean Value Theorem that leads to l'Hôpital's Rule.

We begin this section with some remarks about local maxima and minima of functions.

Definition 10.11 Let f be a function with domain (a, b). A point $C \in (a, b)$ is called a *local maximum* for f (we also say that f has a local maximum at C) if there is a $\delta > 0$ such that $f(t) \leq f(C)$ for all $t \in (C - \delta, C + \delta)$. A point $c \in (a, b)$

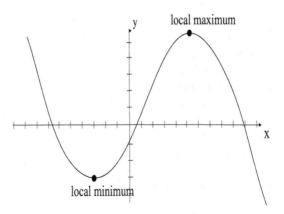

Figure 10.4: Some extrema.

is called a *local minimum* for f (we also say that f has a local minimum at c) if there is a $\delta > 0$ such that $f(t) \geq f(c)$ for all $t \in (c - \delta, c + \delta)$. See Figure 10.4.

Local minima (plural of minimum) and local maxima (plural of maximum) are referred to collectively as *local extrema*.

Proposition 10.12 (Fermat) *If f is a function with domain (a, b), if f has a local extremum at $x \in (a, b)$, and if f is differentiable at x, then $f'(x) = 0$.*

Proof: Suppose that f has a local minimum at x. Then there is a $\delta > 0$ such that $x - \delta < t < x$ implies $f(t) \geq f(x)$. Hence

$$\frac{f(t) - f(x)}{t - x} \leq 0.$$

Letting $t \to x$, it follows that $f'(x) \leq 0$. Similarly, if $x < t < x + \delta$ for suitable δ, then

$$\frac{f(t) - f(x)}{t - x} \geq 0.$$

It follows that $f'(x) \geq 0$. We must conclude that $f'(x) = 0$.

A similar argument applies if f has a local maximum at x. The proof is therefore complete. □

Before going on to mean value theorems, we provide a striking application of Fermat's proposition:

Theorem 10.13 (Darboux's Theorem) *Let f be a differentiable function on an open interval I. Pick points $s < t$ in I and suppose that $f'(s) < \rho < f'(t)$. Then there is a point u between s and t such that $f'(u) = \rho$.*

Proof: Consider the function $g(x) = f(x) - \rho x$. Then $g'(s) < 0$ and $g'(t) > 0$. Assume for simplicity that $s < t$. The sign of the derivative at s shows that $g(\widehat{s}) < g(s)$ for $\widehat{s}$ greater than s and near s. The sign of the derivative at t implies that $g(\widehat{t}) < g(t)$ for $\widehat{t}$ less than t and near t. Thus the minimum of the continuous function g on the compact interval $[s,t]$ must occur at some point u in the interior (s,t). The preceding proposition guarantees that $g'(u) = 0$, or $f'(u) = \rho$ as claimed.   □

If f' were a continuous function then the theorem would just be a special instance of the Intermediate Value Property of continuous functions (see Corollary 9.32). But derivatives need not be continuous, as the example

$$f(x) = \begin{cases} x^2 \cdot \sin(1/x) & \text{if} \quad x \neq 0 \\ 0 & \text{if} \quad x = 0 \end{cases}$$

illustrates. Check for yourself that $f'(0)$ exists and vanishes but $\lim_{x \to 0} f'(x)$ does not exist. This example illustrates the significance of the theorem. Since the theorem says that f' will always satisfy the Intermediate Value Property (even when it is not continuous), its discontinuities cannot be of the first kind. In other words:

Proposition 10.14 *If f is a differentiable function on an open interval I then the discontinuities of f' are all of the second kind.*

Next we turn to the simplest form of the Mean Value Theorem.

Theorem 10.15 (Rolle's Theorem) *Let f be a continuous function on the closed interval $[a,b]$ which is differentiable on (a,b). If $f(a) = f(b) = 0$ then there is a point $\xi \in (a,b)$ such that $f'(\xi) = 0$. See Figure 10.5.*

Proof: If f is a constant function, then any point ξ in the interval will do. So assume that f is nonconstant.

Theorem 9.25 guarantees that f will have both a maximum and a minimum in $[a,b]$. If one of these occurs in (a,b), then Proposition 10.12 guarantees that f' will vanish at that point and we are done. If both occur at the endpoints, then all the values of f lie between 0 and 0. In other words f is constant, contradicting our assumption.   □

Of course the point ξ in Rolle's theorem need not be unique. If $f(x) = x^3 - x^2 - 2x$ on the interval $[-1,2]$ then $f(-1) = f(2) = 0$ and f' vanishes at *two* points of the interval $(-1,2)$. Refer to Figure 10.6.

If you rotate the graph of a function satisfying the hypotheses of Rolle's theorem, the result suggests that, for any continuous function f on an interval $[a,b]$, differentiable on (a,b), we should be able to relate the slope of the chord connecting $(a, f(a))$ and $(b, f(b))$ with the value of f' at some interior point. That is the content of the standard Mean Value Theorem:

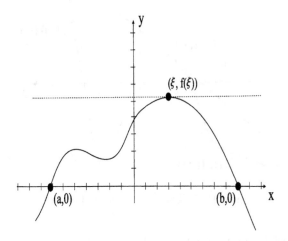

Figure 10.5: Rolle's theorem.

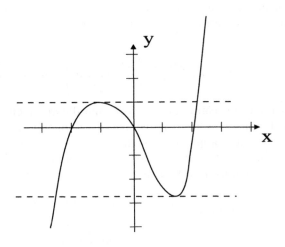

Figure 10.6: An example of Rolle's theorem.

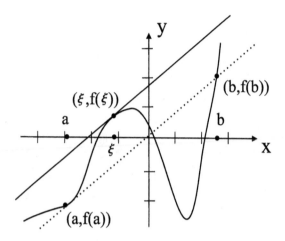

Figure 10.7: The Mean Value Theorem.

Theorem 10.16 (The Mean Value Theorem) *Assume that $a < b$. Let f be a continuous function on the closed interval $[a, b]$ that is differentiable on (a, b). There exists a point $\xi \in (a, b)$ such that*

$$\frac{f(b) - f(a)}{b - a} = f'(\xi).$$

See Figure 10.7.

Proof: Our scheme is to implement the remarks preceding the theorem: we "rotate" the picture to reduce to the case of Rolle's theorem. More precisely, define

$$g(x) = f(x) - \left[f(a) + \frac{f(b) - f(a)}{b - a} \cdot (x - a) \right] \quad \text{if} \quad x \in [a, b].$$

By direct verification, g is continuous on $[a, b]$ and differentiable on (a, b) (after all, g is obtained from f by elementary arithmetic operations). Also $g(a) = g(b) = 0$. Thus we may apply Rolle's theorem to g and we find that there is a $\xi \in (a, b)$ such that $g'(\xi) = 0$. Remembering that x is the variable, we differentiate the formula for g to find that

$$
\begin{aligned}
0 = g'(\xi) &= \left[f'(x) - \frac{f(b) - f(a)}{b - a} \right]\Bigg|_{x=\xi} \\
&= \left[f'(\xi) - \frac{f(b) - f(a)}{b - a} \right].
\end{aligned}
$$

As a result,

$$f'(\xi) = \frac{f(b) - f(a)}{b - a}. \qquad \square$$

POINT OF CONFUSION 10.17 Suppose that you are driving from point A to point B, that these reference points are 100 miles apart, and that the drive takes you 2 hours. We may use the Mean Value Theorem to see that, at some point on the trip, we must have been going *exactly* 50 miles per hour.

Corollary 10.18 If f is a differentiable function on the open interval I and if $f'(x) = 0$ for all $x \in I$ then f is a constant function.

Proof: If s and t are any two elements of I then the theorem tells us that

$$f(s) - f(t) = f'(\xi) \cdot (s - t)$$

for some ξ between s and t. But, by hypothesis, $f'(\xi) = 0$. We conclude that $f(s) = f(t)$. But, since s and t were chosen arbitrarily, we must conclude that f is constant. □

Corollary 10.19 If f is differentiable on an open interval I and $f'(x) \geq 0$ for all $x \in I$, then f is *increasing* on I; that is, if $s < t$ are elements of I, then $f(s) \leq f(t)$.
  If f is differentiable on an open interval I and $f'(x) \leq 0$ for all $x \in I$, then f is *decreasing* on I; that is, if $s < t$ are elements of I, then $f(s) \geq f(t)$.

Proof: Similar to the preceding corollary. □

EXAMPLE 10.20 Let us verify that, if f is a differentiable function on $\mathbb{R}$, and if $|f'(x)| \leq 1$ for all x, then $|f(s) - f(t)| \leq |s - t|$ for all real s and t.
  In fact, for $s \neq t$ there is a ξ between s and t such that

$$\frac{f(s) - f(t)}{s - t} = f'(\xi).$$

But $|f'(\xi)| \leq 1$ by hypothesis hence

$$\left| \frac{f(s) - f(t)}{s - t} \right| \leq 1$$

or

$$|f(s) - f(t)| \leq |s - t|.$$ ∎

EXAMPLE 10.21 Let us verify that

$$\lim_{x \to +\infty} \left(\sqrt{x + 5} - \sqrt{x} \right) = 0.$$

Here the limit operation means that, for any $\epsilon > 0$, there is an $N > 0$ such that $x > N$ implies that the expression in parentheses has absolute value less than ϵ.
  Define $f(x) = \sqrt{x}$ for $x > 0$. Then the expression in parentheses is just $f(x + 5) - f(x)$. By the Mean Value Theorem this equals

$$f'(\xi) \cdot 5$$

for some $x < \xi < x + 5$. But this last expression is

$$\frac{1}{2} \cdot \xi^{-1/2} \cdot 5.$$

By the bounds on ξ, this is

$$\leq \frac{5}{2} x^{-1/2}.$$

Clearly, as $x \to +\infty$, this expression tends to zero. ∎

A powerful tool in analysis is a generalization of the usual Mean Value Theorem that is due to A. L. Cauchy:

Theorem 10.22 (Cauchy's Mean Value Theorem) *Let f and g be continuous functions on the interval $[a, b]$ which are both differentiable on the interval (a, b). Assume that $g' \neq 0$ on the interval. Then there is a point $\xi \in (a, b)$ such that*

$$\frac{f(b) - f(a)}{g(b) - g(a)} = \frac{f'(\xi)}{g'(\xi)}.$$

Proof: Apply the usual Mean Value Theorem to the function

$$h(x) = g(x) \cdot \{f(b) - f(a)\} - f(x) \cdot \{g(b) - g(a)\}. \qquad \square$$

Clearly the usual Mean Value Theorem (Theorem 10.16) is obtained from Cauchy's by taking $g(x)$ to be the function x.

It is a fact that the standard proof of l'Hôpital's Rule (Guillaume François Antoine de l'Hôpital, Marquis de St.-Mesme, 1661–1704) is obtained by way of Cauchy's Mean Value Theorem. This line of reasoning is explored in the next section.

A Look Back

1. What does Fermat's Test say? Why is it intuitively obvious?

2. What does Rolle's theorem say? Why is its justification clear?

3. What does the Mean Value Theorem say? Why is it a generalization of Rolle's theorem?

4. In what sense is Cauchy's Mean Value Theorem a generalization of the usual Mean Value Theorem?

Exercises

1. Let f be a function that is continuous on $[0, \infty)$ and differentiable on $(0, \infty)$. If $f(0) = 0$ and $|f'(x)| \leq |f(x)|$ for all $x > 0$ then prove that $f(x) = 0$ for all x. [This result is often called *Gronwall's inequality*.]

2. Let f be a continuous function on $[a, b]$ that is differentiable on (a, b). Assume that $f(a) = m$ and that $|f'(x)| \leq K$ for all $x \in (a, b)$. What bound can you then put on the magnitude of $f(b)$?

3. Let f be a differentiable function on an open interval I and assume that f has no local minima nor local maxima on I. Prove that f is either increasing or decreasing on I.

4. Let $0 < \alpha \leq 1$. Prove that there is a constant $C_\alpha > 0$ such that, for $0 < x < 1$, it holds that
$$| \ln x | \leq C_\alpha \cdot x^{-\alpha} .$$
Prove that the constant cannot be taken to be independent of α.

5. Let f be a function that is twice differentiable on $(0, \infty)$ and assume that $f''(x) \geq c > 0$ for all x. Prove that f is not bounded from above.

6. Let f be differentiable on an interval I and $f'(x) > 0$ for all $x \in I$. Does it follow that $(f^2)' > 0$ for all $x \in I$? What additional hypothesis on f will make the conclusion true?

7. Use the Mean Value Theorem to say something about the behavior at ∞ of the function $f(x) = \sqrt{x+1} - \sqrt{x}$.

8. Refer to Exercise 7. What can you say about the asymptotics at infinity of $\sqrt{x+1}/\sqrt{x}$?

9. Refer to Exercises 7 and 8. What can you say about the asymptotics at ∞ of $h(x) = \log(x+1) - \log(x)$?

10. Refer to Exercises 7, 8 and 9. What can you say about the asymptotics at ∞ of $k(x) = e^{x+1} - e^x$?

* 11. Answer Exercise 6 with the exponent 2 replaced by any positive integer exponent.

10.3 More on the Theory of Differentiation

Preliminary Remarks

In this section we study l'Hôpital's Rule and related ideas. There are a number of interesting and nontrivial results, and the exercise is worthwhile.

l'Hôpital's Rule (actually due to his teacher J. Bernoulli (1667–1748)) is a useful device for calculating limits, and a nice application of the Cauchy Mean Value Theorem. Here we present a special case of the theorem.

Theorem 10.23 *Suppose that f and g are differentiable functions on an open interval I and that $p \in I$. If $\lim_{x \to p} f(x) = \lim_{x \to p} g(x) = 0$ and if*

$$\lim_{x \to p} \frac{f'(x)}{g'(x)} \qquad (10.23.1)$$

exists and equals a real number ℓ then

$$\lim_{x \to p} \frac{f(x)}{g(x)} = \ell .$$

Proof: Fix a real number $a > \ell$. By (10.23.1) there is a number $q > p$ such that, if $p < x < q$, then

$$\frac{f'(x)}{g'(x)} < a. \tag{10.23.2}$$

But now, if $p < s < t < q$, then

$$\frac{f(t) - f(s)}{g(t) - g(s)} = \frac{f'(x)}{g'(x)}$$

for some $s < x < t$ (by Cauchy's Mean Value Theorem). It follows then from (10.23.2) that

$$\frac{f(t) - f(s)}{g(t) - g(s)} < a.$$

Now let $s \to p$ and invoke the hypothesis about the zero limit of f and g at p to conclude that

$$\frac{f(t)}{g(t)} \leq a$$

when $p < t < q$. Since a is an arbitrary number to the right of ℓ we conclude that

$$\limsup_{t \to p^+} \frac{f(t)}{g(t)} \leq \ell.$$

Similar arguments show that

$$\liminf_{t \to p^+} \frac{f(t)}{g(t)} \geq \ell;$$

$$\limsup_{t \to p^-} \frac{f(t)}{g(t)} \leq \ell;$$

$$\liminf_{t \to p^-} \frac{f(t)}{g(t)} \geq \ell.$$

We conclude that the desired limit exists and equals ℓ. □

EXAMPLE 10.24 Let $f(x) = \sin x$ and $g(x) = x$. Then both functions have limit 0 at $p = 0$. So we can apply l'Hôpital's Rule to the limit

$$\lim_{x \to 0} \frac{\sin x}{x}.$$

We see that this limit is equal to

$$\lim_{x \to 0} \frac{[\sin x]'}{x'} = \lim_{x \to 0} \frac{\cos x}{1} = 1. \qquad \blacksquare$$

A closely related result, with a similar proof, is this:

Theorem 10.25 *Let I be an open interval and $p \in I$. Suppose that f and g are differentiable functions on $I \setminus \{p\}$. If $\lim_{x \to p} f(x) = \lim_{x \to p} g(x) = \pm\infty$ and if*

$$\lim_{x \to p} \frac{f'(x)}{g'(x)} \qquad (10.25.1)$$

exists and equals a real number ℓ then

$$\lim_{x \to p} \frac{f(x)}{g(x)} = \ell .$$

EXAMPLE 10.26 Let

$$f(x) = \big|\ln |x|\big|^{(x^2)} .$$

We wish to determine $\lim_{x \to 0} f(x)$. To do so, we define

$$F(x) = \ln f(x) = x^2 \ln\big|\ln |x|\big| = \frac{\ln |\ln |x||}{1/x^2} .$$

Notice that both the numerator and the denominator tend to $\pm\infty$ as $x \to 0$. So the hypotheses of l'Hôpital's Rule are satisfied and the limit is

$$\lim_{x \to 0} \frac{\ln |\ln |x||}{1/x^2} = \lim_{x \to 0} \frac{1/[x \ln |x|]}{-2/x^3} = \lim_{x \to 0} \frac{-x^2}{2 \ln |x|} = 0 .$$

Since $\lim_{x \to 0} F(x) = 0$ we may calculate that the original limit has value $\lim_{x \to 0} f(x) = 1$. ∎

POINT OF CONFUSION 10.27 Be careful when using l'Hôpital's Rule. You cannot apply it to just any quotient. You must verify that the hypotheses are true (i.e., both the numerator and denominator must tend to 0, or else both the numerator and the denominator must tend to $\pm\infty$).

Now we turn our attention to derivatives of inverse functions.

Proposition 10.28 *Let f be an invertible function on an interval (a, b) with nonzero derivative at a point $x \in (a, b)$. Let $X = f(x)$. Then $\left(f^{-1}\right)'(X)$ exists and equals $1/f'(x)$.*

Proof: Observe that, for $T \neq X$,

$$\frac{f^{-1}(T) - f^{-1}(X)}{T - X} = \frac{1}{(f(t) - f(x))/(t - x)} , \qquad (10.28.1)$$

where $T = f(t)$. Since $f'(x) \neq 0$, the difference quotients for f in the denominator are bounded from zero hence the limit of the formula in (10.28.1) exists. This proves that f^{-1} is differentiable at X and that the derivative at that point equals $1/f'(x)$. □

EXAMPLE 10.29 We know that the function $f(x) = x^k, k$ a positive integer, is one-to-one and differentiable on the interval $(0, 1)$. Moreover the derivative $k \cdot x^{k-1}$ never vanishes on that interval. Therefore the proposition applies and we find for $X \in (0, 1) = f((0, 1))$ that

$$(f^{-1})'(X) = \frac{1}{f'(x)} = \frac{1}{f'(X^{1/k})}$$

$$= \frac{1}{k \cdot X^{1-1/k}} = \frac{1}{k} \cdot X^{\frac{1}{k}-1}.$$

In other words,

$$\left(X^{1/k}\right)' = \frac{1}{k}X^{\frac{1}{k}-1}. \qquad \blacksquare$$

We conclude this section by saying a few words about higher derivatives. If f is a differentiable function on an open interval I then we may ask whether the function f' is differentiable. If it is, then we denote its derivative by

$$f'' \quad \text{or} \quad f^{(2)} \quad \text{or} \quad \frac{d^2}{dx^2}f \quad \text{or} \quad \frac{d^2 f}{dx^2} \quad \text{or} \quad \ddot{f},$$

and call it the second derivative of f. Likewise the derivative of the $(k-1)$th derivative, if it exists, is called the kth derivative and is denoted

$$f''^{\cdots\prime} \quad \text{or} \quad f^{(k)} \quad \text{or} \quad \frac{d^k}{dx^k}f \quad \text{or} \quad \frac{d^k f}{dx^k}.$$

Observe that we cannot even consider whether $f^{(k)}$ exists at a point unless $f^{(k-1)}$ exists in a *neighborhood* of that point.

If f is k times differentiable on an open interval I and if each of the derivatives $f^{(1)}, f^{(2)}, \ldots, f^{(k)}$ is continuous on I then we say that the function f is k *times continuously differentiable* on I. We write $f \in C^k(I)$. Obviously there is some redundancy in this definition since the continuity of $f^{(j-1)}$ follows from the existence of $f^{(j)}$. Thus only the continuity of the last derivative $f^{(k)}$ need be checked. Continuously differentiable functions are useful tools in analysis. We denote the class of k times continuously differentiable functions on I by $C^k(I)$.

For $k = 1, 2, \ldots$ the function

$$f_k(x) = \begin{cases} x^{k+1} & \text{if} \quad x \geq 0 \\ -x^{k+1} & \text{if} \quad x < 0 \end{cases}$$

will be k times continuously differentiable on $\mathbb{R}$ but will fail to be $k + 1$ times differentiable at $x = 0$. More dramatically, an analysis similar to the one we used on the Weierstrass Nowhere Differentiable Function shows that the function

$$g_k(x) = \sum_{j=1}^{\infty} \frac{3^j}{4^{j+jk}} \sin(4^j x)$$

is k times continuously differentiable on $\mathbb{R}$ but will not be $k+1$ times differentiable at any point (this function, with $k = 0$, was Weierstrass's original example).

A more refined notion of smoothness/continuity of functions is that of Hölder continuity or Lipschitz continuity (see Sections 9.2 and 9.3). If f is a function on an open interval I and if $0 < \alpha \leq 1$ then we say that f satisfies a *Lipschitz condition* of order α on I if there is a constant M such that for all $s, t \in I$ we have

$$|f(s) - f(t)| \leq M \cdot |s - t|^\alpha.$$

Such a function is said to be of class $\text{Lip}_\alpha(I)$. Clearly a function of class Lip_α is uniformly continuous on I. For, if $\epsilon > 0$, then we may take $\delta = (\epsilon/M)^{1/\alpha}$: it follows that, for $|s - t| < \delta$, we have

$$|f(s) - f(t)| \leq M \cdot |s - t|^\alpha < M \cdot \epsilon/M = \epsilon.$$

Interestingly, when $\alpha > 1$ the class Lip_α contains only constant functions. For in this instance the inequality

$$|f(s) - f(t)| \leq M \cdot |s - t|^\alpha$$

leads to

$$\left| \frac{f(s) - f(t)}{s - t} \right| \leq M \cdot |s - t|^{\alpha-1}.$$

Because $\alpha - 1 > 0$, letting $s \to t$ yields that $f'(t)$ exists for every $t \in I$ and equals 0. It follows from Corollary 10.18 of the last section that f is constant on I.

Instead of trying to extend the definition of $\text{Lip}_\alpha(I)$ to $\alpha > 1$ it is customary to define classes of functions $C^{k,\alpha}$, for $k = 0, 1, \ldots$ and $0 < \alpha \leq 1$, by the condition that f be of class C^k on I and that $f^{(k)}$ be an element of $\text{Lip}_\alpha(I)$. We leave it as an exercise for you to verify that $C^{k,\alpha} \subset C^{\ell,\beta}$ if either $k > \ell$ or both $k = \ell$ and $\alpha \geq \beta$.

A Look Back

1. Give a precise statement of l'Hôpital's Rule.

2. How is the version of l'Hôpital's Rule for numerator and denominator tending to 0 related to the version of the numerator and denominator tending to ∞?

3. Explain pictorially why the formula for the derivative of an inverse function is valid.

4. Explain concretely and in elementary terms why l'Hôpital's Rule should be true for quotients of polynomials.

Optional APPENDIX: Proof of Theorem 10.7 (The Weierstrass Nowhere Differentiable Function)

Since we have not yet discussed series of functions, we take a moment to understand the definition of f. Fix a real x. Then the series becomes a series of numbers, and the jth summand does not exceed $(3/4)^j$ in absolute value. Thus the series converges absolutely; therefore it converges. So it is clear that the displayed formula defines a function of x.

Step I: f is continuous. To see that f is continuous, pick an $\epsilon > 0$. Choose N so large that

$$\sum_{j=N+1}^{\infty} \left(\frac{3}{4}\right)^j < \frac{\epsilon}{4}$$

(we can of course do this because the series $\sum \left(\frac{3}{4}\right)^j$ converges). Now fix x. Observe that, since ψ is continuous and the graph of ψ is composed of segments of slope ± 1, we have

$$|\psi(s) - \psi(t)| \leq |s - t|$$

for all s and t. Moreover $|\psi(s) - \psi(t)| \leq 1$ for all s, t.

For $j = 1, 2, \ldots, N$, pick $\delta_j > 0$ so that, when $|t - x| < \delta_j$, then

$$\left|\psi\left(4^j t\right) - \psi\left(4^j x\right)\right| < \frac{\epsilon}{8}.$$

Let δ be the minimum of $\delta_1, \ldots \delta_N$.

Now, if $|t - x| < \delta$, then

$$
\begin{aligned}
|f(t) - f(x)| &= \left| \sum_{j=1}^{N} \left(\frac{3}{4}\right)^j \cdot \left(\psi(4^j t) - \psi(4^j x)\right) \right. \\
&\quad \left. + \sum_{j=N+1}^{\infty} \left(\frac{3}{4}\right)^j \cdot \left(\psi(4^j t) - \psi(4^j x)\right) \right| \\
&\leq \sum_{j=1}^{N} \left(\frac{3}{4}\right)^j \left|\left(\psi(4^j t) - \psi(4^j x)\right)\right| \\
&\quad + \sum_{j=N+1}^{\infty} \left(\frac{3}{4}\right)^j \left|\psi(4^j t) - \psi(4^j x)\right| \\
&\leq \sum_{j=1}^{N} \left(\frac{3}{4}\right)^j \cdot \frac{\epsilon}{8} + \sum_{j=N+1}^{\infty} \left(\frac{3}{4}\right)^j.
\end{aligned}
$$

Here we have used the choice of δ to estimate the summands in the first sum. The first sum is thus less than $\epsilon/2$ (just notice that $\sum_{j=1}^{\infty}(3/4)^j < 4$). The second sum is less than $\epsilon/2$ by the choice of N. Altogether then

$$|f(t) - f(x)| < \epsilon$$

whenever $|t - x| < \delta$. Therefore f is continuous, indeed uniformly so.

Step II: f is nowhere differentiable. Fix x. For $\ell = 1, 2, \ldots$ define $t_\ell = x \pm 4^{-\ell}/2$. We will say whether the sign is plus or minus in a moment (this will depend on the position of x relative to the integers). Then

$$\left| \frac{f(t_\ell) - f(x)}{t_\ell - x} \right| = \left| \frac{1}{t_\ell - x} \left[\sum_{j=1}^{\ell} \left(\frac{3}{4} \right)^j \left(\psi(4^j t_\ell) - \psi(4^j x) \right) \right. \right.$$

$$\left. \left. + \sum_{j=\ell+1}^{\infty} \left(\frac{3}{4} \right)^j \left(\psi(4^j t_\ell) - \psi(4^j x) \right) \right] \right|. \qquad (10.7.1)$$

Notice that, when $j \geq \ell + 1$, then $4^j t_\ell$ and $4^j x$ differ by an even integer. Since ψ has period 2, we find that each of the summands in the second sum is 0. Next we turn to the first sum.

We choose the sign—plus or minus—in the definition of t_ℓ so that there is no integer lying between $4^\ell t_\ell$ and $4^\ell x$. We can do this because the two numbers differ by $1/2$. But then the ℓth summand has magnitude

$$(3/4)^\ell \cdot |4^\ell t_\ell - 4^\ell x| = 3^\ell |t_\ell - x|.$$

On the other hand, the first $\ell - 1$ summands add up to not more than

$$\sum_{j=1}^{\ell-1} \left(\frac{3}{4} \right)^j \cdot |4^j t_\ell - 4^j x| = \sum_{j=1}^{\ell-1} 3^j \cdot 4^{-\ell}/2 \leq \frac{3^\ell - 1}{3 - 1} \cdot 4^{-\ell}/2 \leq 3^\ell \cdot 4^{-\ell-1}.$$

It follows that

$$
\left| \frac{f(t_\ell) - f(x)}{t_\ell - x} \right| = \frac{1}{|t_\ell - x|} \cdot \left| \sum_{j=1}^{\ell} \left(\frac{3}{4}\right)^j \left(\psi(4^j t_\ell) - \psi(4^j x)\right) \right|
$$

$$
= \frac{1}{|t_\ell - x|} \cdot \left| \sum_{j=1}^{\ell-1} \left(\frac{3}{4}\right)^j \left(\psi(4^j t_\ell) - \psi(4^j x)\right) \right.
$$

$$
\left. + \left(\frac{3}{4}\right)^\ell \left(\psi(4^\ell t_\ell) - \psi(4^\ell x)\right) \right|
$$

$$
\geq \frac{1}{|t_\ell - x|} \cdot \left| \left(\frac{3}{4}\right)^\ell \psi(4^\ell t_\ell) - \left(\frac{3}{4}\right)^\ell \psi(4^\ell x) \right|
$$

$$
- \frac{1}{|t_\ell - x|} \left| \sum_{j=1}^{\ell-1} \left(\frac{3}{4}\right)^j \left(\psi(4^j t_\ell) - \psi(4^j x)\right) \right|
$$

$$
\geq 3^\ell - \frac{1}{(4^{-\ell}/2)} \cdot 3^\ell \cdot 4^{-\ell-1}
$$

$$
\geq 3^{\ell-1}.
$$

Thus $t_\ell \to x$ but the Newton quotients blow up. Therefore the limit

$$
\lim_{t \to x} \frac{f(t) - f(x)}{t - x}
$$

cannot exist. The function f is not differentiable at x. □

Exercises

1. Suppose that f is a C^2 function on $\mathbb{R}$ and that $|f''(x)| \leq C$ for all x. Prove that

$$
\left| \frac{f(x+h) + f(x-h) - 2f(x)}{h^2} \right| \leq C.
$$

2. Fix a positive integer k. Give an example of two functions f and g neither of which is in C^k but such that $f \cdot g \in C^k$.

3. Fix a positive integer ℓ and define $f(x) = |x|^\ell$. In which class C^k does f lie? In which class $C^{k,\alpha}$ does it lie? [**Hint:** Your answer may depend on the parity of ℓ.]

4. Suppose that f is a differentiable function on an interval I and that $f'(x)$ is never zero. Prove that f is invertible. Then prove that f^{-1} is differentiable. Finally, use the Chain Rule on the identity $f\left(f^{-1}\right) = x$ to derive a formula for $\left(f^{-1}\right)'$.

5. Suppose that a function f on the interval $(0, 1)$ has left derivative equal to zero at every point. What conclusion can you draw?

6. We know that the first derivative can be characterized by the Newton quotient. Find an analogous characterization of second derivatives. What about third derivatives?

* **7.** We know that a continuous function on the interval $[0, 1]$ can be uniformly approximated by polynomials. But if the function f is continuously differentiable on $[0, 1]$, then we can actually say something about the *rate* of approximation. That is, if $\epsilon > 0$ then f can be approximated uniformly within ϵ by a polynomial of degree not greater than $N = N(\epsilon)$. Calculate $N(\epsilon)$.

* **8.** In which class $C^{k,\alpha}$ is the function $x \cdot \ln|x|$ on the interval $[-1/2, 1/2]$? How about the function $x/\ln|x|$?

* **9.** Give an example of a function on $\mathbb{R}$ such that

$$\left| \frac{f(x+h) + f(x-h) - 2f(x)}{h} \right| \leq C$$

for all x and all $h \neq 0$ but f is not in $\mathrm{Lip}_1(\mathbb{R})$.

* **10.** Let f be a differentiable function on an open interval I. Prove that f' is continuous if and only if the inverse image under f' of any point is the intersection of I with a closed set.

* **11.** In the text we give sufficient conditions for the inclusion $C^{k,\alpha} \subset C^{\ell,\beta}$. Show that the inclusion is strict if either $k > \ell$ or $k = \ell$ and $\alpha > \beta$.

12. Prove Theorem 10.25.

Chapter 11

The Integral

11.1 Partitions and the Concept of Integral

Preliminary Remarks

It was Bernhard Riemann who came up with the concept of integral that we use today. His concept is based on the idea of a "partition." That is to say, we address the Riemann integral by breaking up the domain of the function. Interestingly, the more advanced idea of the Lebesgue integral is obtained by breaking up the range of the function.

We learn in calculus that it is often useful to think of an integral as representing area. However, this is but one of many important applications of integration theory. The integral is a generalization of the summation process. That is the point of view that we shall take in the present chapter.

Definition 11.1 Let $[a, b]$ be a closed interval in $\mathbb{R}$. A finite, ordered set of points $\mathcal{P} = \{x_0, x_1, x_2, \ldots, x_{k-1}, x_k\}$ such that

$$a = x_0 \leq x_1 \leq x_2 \leq \cdots \leq x_{k-1} \leq x_k = b$$

is called a *partition* of $[a, b]$. Refer to Figure 11.1.

If $\mathcal{P}$ is a partition of $[a, b]$, then we let I_j denote the interval $[x_{j-1}, x_j]$, $j = 1, 2, \ldots, k$. The symbol Δ_j denotes the *length* of I_j. The *mesh* of $\mathcal{P}$, denoted by $m(\mathcal{P})$, is defined to be max Δ_j.

The points of a partition need not be equally spaced, nor must they be distinct from each other.

EXAMPLE 11.2 The set $\mathcal{P} = \{0, 1, 1, 9/8, 2, 5, 21/4, 23/4, 6\}$ is a partition of the interval $[0, 6]$ with mesh 3 (because $I_5 = [2, 5]$, with length 3, is the longest interval in the partition). See Figure 11.2. ∎

Figure 11.1: A partition.

Figure 11.2: The partition in Example 11.2.

Definition 11.3 Let $[a, b]$ be an interval and let f be a function with domain $[a, b]$. If $\mathcal{P} = \{x_0, x_1, x_2, \ldots, x_{k-1}, x_k\}$ is a partition of $[a, b]$ and if, for each j, s_j is an element of I_j, then the corresponding *Riemann sum* is defined to be

$$\mathcal{R}(f, \mathcal{P}) = \sum_{j=1}^{k} f(s_j)\Delta_j.$$

EXAMPLE 11.4 Let $f(x) = x^2 - x$ and $[a, b] = [1, 4]$. Define the partition $\mathcal{P} = \{1, 3/2, 2, 7/3, 4\}$ of this interval. Then a Riemann sum for this f and $\mathcal{P}$ is

$$\mathcal{R}(f, \mathcal{P}) = (1^2 - 1) \cdot \frac{1}{2} + ((7/4)^2 - (7/4)) \cdot \frac{1}{2}$$
$$+ ((7/3)^2 - (7/3)) \cdot \frac{1}{3} + (3^2 - 3) \cdot \frac{5}{3}$$
$$= \frac{10103}{864}.\qquad\blacksquare$$

POINT OF CONFUSION 11.5 Notice that we have complete latitude in choosing each point s_j from the corresponding interval I_j. While at first confusing, we will find this freedom to be a powerful tool when proving results about the integral.

The first main step in the theory of the Riemann integral is to determine a method for "calculating the limit of the Riemann sums" of a function as the mesh of the partitions tends to zero. There are in fact several means of doing so. We have chosen the simplest one.

Definition 11.6 Let $[a, b]$ be an interval and f a function with domain $[a, b]$. We say that *the Riemann sums of f tend to a limit ℓ as $m(\mathcal{P})$ tends to 0* if, for any $\epsilon > 0$, there is a $\delta > 0$ such that, if $\mathcal{P}$ is any partition of $[a, b]$ with $m(\mathcal{P}) < \delta$, then $|\mathcal{R}(f, \mathcal{P}) - \ell| < \epsilon$ for every choice of $s_j \in I_j$.

It will turn out to be critical for the success of this definition that we require that *every* partition of mesh smaller than δ satisfy the conclusion of the definition. The theory does not work effectively if for every $\epsilon > 0$ there is a $\delta > 0$ and *some* partition $\mathcal{P}$ of mesh less than δ which satisfies the conclusion of the definition.

Definition 11.7 A function f on a closed interval $[a, b]$ is said to be *Riemann integrable* on $[a, b]$ if the Riemann sums of $\mathcal{R}(f, \mathcal{P})$ tend to a finite limit ℓ as $m(\mathcal{P})$ tends to zero.

The value ℓ of the limit, when it exists, is called the *Riemann integral* of f over $[a, b]$ and is denoted by

$$\int_a^b f(x)\, dx.$$

POINT OF CONFUSION 11.8 The value of the integral is well approximated by its Riemann sums. This observation is a useful tool in calculation. We can use Riemann sums to obtain accurate approximations to π and to other important quantities in mathematics. Riemann sums are also powerful devices for studying differential equations and complex analysis.

Remark 11.9 We mention now a useful fact that will be formalized in later sections. Suppose that f is Riemann integrable on $[a, b]$ with the value of the integral being ℓ. Let $\epsilon > 0$. Then, as stated in the definition (with $\epsilon/2$ replacing ϵ), there is a $\delta > 0$ such that, if $\mathcal{Q}$ is a partition of $[a, b]$ of mesh smaller than δ, then $|\mathcal{R}(f, \mathcal{Q}) - \ell| < \epsilon/2$. It follows that, if $\mathcal{P}$ and $\mathcal{P}'$ are partitions of $[a, b]$ of mesh smaller than δ, then

$$|\mathcal{R}(f, \mathcal{P}) - \mathcal{R}(f, \mathcal{P}')| \leq |\mathcal{R}(f, \mathcal{P}) - \ell| + |\ell - \mathcal{R}(f, \mathcal{P}')| < \frac{\epsilon}{2} + \frac{\epsilon}{2} = \epsilon.$$

Note, however, that we may choose $\mathcal{P}'$ to equal the partition $\mathcal{P}$. Also we may for each j choose the points s_j, where f is evaluated for the Riemann sum over $\mathcal{P}$, to be a point where f very nearly assumes its supremum on I_j. Likewise we may for each j choose the points s'_j, where f is evaluated for the Riemann sum over $\mathcal{P}'$, to be a point where f very nearly assumes its infimum on I_j. It easily follows that, when the mesh of $\mathcal{P}$ is less than δ, then

$$\sum_j \left(\sup_{I_j} f - \inf_{I_j} f \right) \Delta_j \leq \epsilon. \tag{11.9.1}$$

This consequence of integrability will prove useful to us in some of the discussions in this and the next section. In the exercises we shall consider in detail the assertion that integrability implies (11.9.1) and the converse as well.

Definition 11.10 If $\mathcal{P}, \mathcal{P}'$ are partitions of $[a, b]$ then their *common refinement* is the union of all the points of $\mathcal{P}$ and $\mathcal{P}'$. See Figure 11.3.

We record now a technical lemma that will be used in several of the proofs that follow:

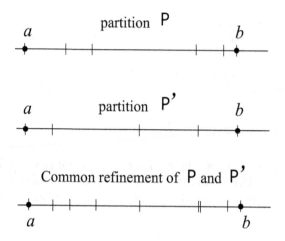

Figure 11.3: The common refinement.

Lemma 11.11 *Let f be a function with domain the closed interval $[a, b]$. The Riemann integral*

$$\int_a^b f(x) \, dx$$

exists if and only if, for every $\epsilon > 0$, there is a $\delta > 0$ such that, if $\mathcal{P}$ and $\mathcal{P}'$ are partitions of $[a, b]$ with $m(\mathcal{P}) < \delta$ and $m(\mathcal{P}') < \delta$, then their common refinement $\mathcal{Q}$ has the property that

$$|\mathcal{R}(f, \mathcal{P}) - \mathcal{R}(f, \mathcal{Q})| < \epsilon$$

and   (11.11.1)

$$|\mathcal{R}(f, \mathcal{P}') - \mathcal{R}(f, \mathcal{Q})| < \epsilon.$$

Proof: If f is Riemann integrable, then the assertion of the lemma follows immediately from the definition of the integral.

For the converse note that (11.11.1) certainly implies that, if $\epsilon > 0$, then there is a $\delta > 0$ such that, if $\mathcal{P}$ and $\mathcal{P}'$ are partitions of $[a, b]$ with $m(\mathcal{P}) < \delta$ and $m(\mathcal{P}') < \delta$, then

$$|\mathcal{R}(f, \mathcal{P}) - \mathcal{R}(f, \mathcal{P}')| < \epsilon$$ (11.11.2)

(just use the triangle inequality).

Now, for each $\epsilon_j = 2^{-j}, j = 1, 2, \ldots$, we can choose a $\delta_j > 0$ as in (11.11.2). Let S_j be the *closure* of the set

$$\{\mathcal{R}(f, \mathcal{P}) : m(\mathcal{P}) < \delta_j\}.$$

By the choice of δ_j (and by (11.11.2)), the set S_j is contained in a closed interval of length not greater than $2\epsilon_j$.

On the one hand,

$$\bigcap_j S_j$$

must be nonempty since it is the decreasing intersection of compact sets. On the other hand, the length estimate implies that the intersection must be contained in a closed interval of length 0—that is, the intersection is a point. That point is then the limit of the Riemann sums, that is, it is the value of the Riemann integral. □

The most important, and perhaps the simplest, fact about the Riemann integral is that a large class of familiar functions is Riemann integrable.

Theorem 11.12 *Let f be a continuous function on a nontrivial closed, bounded interval $I = [a, b]$. Then f is Riemann integrable on $[a, b]$.*

Proof: We use the lemma. Given $\epsilon > 0$, choose (by the uniform continuity of f on I—Theorem 5.30) a $\delta > 0$ such that, whenever $|s - t| < \delta$ then

$$|f(s) - f(t)| < \frac{\epsilon}{b - a}. \tag{11.12.1}$$

Let $\mathcal{P}$ and $\mathcal{P}'$ be any two partitions of $[a, b]$ of mesh smaller than δ. Let $\mathcal{Q}$ be the common refinement of $\mathcal{P}$ and $\mathcal{P}'$.

Now we let I_j denote the intervals arising in the partition $\mathcal{P}$ (and having length Δ_j) and $\widetilde{I}_\ell$ the intervals arising in the partition $\mathcal{Q}$ (and having length $\widetilde{\Delta}_\ell$). Since the partition $\mathcal{Q}$ contains every point of $\mathcal{P}$, plus some additional points as well, every $\widetilde{I}_\ell$ is contained in some I_j. Fix j and consider the expression

$$\left| f(s_j)\Delta_j - \sum_{\widetilde{I}_\ell \subset I_j} f(t_\ell)\widetilde{\Delta}_\ell \right|. \tag{11.12.2}$$

We write

$$\Delta_j = \sum_{\widetilde{I}_\ell \subset I_j} \widetilde{\Delta}_\ell.$$

This equality enables us to rearrange (11.12.2) as

$$\left| f(s_j) \cdot \sum_{\widetilde{I}_\ell \subset I_j} \widetilde{\Delta}_\ell - \sum_{\widetilde{I}_\ell \subset I_j} f(t_\ell)\widetilde{\Delta}_\ell \right|$$

$$= \left| \sum_{\widetilde{I}_\ell \subset I_j} [f(s_j) - f(t_\ell)]\widetilde{\Delta}_\ell \right|$$

$$\leq \sum_{\widetilde{I}_\ell \subset I_j} |f(s_j) - f(t_\ell)|\widetilde{\Delta}_\ell.$$

But each of the points t_ℓ is in the interval I_j, as is s_j. So they differ by less than δ. Therefore, by (11.12.1), the last expression is less than

$$\sum_{\tilde{I}_\ell \subset I_j} \frac{\epsilon}{b-a} \tilde{\Delta}_\ell = \frac{\epsilon}{b-a} \sum_{\tilde{I}_\ell \subset I_j} \tilde{\Delta}_\ell$$

$$= \frac{\epsilon}{b-a} \cdot \Delta_j.$$

Now we conclude the argument by writing

$$|\mathcal{R}(f,\mathcal{P}) - \mathcal{R}(f,\mathcal{Q})| = \left| \sum_j f(s_j)\Delta_j - \sum_\ell f(t_\ell)\tilde{\Delta}_\ell \right|$$

$$\leq \sum_j \left| f(s_j)\Delta_j - \sum_{\tilde{I}_\ell \subset I_j} f(t_\ell)\tilde{\Delta}_\ell \right|$$

$$< \sum_j \frac{\epsilon}{b-a} \cdot \Delta_j$$

$$= \frac{\epsilon}{b-a} \cdot \sum_j \Delta_j$$

$$= \frac{\epsilon}{b-a} \cdot (b-a)$$

$$= \epsilon.$$

The estimate for $|\mathcal{R}(f,\mathcal{P}') - \mathcal{R}(f,\mathcal{Q})|$ is identical and we omit it. The result now follows from Lemma 11.11. □

In the exercises we will ask you to extend the theorem to the case of functions f on $[a,b]$ that are bounded and have finitely many, or even countably many, discontinuities.

We conclude this section by noting an important fact about Riemann integrable functions. A Riemann integrable function on an interval $[a,b]$ *must be bounded*. If it were not, then one could choose the points s_j in the construction of $\mathcal{R}(f,\mathcal{P})$ so that $f(s_j)$ is arbitrarily large, and the Riemann sums would become arbitrarily large, hence cannot converge. You will be asked in the exercises to work out the details of this assertion.

A Look Back

1. What is a partition?

2. What is the mesh of a partition?

3. What is the common refinement of two partitions?

4. What does it mean for the integral of a function f to exist?

5. What is a fairly large class of functions that are Riemann integrable?

Exercises

1. If f is a Riemann integrable function on $[a, b]$, then show that f must be a bounded function.

2. Define the *Dirichlet function* to be

$$f(x) = \begin{cases} 1 & \text{if} \quad x \text{ is rational} \\ 0 & \text{if} \quad x \text{ is irrational.} \end{cases}$$

Prove that the Dirichlet function is not Riemann integrable on the interval $[a, b]$.

3. Define

$$g(x) = \begin{cases} x \cdot \sin(1/x) & \text{if} \quad x \neq 0 \\ 0 & \text{if} \quad x = 0. \end{cases}$$

Is g Riemann integrable on the interval $[-1, 1]$?

4. Provide the details of the assertion that, if f is Riemann integrable on the interval $[a, b]$ then, for any $\epsilon > 0$, there is a $\delta > 0$ such that, if $\mathcal{P}$ is a partition of mesh less than δ, then

$$\sum_j \left(\sup_{I_j} f - \inf_{I_j} f \right) \Delta_j < \epsilon.$$

[**Hint:** Follow the scheme presented in Remark 11.9. Given $\epsilon > 0$, choose $\delta > 0$ as in the definition of the integral. Fix a partition $\mathcal{P}$ with mesh smaller than δ. Let $K + 1$ be the number of points in $\mathcal{P}$. Choose points $t_j \in I_j$ so that $|f(t_j) - \sup_{I_j} f| < \epsilon/(2(K + 1))$; also choose points $t'_j \in I_j$ so that $|f(t'_j) - \inf_{I_j} f| < \epsilon/(2(K + 1))$. By applying the definition of the integral to this choice of t_j and t'_j we find that

$$\sum_j \left(\sup_{I_j} f - \inf_{I_j} f \right) \Delta_j < 2\epsilon.$$

The result follows.]

5. To what extent is the following statement true? If f is Riemann integrable and bounded from 0 on $[a, b]$ then $1/f$ is Riemann integrable on $[a, b]$.

6. Prove the converse of the statement in Exercise 4. [**Hint:** Note that any Riemann sum over a sufficiently fine partition $\mathcal{P}$ is trapped between the sum in which the infimum is always chosen and the sum in which the supremum is always chosen.]

7. Give an example of a function f such that f^2 is Riemann integrable but f is not.

8. If f is Riemann integrable on the interval $[a, b]$ and if $\mu : [\alpha, \beta] \to [a, b]$ is continuously differentiable then prove that $f \circ \mu$ is Riemann integrable on $[\alpha, \beta]$.

9. Prove that, if f is continuous on the interval $[a, b]$ except at finitely many points and is bounded, then f is Riemann integrable on $[a, b]$.

* 10. Do Exercise 9 with the phrase "finitely many" replaced by "countably many."

* 11. Prove that the Dirichlet function (see Exercise 2) is the pointwise limit of Riemann integrable functions.

* 12. Show that any Riemann integrable function is the pointwise limit of continuous functions.

* 13. Give an example to show that the composition of Riemann integrable functions need not be Riemann integrable.

11.2   Properties of the Riemann Integral

Preliminary Remarks

Of course the integral is a linear operator on functions, and enjoys thereby a number of useful properties. These include ways in which the integral respects arithmetic operations. We explore these in the present section.

We begin this section with a few elementary properties of the integral that reflect its linear nature.

Theorem 11.13 Let $[a, b]$ be a nonempty interval, let f and g be Riemann integrable functions on the interval, and let α be a real number. Then $f \pm g$ and $\alpha \cdot f$ are integrable and we have

(a) $\int_a^b f(x) \pm g(x)\, dx = \int_a^b f(x)\, dx \pm \int_a^b g(x)\, dx$;

(b) $\int_a^b \alpha \cdot f(x)\, dx = \alpha \cdot \int_a^b f(x)\, dx$.

Proof: For (a), let

$$A = \int_a^b f(x)\, dx$$

and

$$B = \int_a^b g(x)\, dx\,.$$

Let $\epsilon > 0$. Choose a $\delta_1 > 0$ such that, if $\mathcal{P}$ is a partition of $[a, b]$ with mesh less than δ_1, then

$$|\mathcal{R}(f, \mathcal{P}) - A| < \frac{\epsilon}{2}\,.$$

Similarly choose a $\delta_2 > 0$ such that, if $\mathcal{P}'$ is a partition of $[a, b]$ with mesh less than δ_2, then

$$|\mathcal{R}(g, \mathcal{P}') - B| < \frac{\epsilon}{2}\,.$$

Let $\delta = \min\{\delta_1, \delta_2\}$. If $\mathcal{P}''$ is any partition of $[a, b]$ with $m(\mathcal{P}'') < \delta$ then

$$
\begin{aligned}
|\mathcal{R}(f \pm g, \mathcal{P}'') - (A \pm B)| &= |\mathcal{R}(f, \mathcal{P}'') \pm \mathcal{R}(g, \mathcal{P}'') - (A \pm B)| \\
&\leq |\mathcal{R}(f, \mathcal{P}'') - A| + |\mathcal{R}(g, \mathcal{P}'') - B| \\
&< \frac{\epsilon}{2} + \frac{\epsilon}{2} \\
&= \epsilon.
\end{aligned}
$$

This means that the integral of $f \pm g$ exists and equals $A \pm B$, as we were required to prove.

The proof of (b) follows similar lines but is much easier and we leave it as an exercise for you.   □

Theorem 11.14 *If c is a point of the interval $[a, b]$ and if f is Riemann integrable on both $[a, c]$ and $[c, b]$ then f is integrable on $[a, b]$ and $\int_a^c f(x)\, dx + \int_c^b f(x)\, dx = \int_a^b f(x)\, dx.$*

Proof: Let us write

$$A = \int_a^c f(x)\, dx$$

and

$$B = \int_c^b f(x)\, dx.$$

Now pick $\epsilon > 0$. There is a $\delta_1 > 0$ such that, if $\mathcal{P}$ is a partition of $[a, c]$ with mesh less than δ_1, then

$$|\mathcal{R}(f, \mathcal{P}) - A| < \frac{\epsilon}{3}.$$

Similarly, choose $\delta_2 > 0$ such that, if $\mathcal{P}'$ is a partition of $[c, b]$ with mesh less than δ_2, then

$$|\mathcal{R}(f, \mathcal{P}') - B| < \frac{\epsilon}{3}.$$

Let M be an upper bound for $|f|$ (recall, from the remark at the end of Section 11.1, that a Riemann integrable function must be bounded). Set $\delta = \min\{\delta_1, \delta_2, \epsilon/(6(M + 1))\}$. Now let $\mathcal{V} = \{v_1, \ldots, v_k\}$ be any partition of $[a, b]$ with mesh less than δ. There is a last point v_n which is in $[a, c]$ and a first point v_{n+1} in $[c, b]$. Observe that $\mathcal{P} = \{v_0, \ldots, v_n, c\}$ is a partition of $[a, c]$ with mesh smaller than δ_1 and $\mathcal{P}' = \{c, v_{n+1}, \ldots, v_k\}$ is a partition of $[c, b]$ with mesh smaller than δ_2. Let us rename the elements of $\mathcal{P}$ as $\{p_0, \ldots, p_n\}$ and the elements of $\mathcal{P}'$ as $\{p'_0, \cdots p'_{k-n}\}$. Notice that $p_n = p'_0 = c$. For each j let s_j be a point chosen in the interval $I_j = [v_{j-1}, v_j]$ from the partition $\mathcal{V}$.

Then we have

$$\left| \mathcal{R}(f, \mathcal{V}) - \left[A + B\right] \right|$$

$$= \left| \left(\sum_{j=1}^n f(s_j)\Delta_j - A \right) + f(s_{n+1})\Delta_{n+1} + \left(\sum_{j=n+2}^k f(s_j)\Delta_j - B \right) \right|$$

$$= \left| \left(\sum_{j=1}^n f(s_j)\Delta_j + f(c) \cdot (c - v_n) - A \right) \right.$$

$$+ \left(f(c) \cdot (v_{n+1} - c) + \sum_{j=n+2}^k f(s_j)\Delta_j - B \right)$$

$$+ \left. \left(f(s_{n+1}) - f(c) \right) \cdot (c - v_n) + \left(f(s_{n+1}) - f(c) \right) \cdot (v_{n+1} - c) \right|$$

$$\leq \left| \left(\sum_{j=1}^{n} f(s_j)\Delta_j + f(c) \cdot (c - v_n) - A \right) \right|$$

$$+ \left| \left(f(c) \cdot (v_{n+1} - c) + \sum_{j=n+2}^{k} f(s_j)\Delta_j - B \right) \right|$$

$$+ \left| (f(s_{n+1}) - f(c)) \cdot (v_{n+1} - v_n) \right|$$

$$= \left| \mathcal{R}(f, \mathcal{P}) - A \right| + \left| \mathcal{R}(f, \mathcal{P}') - B \right|$$

$$+ \left| (f(s_{n+1}) - f(c)) \cdot (v_{n+1} - v_n) \right|$$

$$< \frac{\epsilon}{3} + \frac{\epsilon}{3} + 2M \cdot \delta$$

$$\leq \epsilon$$

by the choice of δ.

This shows that f is integrable on the entire interval $[a, b]$ and the value of the integral is

$$A + B = \int_a^c f(x)\,dx + \int_c^b f(x)\,dx. \qquad \square$$

Remark 11.15 The last proof illustrates why it is useful to be able to choose the $s_j \in I_j$ arbitrarily. The nub of the proof is to be able to express the integral of f on $[a, b]$, and thus a Riemann sum for that integral, in terms of integrals (and hence Riemann sums) on the two subintervals.

POINT OF CONFUSION 11.16 If we adopt the convention that

$$\int_b^a f(x)\,dx = - \int_a^b f(x)\,dx$$

(which is consistent with the way that the integral was defined in the first place), then Theorem 11.14 is true even when c is not an element of $[a, b]$. For instance, suppose that $c < a < b$. Then, by Theorem 11.14,

$$\int_c^a f(x)\,dx + \int_a^b f(x)\,dx = \int_c^b f(x)\,dx.$$

But this may be rearranged to read

$$\int_a^b f(x)\,dx = - \int_c^a f(x)\,dx + \int_c^b f(x)\,dx = \int_a^c f(x)\,dx + \int_c^b f(x)\,dx.$$

One of the basic tools of analysis is to perform estimates. Thus we require certain fundamental inequalities about integrals. These are recorded in the next theorem.

Theorem 11.17 *Let f and g be integrable functions on a nonempty interval $[a, b]$. Then*

(i) $$\left| \int_a^b f(x)\, dx \right| \le \int_a^b |f(x)|\, dx;$$

(ii) *If $f(x) \le g(x)$ for all $x \in [a, b]$ then* $\int_a^b f(x)\, dx \le \int_a^b g(x)\, dx.$

Proof: If $\mathcal{P}$ is any partition of $[a, b]$ then

$$|\mathcal{R}(f, \mathcal{P})| \le \mathcal{R}(|f|, \mathcal{P}).$$

Assertion **(i)** follows.

Next, for part **(ii)**,
$$\mathcal{R}(f, \mathcal{P}) \le \mathcal{R}(g, \mathcal{P}).$$

This inequality implies the second assertion. $\qquad\qquad\square$

Another fundamental operation in the theory of the integral is "change of variable" (sometimes called the "u-substitution" in calculus books). We next turn to a careful formulation and proof of this operation. First we need a lemma:

Lemma 11.18 *If f is Riemann integrable on an interval $[a, b]$ with values in $[c, d]$, and if $\phi : [c, d] \to \mathbb{R}$ is continuously differentiable, then $\phi \circ f$ is Riemann integrable.*

Proof: The proof is complicated, and we omit the details. See [KRA5, Ch. 7] for the full story. $\qquad\qquad\square$

Corollary 11.19 *If f and g are Riemann integrable on $[a, b]$, then so is the function $f \cdot g$.*

Proof: By Theorem 11.13, $f + g$ is integrable. By the lemma, $(f + g)^2 = f^2 + 2f \cdot g + g^2$ is integrable. But the lemma also implies that f^2 and g^2 are integrable (here we use the function $\phi(x) = x^2$). It results, by subtraction, that $2 \cdot f \cdot g$ is integrable. Hence $f \cdot g$ is integrable. $\qquad\qquad\square$

Theorem 11.20 *Let f be an integrable function on an interval $[a, b]$ of positive length. Let ψ be a continuously differentiable function from another interval $[\alpha, \beta]$ of positive length into $[a, b]$. Assume that ψ is increasing, one-to-one, and onto. Then*

$$\int_a^b f(x)\, dx = \int_\alpha^\beta f(\psi(x)) \cdot \psi'(x)\, dx.$$

Proof: Since f is integrable, its absolute value is bounded by some number M. Fix $\epsilon > 0$. Since ψ' is continuous on the compact interval $[\alpha, \beta]$, it is uniformly continuous (Theorem 9.30). Hence we may choose $\delta > 0$ so small that if $|s-t| < \delta$ then $|\psi'(s) - \psi'(t)| < \epsilon/(M \cdot (\beta - \alpha))$. If $\mathcal{P} = \{p_0, \ldots, p_k\}$ is any partition of $[a, b]$ then there is an associated partition $\widetilde{\mathcal{P}} = \{\psi^{-1}(p_0), \ldots, \psi^{-1}(p_k)\}$ of $[\alpha, \beta]$. For simplicity denote the points of $\widetilde{\mathcal{P}}$ by $\widetilde{p}_j$. Let us choose the partition $\mathcal{P}$ so fine that the mesh of $\widetilde{\mathcal{P}}$ is less than δ. If t_j are points of $I_j = [p_{j-1}, p_j]$ then there are corresponding points $s_j = \psi^{-1}(t_j)$ of $\widetilde{I}_j = [\widetilde{p}_{j-1}, \widetilde{p}_j]$. Then we have

$$
\begin{aligned}
\sum_{j=1}^{k} f(t_j) \Delta_j &= \sum_{j=1}^{k} f(t_j)(p_j - p_{j-1}) \\
&= \sum_{j=1}^{k} f(\psi(s_j))(\psi(\widetilde{p}_j) - \psi(\widetilde{p}_{j-1})) \\
&= \sum_{j=1}^{k} f(\psi(s_j))\psi'(u_j)(\widetilde{p}_j - \widetilde{p}_{j-1}),
\end{aligned}
$$

where we have used the Mean Value Theorem in the last line to find each u_j. Our problem at this point is that $f \circ \psi$ and ψ' are evaluated at different points. So we must do some estimation to correct that problem.

The last displayed line equals

$$
\sum_{j=1}^{k} f(\psi(s_j))\psi'(s_j)(\widetilde{p}_j - \widetilde{p}_{j-1}) + \sum_{j=1}^{k} f(\psi(s_j)) \left(\psi'(u_j) - \psi'(s_j)\right) (\widetilde{p}_j - \widetilde{p}_{j-1}).
$$

The first sum is a Riemann sum for $f(\psi(x) \cdot \psi'(x))$ and the second sum is an error term. Since the points u_j and s_j are elements of the same interval $\widetilde{I}_j$ of length less than δ, we conclude that $|\psi'(u_j) - \psi'(s_j)| < \epsilon/(M \cdot |\beta - \alpha|)$. Thus the error term in absolute value does not exceed

$$
\sum_{j=1}^{k} M \cdot \frac{\epsilon}{M \cdot |\beta - \alpha|} \cdot (\widetilde{p}_j - \widetilde{p}_{j-1}) = \frac{\epsilon}{\beta - \alpha} \sum_{j=0}^{k} (\widetilde{p}_j - \widetilde{p}_{j-1}) = \epsilon.
$$

This shows that every Riemann sum for f on $[a, b]$ with sufficiently small mesh corresponds to a Riemann sum for $f(\psi(x)) \cdot \psi'(x)$ on $[\alpha, \beta]$ plus an error term of size less than ϵ. A similar argument shows that every Riemann sum for $f(\psi(x)) \cdot \psi'(x)$ on $[\alpha, \beta]$ with sufficiently small mesh corresponds to a Riemann sum for f on $[a, b]$ plus an error term of magnitude less than ϵ. The conclusion is then that the integral of f on $[a, b]$ (which exists by hypothesis) and the integral of $f(\psi(x)) \cdot \psi'(x)$ on $[\alpha, \beta]$ (which exists by the corollary to the lemma) agree. $\square$

We conclude this section with the very important Fundamental Theorem of Calculus.

Theorem 11.21 (The Fundamental Theorem of Calculus) *Let f be a continuous function on the interval $[a, b]$. For $x \in [a, b]$ we define*

$$F(x) = \int_a^x f(s) \, ds \,.$$

For any $x \in (a, b)$ we then have

$$F'(x) = f(x) \,.$$

Proof: Fix $x \in (a, b)$. Let $\epsilon > 0$. Choose, by the continuity of f at x, a $\delta > 0$ such that $|s - x| < \delta$ implies $|f(s) - f(x)| < \epsilon$. We may assume that $\delta < \min\{x - a, b - x\}$. If $|t - x| < \delta$ then

$$
\begin{aligned}
\left| \frac{F(t) - F(x)}{t - x} - f(x) \right|
&= \left| \frac{\int_a^t f(s)\,ds - \int_a^x f(s)\,ds}{t - x} - f(x) \right| \\
&= \left| \frac{\int_x^t f(s)\,ds}{t - x} - \frac{\int_x^t f(x)\,ds}{t - x} \right| \\
&= \left| \frac{\int_x^t (f(s) - f(x))\,ds}{t - x} \right| \,.
\end{aligned}
$$

Notice that we rewrote $f(x)$ as the integral with respect to a dummy variable s over an interval of length $|t - x|$ divided by $(t - x)$. Assume for the moment that $t > x$. Then the last line is dominated by

$$
\frac{\int_x^t |f(s) - f(x)|\,ds}{t - x} \;\leq\; \frac{\int_x^t \epsilon\,ds}{t - x} \\
= \epsilon \,.
$$

A similar estimate holds when $t < x$ (simply reverse the limits of integration). This shows that

$$\lim_{t \to x} \frac{F(t) - F(x)}{t - x}$$

exists and equals $f(x)$. Thus $F'(x)$ exists and equals $f(x)$. $\qquad \square$

In the exercises we shall consider how to use the theory of one-sided limits to make the conclusion of the Fundamental Theorem true on the entire interval $[a, b]$. We conclude with

Corollary 11.22 *If f is a continuous function on $[a, b]$ and if G is any continuously differentiable function on $[a, b]$ whose derivative equals f on (a, b) then*

$$\int_a^b f(x) \, dx = G(b) - G(a) \,.$$

Proof: Define F as in the theorem. Since F and G have the same derivative on (a, b), they differ by a constant (Corollary 10.18). Then

$$\int_a^b f(x)\, dx = F(b) = F(b) - F(a) = G(b) - G(a)$$

as desired.   □

A Look Back

1. How does change of variable in an integral work?

2. How are we allowed to break up the domain of integration of a function?

3. Why is the Fundamental Theorem of Calculus important?

4. Why are there two different statements of the Fundamental Theorem of Calculus?

Exercises

1. Imitate the proof of the Fundamental Theorem of Calculus in this section to show that, if f is continuous on $[a, b]$ and if we define

$$F(x) = \int_a^x f(t)\, dt \,,$$

then the one-sided derivative $F'(a)$ exists and equals $f(a)$ in the sense that

$$\lim_{t \to a^+} \frac{F(t) - F(a)}{t - a} = f(a)\,.$$

Formulate and prove an analogous statement for the one-sided derivative of F at b.

2. Let f be a bounded function on an unbounded interval of the form $[A, \infty)$. We say that f is integrable on $[A, \infty)$ if f is integrable on every compact subinterval of $[A, \infty)$ and

$$\lim_{B \to +\infty} \int_A^B f(x)\, dx$$

exists and is finite.

Assume that f is Riemann integrable on $[1, N]$ for every $N > 1$ and that f is decreasing. Show that f is Riemann integrable on $[1, \infty)$ if and only if $\sum_{j=1}^\infty f(j)$ is finite.

Suppose that g is nonnegative and integrable on $[1, \infty)$. If $0 \le |f(x)| \le g(x)$ for $x \in [1, \infty)$ and f is integrable on compact subintervals of $[1, \infty)$, then prove that f is integrable on $[1, \infty)$.

3. Let f be a function on an interval of the form $(a, b]$ such that f is integrable on compact subintervals of $(a, b]$. If

$$\lim_{\epsilon \to 0^+} \int_{a+\epsilon}^{b} f(x) \, dx$$

exists and is finite then we say that f is integrable on $(a, b]$. Prove that, if we restrict attention to bounded f, then in fact this definition gives rise to no new integrable functions. However, there are unbounded functions that can now be integrated. Give an example.

Give an example of a function g that is integrable by the definition in the preceding paragraph but is such that $|g|$ is not integrable.

4. Suppose that f is a continuous, nonnegative function on the interval $[0, 1]$. Let M be the maximum of f on the interval. Prove that

$$\lim_{n \to \infty} \left[\int_0^1 f(t)^n \, dt \right]^{1/n} = M \, .$$

5. Let f be a continuously differentiable function on the interval $[0, 2\pi]$. Further assume that $f(0) = f(2\pi)$ and $f'(0) = f'(2\pi)$. For $n \in \mathbb{N}$ define

$$\widehat{f}(n) = \frac{1}{2\pi} \int_0^{2\pi} f(x) \sin nx \, dx \, .$$

Prove that

$$\sum_{n=1}^{\infty} |\widehat{f}(n)|^2$$

converges. [**Hint:** Use integration by parts to obtain a favorable estimate on $|\widehat{f}(n)|$.]

6. Prove part **(b)** of Theorem 11.13.

* 7. Let $f_1, f_2, \ldots$ be Riemann integrable functions on $[0, 1]$. Suppose that $f_1(x) \geq f_2(x) \geq \cdots$ for every x and that $\lim_{j \to \infty} f_j(x) \equiv f(x)$ exists and is finite for every x. Is it the case that f is Riemann integrable?

* 8. Prove that

$$\lim_{\eta \to 0^+} \int_{\eta}^{1/\eta} \frac{\cos(2r) - \cos r}{r} \, dr$$

exists and is finite.

* 9. Suppose that f is a Riemann integrable function on the interval $[0, 1]$. Let $\epsilon > 0$. Show that there is a polynomial p so that

$$\int_0^1 |f(x) - p(x)| \, dx < \epsilon \, .$$

* 10. Refer to Exercise 5 for terminology. Prove that $\lim_{n \to \pm\infty} |\widehat{f}(n)| = 0$.

* 11. Refer to Exercise 10. Prove that if f is twice continuously differentiable on $\mathbb{R}$ and $f \equiv 0$ off the interval $[0, 2\pi]$, then

$$|\widehat{f}(n)| \leq C \cdot n^{-2} \, .$$

Chapter 12

Sequences and Series of Functions

12.1 Convergence of a Sequence of Functions

Preliminary Remarks

Sequences and series of functions play a pivotal role in modern mathematics, and also in engineering and physics. The theory of Fourier series, just as an example, was invented in the early nineteenth century as a means of decomposing a fairly arbitrary function into simple units (namely sines and cosines). The more modern theory of wavelets takes the Fourier theory to new heights of beauty and power.

A *sequence of functions* is usually written

$$f_1, f_2, \ldots \quad \text{or} \quad \{f_j\}_{j=1}^{\infty} .$$

We will generally assume that the functions f_j all have the same domain S.

Definition 12.1 A sequence of functions $\{f_j\}_{j=1}^{\infty}$ with domain $S \subset \mathbb{R}$ is said to *converge pointwise* to a limit function f on S if, for each $x \in S$, the sequence of numbers $\{f_j(x)\}$ converges to $f(x)$.

EXAMPLE 12.2 Define $f_j(x) = x^j$ with domain $S = \{x : 0 \leq x \leq 1\}$. If $0 \leq x < 1$ then $f_j(x) \to 0$. However, $f_j(1) \to 1$. Therefore the sequence f_j converges pointwise to the function

$$f(x) = \begin{cases} 0 & \text{if} \quad 0 \leq x < 1 \\ 1 & \text{if} \quad x = 1 . \end{cases}$$

See Figure 12.1. We see that, even though the f_j are each continuous, the limit function f is not. ∎

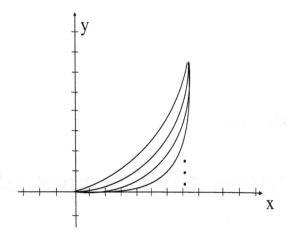

Figure 12.1: The sequence $\{x^j\}$.

Here are some of the basic questions that we must ask about a sequence of functions f_j that converges to a function f on a domain S:

(1) If the functions f_j are continuous, then is f continuous?

(2) If the functions f_j are integrable on an interval I, then is f integrable on I?

(3) If f is integrable on I, then does the sequence $\int_I f_j(x)\,dx$ converge to $\int_I f(x)\,dx$?

(4) If the functions f_j are differentiable, then is f differentiable?

(5) If f is differentiable, then does the sequence f_j' converge to f'?

We see from Example 12.2 that the answer to the first question is "no": Each of the f_j is continuous but f certainly is not. It turns out that, in order to obtain a favorable answer to our questions, we must consider a stricter notion of convergence of functions. This motivates the next definition.

Definition 12.3 Let f_j be a sequence of functions on a domain S. We say that the functions f_j *converge uniformly* to f if, given $\epsilon > 0$, there is an $N > 0$ such that, for any $j > N$ and any $x \in S$, it holds that $|f_j(x) - f(x)| < \epsilon$.

POINT OF CONFUSION 12.4 Notice that the special feature of uniform convergence is that the rate at which $f_j(x)$ converges is independent of $x \in S$. In Example 12.2, $f_j(x)$ is converging very rapidly to zero for x near zero but arbitrarily slowly to zero for x near 1—see Figure 12.1. In the next example we shall prove this assertion rigorously.

EXAMPLE 12.5 The sequence $f_j(x) = x^j$ does *not* converge uniformly to the limit function

$$f(x) = \begin{cases} 0 & \text{if } 0 \le x < 1 \\ 1 & \text{if } x = 1 \end{cases}$$

on the domain $S = [0, 1]$. In fact it does not even do so on the smaller domain $[0, 1)$. To see this, notice that no matter how large j is, we have, by the Mean Value Theorem, that

$$f_j(1) - f_j(1 - 1/(2j)) = \frac{1}{2j} \cdot f_j'(\xi)$$

for some ξ between $1 - 1/(2j)$ and 1. But $f_j'(x) = j \cdot x^{j-1}$ hence $|f_j'(\xi)| < j$ and we conclude that

$$|f_j(1) - f_j(1 - 1/(2j))| < \frac{1}{2}$$

or

$$f_j(1 - 1/(2j)) > f_j(1) - \frac{1}{2} = \frac{1}{2}.$$

In conclusion, no matter how large j, there will be values of x (namely, $x = 1 - 1/(2j)$) at which $f_j(x)$ is at least distance $1/2$ from the limit 0. We conclude that the convergence is not uniform. ∎

Theorem 12.6 *If f_j are continuous functions on a set S that converge uniformly on S to a function f then f is also continuous.*

Proof: Let $\epsilon > 0$. Fix an integer N so large that, if $j > N$, then $|f_j(x) - f(x)| < \epsilon/3$ for all $x \in S$. Fix $c \in S$. Choose $\delta > 0$ so small that if $x \in S$ and $|x - c| < \delta$ then $|f_{N+1}(x) - f_{N+1}(c)| < \epsilon/3$. For such x we have

$$|f(x) - f(c)| \le |f(x) - f_{N+1}(x)| + |f_{N+1}(x) - f_{N+1}(c)| + |f_{N+1}(c) - f(c)|$$
$$< \frac{\epsilon}{3} + \frac{\epsilon}{3} + \frac{\epsilon}{3}$$

by the way that we chose N and δ. But the last line sums to ϵ, proving that f is continuous at c. Since $c \in S$ was chosen arbitrarily, we are done. □

EXAMPLE 12.7 Define functions

$$f_j(x) = \begin{cases} 0 & \text{if } x = 0 \\ j & \text{if } 0 < x \le 1/j \\ 0 & \text{if } 1/j < x \le 1 \end{cases}$$

for $j = 2, 3, \ldots$. Then $\lim_{j \to \infty} f_j(x) = 0 \equiv f(x)$ for all x in the interval $I = [0, 1]$. However

$$\int_0^1 f_j(x)\, dx = \int_0^{1/j} j\, dx = 1$$

for every j. Thus the f_j converge to the integrable limit function $f(x) \equiv 0$ (with integral 0), but their integrals $\int f_j(x)\, dx = 1$ do not converge to the integral $\int f(x)\, dx = 0$ of f. ∎

EXAMPLE 12.8 Let $q_1, q_2, \ldots$ be an enumeration of the rationals in the interval $I = [0, 1]$. Define functions

$$f_j(x) = \begin{cases} 1 & \text{if} \quad x \in \{q_1, q_2, \ldots, q_j\} \\ 0 & \text{if} \quad x \notin \{q_1, q_2, \ldots, q_j\}. \end{cases}$$

Then the functions f_j converge pointwise to the Dirichlet function f which is equal to 1 on the rationals and 0 on the irrationals. Each of the functions f_j has integral 0 on I. But the function f is not integrable on I.   ∎

The last two examples show that something more than pointwise convergence is needed in order for the integral to respect the limit process.

Theorem 12.9 *Let f_j be integrable functions on a nontrivial bounded interval $[a, b]$ and suppose that the functions f_j converge uniformly to the limit function f. Then f is integrable on $[a, b]$ and*

$$\lim_{j \to \infty} \int_a^b f_j(x)\, dx = \int_a^b f(x)\, dx.$$

Proof: Pick $\epsilon > 0$. Choose N so large that, if $j > N$, then $|f_j(x) - f(x)| < \epsilon/[6(b - a)]$ for all $x \in [a, b]$. Notice that, if $j, k > N$, then

$$\left| \int_a^b f_j(x)\, dx - \int_a^b f_k(x)\, dx \right| \le \int_a^b |f_j(x) - f_k(x)|\, dx. \tag{12.9.1}$$

But $|f_j(x) - f_k(x)| \le |f_j(x) - f(x)| + |f(x) - f_k(x)| < \epsilon/[3(b-a)]$. Therefore line (12.9.1) does not exceed

$$\int_a^b \frac{\epsilon}{3(b-a)}\, dx = \frac{\epsilon}{3}.$$

Thus the numbers $\int_a^b f_j(x)\, dx$ form a Cauchy sequence. Let the limit of this sequence be called A. Notice that, if we let $k \to \infty$ in the inequality

$$\left| \int_a^b f_j(x)\, dx - \int_a^b f_k(x)\, dx \right| \le \frac{\epsilon}{3},$$

then we obtain

$$\left| \int_a^b f_j(x)\, dx - A \right| \le \frac{\epsilon}{3}$$

for all $j > N$. This estimate will be used below.

By hypothesis there is a $\delta > 0$ such that, if $\mathcal{P} = \{p_1, \ldots, p_k\}$ is a partition of $[a, b]$ with $m(\mathcal{P}) < \delta$, then

$$\left| \mathcal{R}(f_{N+1}, \mathcal{P}) - \int_a^b f_{N+1}(x)\, dx \right| < \frac{\epsilon}{3}.$$

But then, for such a partition, we have

$$|R(f, P) - A| \leq \left|R(f, P) - R(f_{N+1}, P)\right| + \left|R(f_{N+1}, P) - \int_a^b f_{N+1}(x)\, dx\right|$$
$$+ \left|\int_a^b f_{N+1}(x)\, dx - A\right|.$$

We have already noted that, by the choice of N, the third term on the right does not exceed $\epsilon/3$. The second term is smaller than $\epsilon/3$ by the way that we chose the partition P. It remains to examine the first term. Now

$$\left|R(f, P) - R(f_{N+1}, P)\right| = \left|\sum_{j=1}^k f(s_j)\Delta_j - \sum_{j=1}^k f_{N+1}(s_j)\Delta_j\right|$$
$$\leq \sum_{j=1}^k \left|f(s_j) - f_{N+1}(s_j)\right|\Delta_j$$
$$< \sum_{j=1}^k \frac{\epsilon}{6(b-a)}\Delta_j$$
$$= \frac{\epsilon}{6(b-a)}\sum_{j=1}^k \Delta_j$$
$$= \frac{\epsilon}{6}.$$

Therefore $|R(f, P) - A| < \epsilon$ when $m(P) < \delta$. This shows that the function f is integrable on $[a, b]$ and has integral with value A. $\square$

POINT OF CONFUSION 12.10 We have seen a few instances now in which the concept of uniform convergence addresses significant issues of convergence of a sequence of functions. This is an important idea for you to master and to be able to use easily.

We have succeeded in answering questions (1) and (2) that were raised at the beginning of the section. In the next section we will answer questions (3), (4), and (5).

A Look Back

1. What does it mean for a sequence of functions to converge pointwise?

2. What does it mean for a sequence of functions to converge uniformly?

3. What condition on a sequence of integrable functions will guarantee that their limit function will be integrable?

4. Is it true that the limit of a sequence of continuous functions is continuous?

Exercises

1. If $f_j \to f$ uniformly on a domain S and if f_j, f never vanish on S then does it follow that the functions $1/f_j$ converge uniformly to $1/f$ on S?

2. Write out the first five partial sums for the series

$$\sum_{j=1}^{\infty} \frac{\sin^3 j}{j^2}.$$

3. Write a series of polynomials that converges to $f(x) = \sin x^2$. Can you prove that it converges?

4. Write a series of trigonometric functions that converges to $f(x) = x$. Can you prove that it converges?

5. Write a series of piecewise linear functions that converges to $f(x) = x^2$ on the interval $[0, 1]$. Can you prove that it converges?

6. Write a series of functions that converges pointwise on $[0, 1]$ but does not converge uniformly on any proper subinterval. [**Hint:** First consider a sequence.]

7. Show that if $\sum_j f_j'$ converges uniformly on $[0, 1]$ (where the prime stands for the derivative), and if $f_j(0) = 0$ for all j, then $\sum_j f_j$ converges uniformly on $[0, 1]$.

8. TRUE or FALSE: If $\sum_j f_j$ converges absolutely and uniformly and $\sum_j g_j$ converges absolutely and uniformly on a compact interval $[a, b]$, then so does $\sum_j f_j g_j$.

9. If a power series $\sum a_j x^j$ converges at each point $x = 1$, $x = 2$, $x = 3$, etc., then show that the series converges uniformly on each interval of the form $[-N, N]$.

* **10.** Give an example of a Taylor series that converges uniformly on compact sets to its limit function.

* **11.** A Taylor series will never converge only pointwise. Explain.

12.2   More on Uniform Convergence

Preliminary Remarks

As we have seen, uniform convergence is a very powerful idea for guaranteeing that the limit of a sequence of functions is well behaved. In the present section we will see some very explicit and useful applications and extensions of this idea.

In general, limits do not commute. Since the integral is defined with a limit, and since we saw in the last section that integrals do not always respect limits of functions, we know some concrete instances of non-commutation of limits. The fact that continuity is defined with a limit, and that the limit of continuous functions need not be continuous, gives even more examples of situations in which limits do not commute. Let us now turn to a situation in which limits *do* commute:

Theorem 12.11 *Fix a set S and a point $s \in S$. Assume that the functions f_j converge uniformly on the domain $S \setminus \{s\}$ to a limit function f. Suppose that each function $f_j(x)$ has a limit as $x \to s$. Then f itself has a limit as $x \to s$ and*

$$\lim_{x \to s} f(x) = \lim_{j \to \infty} \lim_{x \to s} f_j(x) \,.$$

Because of the way that f is defined, we may rewrite this conclusion as

$$\lim_{x \to s} \lim_{j \to \infty} f_j(x) = \lim_{j \to \infty} \lim_{x \to s} f_j(x) \,.$$

In other words, the limits $\lim_{x \to s}$ and $\lim_{j \to \infty}$ commute.

Proof: Let $\alpha_j = \lim_{x \to s} f_j(x)$. Let $\epsilon > 0$. There is a number $N > 0$ (independent of $x \in S \setminus \{s\}$) such that $j > N$ implies that $|f_j(x) - f(x)| < \epsilon/4$. Fix $j, k > N$. Choose $\delta > 0$ such that $0 < |x - s| < \delta$ implies both that $|f_j(x) - \alpha_j| < \epsilon/4$ and $|f_k(x) - \alpha_k| < \epsilon/4$. Fix such an x. Then

$$|\alpha_j - \alpha_k| \leq |\alpha_j - f_j(x)| + |f_j(x) - f(x)| + |f(x) - f_k(x)| + |f_k(x) - \alpha_k| \,.$$

The first and last expressions are less than $\epsilon/4$ by the choice of x. The middle two expressions are less than $\epsilon/4$ by the choice of $j, k > N$. We conclude that $|\alpha_j - \alpha_k| < \epsilon$ for $j, k > N$, so the sequence $\{\alpha_j\}$ is Cauchy.[1] Let α be the limit of that sequence.

Letting $k \to \infty$ in the inequality

$$|\alpha_j - \alpha_k| < \epsilon$$

that we obtained above yields

$$|\alpha_j - \alpha| \leq \epsilon$$

for $j > N$. Now, with δ as above and $0 < |x - s| < \delta$, we have

$$|f(x) - \alpha| \leq |f(x) - f_j(x)| + |f_j(x) - \alpha_j| + |\alpha_j - \alpha| \,.$$

By the choices we have made, the first term is less than $\epsilon/4$, the second is less than $\epsilon/4$, and the third is less than or equal to ϵ. Altogether, if $0 < |x - s| < \delta$ then $|f(x) - \alpha| < 2\epsilon$. This is the desired conclusion. $\square$

POINT OF CONFUSION 12.12 *Once again we see that uniform convergence is the key to understanding a tricky situation involving limits. It will be a recurring idea in the rest of this text.*

Parallel with our notion of Cauchy sequence of numbers, we have a concept of Cauchy sequence of functions in the uniform sense:

---

[1] The argument to get this estimate entailed choosing an x that depended on j and k. But x has now faded into the background and the Cauchy property of $\{\alpha_j\}$ stands.

Definition 12.13 A sequence of functions f_j on a domain S is called *a uniformly Cauchy sequence* if, for each $\epsilon > 0$, there is an $N > 0$ such that, if $j, k > N$, then

$$|f_j(x) - f_k(x)| < \epsilon \quad \forall x \in S.$$

POINT OF CONFUSION 12.14 We see that a uniformly Cauchy sequence is one for which the functions get closer together at a uniform rate across the common domain of the functions. This is consistent with our notion of uniform convergence discussed earlier.

Proposition 12.15 *A sequence of functions f_j is uniformly Cauchy on a domain S if and only if the sequence converges uniformly to a limit function f on the domain S.*

Proof: The proof is straightforward and is assigned as an exercise.   □

We will use the last two results in our study of the limits of differentiable functions. First we consider an example.

EXAMPLE 12.16 Define the function

$$f_j(x) = \begin{cases} 0 & \text{if} \quad x \le 0 \\ jx^2 & \text{if} \quad 0 < x \le 1/(2j) \\ x - 1/(4j) & \text{if} \quad 1/(2j) < x < \infty. \end{cases}$$

We leave it as an exercise for you to check that the functions f_j converge uniformly on the entire real line to the function

$$f(x) = \begin{cases} 0 & \text{if} \quad x \le 0 \\ x & \text{if} \quad x > 0 \end{cases}$$

(draw a sketch to help you see this). Notice that each of the functions f_j is continuously differentiable on the entire real line, but f is not differentiable at 0.   ∎

It turns out that we must strengthen our convergence hypotheses if we want the limit process to respect differentiation. The basic result is:

Theorem 12.17 *Suppose that a sequence f_j of differentiable functions on an open interval I converges pointwise to a limit function f. Suppose further that the sequence f_j' converges uniformly on I to a limit function g. Then the limit function f is differentiable on I and $f'(x) = g(x)$ for all $x \in I$.*

Proof: Let $\epsilon > 0$. The sequence $\{f_j'\}$ is uniformly Cauchy. Therefore we may choose N so large that $j, k > N$ implies that

$$|f_j'(x) - f_k'(x)| < \frac{\epsilon}{2} \quad \forall x \in I. \tag{12.17.1}$$

Fix a point $c \in I$. Define

$$\mu_j(x) = \frac{f_j(x) - f_j(c)}{x - c}$$

for $x \in I, x \neq c$. It is our intention to apply Theorem 12.11 above to the functions μ_j.

First notice that, for each j, we have

$$\lim_{x \to c} \mu_j(x) = f_j'(c).$$

Thus

$$\lim_{j \to \infty} \lim_{x \to c} \mu_j(x) = \lim_{j \to \infty} f_j'(c) = g(c).$$

That calculates the limits in one order.

On the other hand,

$$\lim_{j \to \infty} \mu_j(x) = \frac{f(x) - f(c)}{x - c} \equiv \mu(x)$$

for $x \in I \setminus \{c\}$. If we can show that this convergence is uniform then Theorem 12.11 applies and we may conclude that

$$\lim_{x \to c} \mu(x) = \lim_{j \to \infty} \lim_{x \to c} \mu_j(x) = \lim_{j \to \infty} f_j'(c) = g(c).$$

But this just says that f is differentiable at c and the derivative equals g. That is the desired result.

To verify the uniform convergence of the μ_j, we apply the Mean Value Theorem to the function $f_j - f_k$. For $x \neq c$ we have

$$
\begin{aligned}
|\mu_j(x) - \mu_k(x)| &= \frac{1}{|x - c|} \cdot |(f_j(x) - f_k(x)) - (f_j(c) - f_k(c))| \\
&= \frac{1}{|x - c|} \cdot |x - c| \cdot |(f_j - f_k)'(\xi)| \\
&= |(f_j - f_k)'(\xi)|
\end{aligned}
$$

for some ξ between x and c. But line (12.17.1) guarantees that the last line does not exceed $\epsilon/2$. That shows that the μ_j converge uniformly and concludes the proof. □

Remark 12.18 A little additional effort shows that we need only assume in the theorem that the functions f_j converge at a single point x_0 in the domain. One of the exercises asks you to prove this assertion.

Notice further that, if we make the additional assumption that each of the functions f_j' is continuous, then the proof of the theorem becomes much easier. For then

$$f_j(x) = f_j(x_0) + \int_{x_0}^{x} f_j'(t)\, dt$$

by the Fundamental Theorem of Calculus. The hypothesis that the f_j' converge uniformly then implies, by Theorem 12.9, that the integrals converge to

$$\int_{x_0}^{x} g(t)\,dt\,.$$

The hypothesis that the functions f_j converge at x_0 then allows us to conclude that the sequence $f_j(x)$ converges for every x to $f(x)$ and

$$f(x) = f(x_0) + \int_{x_0}^{x} g(t)\,dt\,.$$

The Fundamental Theorem of Calculus then yields that $f' = g$ as desired.

A Look Back

1. Give a verbal description of what "uniformly Cauchy" means.

2. What is the role of uniform convergence in addressing the issue of commutation of limits?

3. What do we need to assume about a sequence of functions in order to guarantee that their derivatives converge in a natural fashion?

4. What would be a uniform Cauchy condition for series?

Exercises

1. Prove that, if a series of continuous functions converges uniformly, then the sum function is also continuous.

2. Prove Proposition 12.15. Refer to the parallel result in Chapter 6 for some hints.

3. Prove the assertion made in Remark 12.18 that Theorem 12.17 is still true if the functions f_j are assumed to converge at just one point (and also that the derivatives f_j' converge uniformly).

4. If a sequence of functions f_j on a domain $S \subset \mathbb{R}$ has the property that $f_j \to f$ uniformly on S, then does it follow that $(f_j)^2 \to f^2$ uniformly on S? What simple additional hypothesis will make your answer affirmative?

5. Assume that f_j are continuous functions on the interval $[0, 1]$. Suppose that $\lim_{j \to \infty} f_j(x)$ exists for each $x \in [0, 1]$ and defines a function f on $[0, 1]$. Further suppose that $f_1 \leq f_2 \leq \cdots$. Can you conclude that f is continuous?

6. Let $f : \mathbb{R} \to \mathbb{R}$ be a function. We say that f is *piecewise constant* if the real line can be written as the infinite disjoint union of intervals and f is constant on each of those intervals. Now let φ be a continuous function on $[a, b]$. Show that φ can be uniformly approximated by piecewise constant functions.

7. Give an example of a sequence of polynomials on $[0, 1]$ that converges uniformly to $\sin x$.

*   8. Refer to Exercise 6 for terminology. Let f be a piecewise constant function. Show that f is the pointwise limit of polynomials.

* **9.** A function is called "piecewise linear" if it is **(i)** continuous and **(ii)** its graph consists of finitely many linear segments. Prove that a continuous function on an interval $[a, b]$ is the uniform limit of a sequence of piecewise linear functions.

* **10.** Let f_j be a uniformly convergent sequence of functions on a common domain S. What would be suitable conditions on a function ϕ to guarantee that $\phi \circ f_j$ converges uniformly on S?

* **11.** Construct a sequence of continuous functions $f_j(x)$ that has the property that $f_j(q)$ increases monotonically to $+\infty$ for each rational q but such that, at each irrational x, $|f_j(x)| \leq 1$ for infinitely many j.

* **12.** Prove that a sequence $\{f_j\}$ of functions converges pointwise if and only if the series

$$f_1 + \sum_{j=2}^{\infty}(f_j - f_{j-1})$$

converges pointwise. Prove the same result for uniform convergence.

12.3 Series of Functions

Preliminary Remarks

Of course a series of functions is understood by studying the sequence of its partial sums. So, in some sense, the theory of sequences of functions and the theory of series of functions are equivalent. But series are useful because they explicitly represent the idea of decomposing an arbitrary function into elemental pieces.

Definition 12.19 The formal expression

$$\sum_{j=1}^{\infty} f_j(x),$$

where the f_j are functions on a common domain S, is called a *series of functions*. For $N = 1, 2, 3, \ldots$ the expression

$$S_N(x) = \sum_{j=1}^{N} f_j(x) = f_1(x) + f_2(x) + \cdots + f_N(x)$$

is called the Nth *partial sum* for the series. In case

$$\lim_{N \to \infty} S_N(x)$$

exists and is finite then we say that the series *converges* at x. Otherwise we say that the series *diverges* at x.

POINT OF CONFUSION 12.20 Notice that the question of convergence of a series of functions, which should be thought of as an *addition process*, reduces to a question about the *sequence* of partial sums. Sometimes, as in the next example, it is convenient to begin the series at some index other than $j = 1$.

EXAMPLE 12.21 Consider the series

$$\sum_{j=0}^{\infty} x^j .$$

This is the geometric series from Proposition 8.21. It converges absolutely for $|x| < 1$ and diverges otherwise.

By the formula for the partial sums of a geometric series,

$$S_N(x) = \sum_{j=0}^{N} x^j = \frac{1 - x^{N+1}}{1 - x}$$

for $x \neq 1$. When $|x| < 1$ we see that

$$S_N(x) \to \frac{1}{1 - x}.$$ ∎

Definition 12.22 Let

$$\sum_{j=1}^{\infty} f_j(x)$$

be a series of functions on a domain S. If the partial sums $S_N(x)$ converge uniformly on S to a limit function $g(x)$, then we say that the series *converges uniformly* on S. We write

$$\sum_{j=1}^{\infty} f_j(x) = g(x)$$

for $x \in S$.

Of course all of our results about uniform convergence of *sequences* of functions translate, via the sequence of partial sums of a series, to results about uniformly convergent series of functions. For example:

(a) If f_j are continuous functions on a domain S and if the series

$$\sum_{j=1}^{\infty} f_j(x)$$

converges uniformly on S to a limit function f, then f is also continuous on S.

(b) If f_j are integrable functions on $[a, b]$ and if

$$\sum_{j=1}^{\infty} f_j(x)$$

converges uniformly on $[a, b]$ to a limit function f, then f is also integrable on $[a, b]$ and

$$\int_a^b f(x)\, dx = \sum_{j=1}^{\infty} \int_a^b f_j(x)\, dx.$$

You will be asked to provide details of these assertions, as well as a statement and proof of a result about derivatives of series, in the exercises. Meanwhile we turn to an elegant test for uniform convergence that is due to Weierstrass.

POINT OF CONFUSION 12.23 Notice that, in the last displayed equation, the entity on the left is a number and the entity on the right is a sum of real numbers. This is a statement about the value of an integral.

Theorem 12.24 (The Weierstrass M-Test) *Let $\{f_j\}_{j=1}^{\infty}$ be functions on a common domain S. Assume that each $|f_j|$ is bounded on S by a constant M_j and that*

$$\sum_{j=1}^{\infty} M_j < \infty.$$

Then the series

$$\sum_{j=1}^{\infty} f_j \qquad\qquad (12.24.1)$$

converges uniformly and absolutely on the set S.

Proof: By hypothesis, the sequence T_N of partial sums of the series $\sum_{j=1}^{\infty} M_j$ is Cauchy. Given $\epsilon > 0$ there is therefore a number K so large that $q > p > K$ implies that

$$\sum_{j=p+1}^{q} M_j = |T_q - T_p| < \epsilon.$$

We may conclude that the partial sums S_N of the original series $\sum f_j$ satisfy, for $q > p > K$,

$$|S_q(x) - S_p(x)| = \left| \sum_{j=p+1}^{q} f_j(x) \right|$$

$$\leq \sum_{j=p+1}^{q} |f_j(x)| \leq \sum_{j=p+1}^{q} M_j < \epsilon.$$

Thus the partial sums $S_N(x)$ of the series (12.24.1) are uniformly Cauchy. The series (12.24.1) therefore converges uniformly. The same estimates apply to the partial sums of the series $\sum_j |f_j|$. Therefore the series must also converge absolutely. $\qquad\square$

EXAMPLE 12.25 Let us consider the series

$$f(x) = \sum_{j=1}^{\infty} 2^{-j} \sin\left(2^j x\right) .$$

The sine terms oscillate so erratically that it would be difficult to calculate partial sums for this series. However, noting that the jth summand $f_j(x) = 2^{-j} \sin(2^j x)$ is dominated in absolute value by 2^{-j}, we see that the Weierstrass M-Test applies to this series. We conclude that the series converges uniformly and absolutely on the entire real line.

By property (a) of uniformly convergent series of continuous functions that was noted above, we may conclude that the function f defined by our series is continuous. It is also 2π-periodic: $f(x + 2\pi) = f(x)$ for every x since this assertion is true for each summand. Since the continuous function f restricted to the compact interval $[0, 2\pi]$ is uniformly continuous (Theorem 5.30), we may conclude that f is uniformly continuous on the entire real line.

However, it turns out that f is nowhere differentiable. The proof of this assertion follows lines similar to the treatment of nowhere differentiable functions in Theorem 10.7. The details will be covered in an exercise. ∎

A Look Back

1. Say in words what the Weierstrass M-Test says.

2. What does it mean for a series of functions to converge uniformly?

3. What does it mean for a series of functions to converge absolutely?

4. What does it mean for a series of functions to converge pointwise?

5. Why does every question about series reduce to a question about a sequence?

Exercises

1. Formulate and prove a result about the derivative of the sum of a convergent series of differentiable functions.

2. Prove Dini's theorem: If f_j are continuous functions on a compact set K, $f_1(x) \leq f_2(x) \leq \ldots$ for all $x \in K$, and the f_j converge to a continuous function f on K then in fact the f_j converge *uniformly* to f on K.

3. Use the concept of boundedness of a function to show that the functions $\sin x$ and $\cos x$ cannot be polynomials.

4. Prove that, if p is any polynomial, then there is an N large enough that $e^x > |p(x)|$ for $x > N$. Conclude that the function e^x is not a polynomial.

5. Find a way to prove that $\tan x$ and $\ln x$ are not polynomials.

6. Prove that the series

$$\sum_{j=1}^{\infty} \frac{\sin jx}{j}$$

converges uniformly on compact intervals that do not contain odd multiples of $\pi/2$. (**Hint:** Sum by parts.)

7. Suppose that the sequence $f_j(x)$ on the interval $[0, 1]$ satisfies

$$|f_j(s) - f_j(t)| \leq |s - t| \qquad (*)$$

for all $s, t \in [0, 1]$. Further assume that the f_j converge pointwise to a limit function f on the interval $[0, 1]$. Does the limit function f satisfy $(*)$?

8. Prove a comparison test for uniform convergence of series: if f_j, g_j are functions and $0 \leq f_j \leq g_j$ and the series $\sum g_j$ converges uniformly then so also does the series $\sum f_j$.

9. Show by giving an example that the converse of the Weierstrass M-Test is false.

10. Prove that if f_j are continuous functions on a domain S and if the series

$$\sum_{j=1}^{\infty} f_j(x)$$

converges uniformly on S to a limit function f, then f is also continuous on S.

11. Prove that if a series $\sum_{j=1}^{\infty} f_j$ of integrable functions on an interval $[a, b]$ is uniformly convergent on $[a, b]$ then the sum function f is integrable and

$$\int_a^b f(x)\, dx = \sum_{j=1}^{\infty} \int_a^b f_j(x)\, dx\,.$$

* **12.** Let $0 < \alpha \leq 1$. Prove that the series

$$\sum_{j=1}^{\infty} 2^{-j\alpha} \sin\left(2^j x\right)$$

defines a function f that is nowhere differentiable. To achieve this end, follow the scheme that was used to prove Theorem 10.7: **(a)** Fix x; **(b)** for h small, choose M such that 2^{-M} is approximately equal to $|h|$; **(c)** break the series up into the sum from 1 to $M - 1$, the single summand $j = M$, and the sum from $j = M + 1$ to ∞. The middle term has very large Newton quotient and the first and last terms are relatively small.

* **13.** Prove that the sequence of functions $f_j(x) = \sin(jx)$ has no subsequence that converges at every x.

12.4 The Weierstrass Approximation Theorem

Preliminary Remarks

One of the most powerful and astonishing theorems of nineteenth-century mathematics is the Weierstrass theorem that we study in this section. It tells us that absolutely any continuous function on a closed, bounded interval can be uniformly approximated by a polynomial. This fact has practical significance today, because we cannot program an arbitrary function onto a computer, but we certainly can program a polynomial function.

The name Weierstrass has occurred frequently in this chapter. In fact Karl Weierstrass (1815–1897) revolutionized analysis with his examples and theorems. This section is devoted to one of his most striking results. We introduce it with a motivating discussion.

It is natural to wonder whether the usual functions of calculus—$\sin x$, $\cos x$, and e^x, for instance—are actually polynomials of some very high degree. Since polynomials are so much easier to understand than these transcendental functions, an affirmative answer to this question would certainly simplify mathematics. Of course a moment's thought shows that this wish is impossible: a polynomial of degree k has at most k real roots. Since sine and cosine have infinitely many real roots they cannot be polynomials. A polynomial of degree k has the property that, if it is differentiated enough times (namely, $k + 1$ times), then the derivative is zero. Since this is not the case for e^x, we conclude that e^x cannot be a polynomial. Similarly for $\log x$. The exercises of the last section discuss other means for distinguishing the familiar transcendental functions of calculus from polynomial functions.

In calculus we learned of a formal procedure, called Taylor series, for associating polynomials with a given function f. In some instances these polynomials form a sequence that converges back to the original function. Of course the method of the Taylor expansion has no hope of working unless f is infinitely differentiable. Even then, it turns out that the Taylor series rarely converges back to the original function. Nevertheless, Taylor's theorem with remainder might cause us to speculate that any reasonable function can be approximated in some fashion by polynomials. In fact the theorem of Weierstrass gives a spectacular affirmation of this speculation:

Theorem 12.26 (The Weierstrass Approximation Theorem) *Let f be a continuous function on an interval $[a, b]$. Then there is a sequence of polynomials $p_j(x)$ with the property that the sequence p_j converges uniformly on $[a, b]$ to f.*

We prove this theorem in detail in the optional Appendix to this chapter. For now, let us consider some of its consequences. A restatement of the theorem would be that, given a continuous function f on $[a, b]$ and an $\epsilon > 0$, there is a polynomial p such that

$$|f(x) - p(x)| < \epsilon$$

for every $x \in [a, b]$. If one were programming a computer to calculate values of a fairly wild function f, the theorem guarantees that, up to a given degree of accuracy, one could use a polynomial instead (which would in fact be much easier for the computer to handle). Advanced techniques can even tell what degree of polynomial is needed to achieve a given degree of accuracy. The proof that we shall present also suggests how this might be done.

Let f be the Weierstrass Nowhere Differentiable Function. The theorem guarantees that, on any compact interval, f is the uniform limit of polynomials. Thus even the uniform limit of infinitely differentiable functions need not be differentiable—even at one point. This explains why the hypotheses of Theorem 12.17 needed to be so stringent.

A Look Back

1. State verbally what the Weierstrass Approximation Theorem says.

2. Why is Weierstrass's theorem surprising?

3. Is it possible to approximate a continuous function with something other than polynomials?

4. Is there a converse to Weierstrass's theorem?

Optional APPENDIX: Proof of the Weierstrass Approximation Theorem

We break up the proof of the Weierstrass Approximation Theorem into a sequence of lemmas.

Lemma 12.27 *Let ψ_j be a sequence of continuous functions on the interval $I = [-1, 1]$ with the following properties:*

(i) *$\psi_j(x) \geq 0$ for all x;*

(ii) *$\int_{-1}^{1} \psi_j(x)\,dx = 1$ for each j;*

(iii) *For any $\delta > 0$ we have*

$$\lim_{j \to \infty} \int_{\delta \leq |x| \leq 1} \psi_j(x)\,dx = 0.$$

If f is a continuous function on the real line which is identically zero off the interval $[0, 1]$ then the functions

$$f_j(x) = \int_{-1}^{1} \psi_j(t) f(x - t)\,dt$$

converge uniformly on the interval $[0, 1]$ to $f(x)$.

Proof: By multiplying f by a constant we may assume that $\sup |f| = 1$. Let $\epsilon > 0$. Since f is uniformly continuous on the interval $[0, 1]$ we may choose a $\delta > 0$ such that if $x, t \in I$ $|x - t| < \delta$ then $|f(x) - f(t)| < \epsilon/2$. By property **(iii)** above we may choose an N so large that $j > N$ implies that $|\int_{\delta \leq |t| \leq 1} \psi_j(t)\,dt| < \epsilon/4$. Then, for any $x \in [0, 1]$, we have

$$
\begin{aligned}
|f_j(x) - f(x)| &= \left| \int_{-1}^{1} \psi_j(t) f(x - t)\,dt - f(x) \right| \\
&= \left| \int_{-1}^{1} \psi_j(t) f(x - t)\,dt - \int_{-1}^{1} \psi_j(t) f(x)\,dt \right|.
\end{aligned}
$$

Notice that, in the last line, we have used fact **(ii)** about the functions ψ_j to multiply the term $f(x)$ by 1 in a clever way. Now we may combine the two integrals to find that the last line

$$= \left| \int_{-1}^{1} (f(x-t) - f(x))\psi_j(t)\, dt \right|$$

$$\leq \int_{-\delta}^{\delta} |f(x-t) - f(x)|\psi_j(t)\, dt$$

$$+ \int_{\delta \leq |t| \leq 1} |f(x-t) - f(x)|\psi_j(t)\, dt$$

$$= A + B.$$

To estimate term A, we recall that, for $|t| < \delta$, we have $|f(x-t) - f(x)| < \epsilon/2$; hence

$$A \leq \int_{-\delta}^{\delta} \frac{\epsilon}{2} \psi_j(t)\, dt \leq \frac{\epsilon}{2} \cdot \int_{-1}^{1} \psi_j(t)\, dt = \frac{\epsilon}{2}.$$

For B we write

$$B \leq \int_{\delta \leq |t| \leq 1} 2 \cdot \sup |f| \cdot \psi_j(t)\, dt$$

$$\leq 2 \cdot \int_{\delta \leq |t| \leq 1} \psi_j(t)\, dt$$

$$< 2 \cdot \frac{\epsilon}{4} = \frac{\epsilon}{2},$$

where in the penultimate line we have used the choice of j. Adding together our estimates for A and B, and noting that these estimates are independent of the choice of x, yields the result. $\qquad\square$

Lemma 12.28 Define $\psi_j(t) = k_j \cdot (1 - t^2)^j$, where the positive constants k_j are chosen so that $\int_{-1}^{1} \psi_j(t)\, dt = 1$. Then the functions ψ_j satisfy the properties **(i)–(iii)** of the last lemma.

Proof: Of course property **(ii)** is true by design. Property **(i)** is obvious. In order to verify property **(iii)**, we need to estimate the size of k_j.

Notice that

$$\int_{-1}^{1} (1 - t^2)^j\, dt = 2 \cdot \int_{0}^{1} (1 - t^2)^j\, dt$$

$$\geq 2 \cdot \int_{0}^{1/\sqrt{j}} (1 - t^2)^j\, dt$$

$$\geq 2 \cdot \int_{0}^{1/\sqrt{j}} (1 - jt^2)\, dt,$$

where we have used the Binomial Theorem and $0 \le t \le 1/\sqrt{j}$ to obtain the last inequality. But this last integral is easily evaluated and equals $4/(3\sqrt{j})$. We conclude that

$$\int_{-1}^{1} (1 - t^2)^j \, dt > \frac{1}{\sqrt{j}}.$$

As a result, $k_j < \sqrt{j}$.

Now, to verify property (iii) of the lemma, we notice that, for $\delta > 0$ fixed and $\delta \le |t| \le 1$, it holds that

$$|\psi_j(t)| \le k_j \cdot (1 - \delta^2)^j \le \sqrt{j} \cdot (1 - \delta^2)^j$$

and this expression tends to 0 as $j \to \infty$. Thus $\psi_j \to 0$ uniformly on $\{t : \delta \le |t| \le 1\}$. It follows that the ψ_j satisfy property (iii) of the lemma. □

Proof of the Theorem: We may assume without loss of generality (just by changing coordinates) that f is a continuous function on the interval $[0, 1]$. After adding a linear function (which is a polynomial) to f, we may assume that $f(0) = f(1) = 0$. Thus f may be continued to be a continuous function which is identically zero on $(-\infty, 0]$ and $[1, \infty)$.

Let ψ_j be as in Lemma 12.28 and form f_j as in Lemma 12.27. Then we know that the f_j converge uniformly on $[0, 1]$ to f. Finally,

$$\begin{aligned}
f_j(x) &= \int_{-1}^{1} \psi_j(t) f(x - t) \, dt \\
&= \int_{0}^{1} \psi_j(x - t) f(t) \, dt \\
&= k_j \int_{0}^{1} (1 + (x - t)^2)^j f(t) \, dt.
\end{aligned}$$

In the second equality we performed a simple change of variable.

But multiplying out the expression $(1 + (x - t)^2)^j$ in the integrand then shows that f_j is a polynomial of degree at most $2j$ in x. Thus we have constructed a sequence of polynomials f_j that converges uniformly to the function f on the interval $[0, 1]$.

Exercises

1. Let $\{f_j\}$ be a sequence of continuous functions on the real line. Suppose that the f_j converge uniformly to a function f. Prove that

$$\lim_{j \to \infty} f_j(x + 1/j) = f(x)$$

uniformly on any bounded interval.

Can any of these hypotheses be weakened?

2. Prove that the Weierstrass Approximation Theorem fails if we restrict attention to polynomials of degree less than or equal to 1000.

3. Is the Weierstrass Approximation Theorem true if we restrict ourselves to only using polynomials of even degree?

4. Is the Weierstrass Approximation Theorem true if we restrict ourselves to only using polynomials with coefficients of size not exceeding 1?

5. TRUE or FALSE: If a sequence of polynomials $\{p_j\}$ converges uniformly to 0 on the interval $[-2, 2]$, then the sequence of derivatives $\{p_j'\}$ converges uniformly to 0 on $[-1, 1]$.

6. TRUE or FALSE: If a sequence of polynomials $\{p_j\}_{j=1}^{\infty}$ converges uniformly to 0 on the interval $[-2, 2]$, then the sequence of antiderivatives $\{\int_0^x p_j(t)\, dt\}_{j=1}^{\infty}$ converges uniformly on $[-1, 1]$.

7. Refer to Exercise 9 below for terminology. Prove that the sum of a series of step functions need not be a step function.

* 8. Use the Weierstrass Approximation Theorem and Mathematical Induction to prove that, if f is k times continuously differentiable on an interval $[a, b]$, then there is a sequence of polynomials p_j with the property that

$$p_j \to f$$

uniformly on $[a, b]$,

$$p_j' \to f'$$

uniformly on $[a, b]$,

$$\cdots$$

$$p_j^{(k)} \to f^{(k)}$$

uniformly on $[a, b]$.

* 9. Let $a < b$ be real numbers. Call a function of the form

$$f(x) = \begin{cases} 1 & \text{if} \quad a \le x \le b \\ 0 & \text{if} \quad x < a \text{ or } x > b \end{cases}$$

a *characteristic function* for the interval $[a, b]$. Then a function of the form

$$g(x) = \sum_{j=1}^{k} a_j \cdot f_j(x),$$

with the f_j characteristic functions of intervals $[a_j, b_j]$, is called a *step function*. Prove that any continuous function on an interval $[c, d]$ is the uniform limit of a sequence of step functions. (**Hint:** The proof of this assertion is conceptually easy; do *not* imitate the proof of the Weierstrass Approximation Theorem.)

* 10. If f is a continuous function on the interval $[a, b]$ and if

$$\int_a^b f(x)p(x)\, dx = 0$$

for every polynomial p, then prove that f must be the zero function. (**Hint:** Use Weierstrass's Approximation Theorem.)

* **11.** Define a *trigonometric polynomial* to be a function of the form

$$\sum_{j=1}^{k} a_j \cdot \cos jx + \sum_{j=1}^{\ell} b_j \cdot \sin jx .$$

Prove a version of the Weierstrass Approximation Theorem on the interval $[0, 2\pi]$ for 2π-periodic continuous functions and with the phrase "trigonometric polynomial" replacing "polynomial." (**Hint:** Prove that

$$\sum_{\ell=-j}^{j} \left(1 - \frac{|\ell|}{j+1} \right) (\cos \ell t) =$$

$$\frac{1}{j+1} \left(\frac{\sin \frac{j+1}{2}t}{\sin \frac{1}{2}t} \right)^2 .$$

Use these functions as the functions ψ_j in the proof of Weierstrass's theorem.)

Chapter 13

Elementary Transcendental Functions

13.1 Power Series

Preliminary Remarks

When we learn about power series, and especially Taylor's formula, in calculus class we generally come away with the impression that most any function can be expanded in a power series. Unfortunately this is not the case. Functions that have convergent power series expansions are called real analytic functions, and have many special properties. We shall learn about them in the present section.

A series of the form

$$\sum_{j=0}^{\infty} a_j (x - c)^j$$

is called a *power series* expanded about the point c. Our first task is to determine the nature of the set on which a power series converges.

Proposition 13.1 *Assume that the power series*

$$\sum_{j=0}^{\infty} a_j (x - c)^j.$$

converges at the value $x = d$ with $d \neq c$. Let $r = |d - c|$. Then the series converges uniformly and absolutely on compact subsets of $\mathcal{I} = \{x : |x - c| < r\}$.

Proof: We may take the compact subset of $\mathcal{I}$ to be $K = [c - s, c + s]$ for some number $0 < s < r$. For $x \in K$ it then holds that

$$\sum_{j=0}^{\infty} |a_j(x-c)^j| = \sum_{j=0}^{\infty} |a_j(d-c)^j| \cdot \left|\frac{x-c}{d-c}\right|^j .$$

In the sum on the right, the first expression in absolute values is bounded by some constant C (by the convergence hypothesis). The quotient in absolute values is majorized by $L = s/r < 1$. The series on the right is thus dominated by

$$\sum_{j=0}^{\infty} C \cdot L^j .$$

This geometric series converges. By the Weierstrass M-Test, the original series converges absolutely and uniformly on K. $\qquad\square$

An immediate consequence of the proposition is that the set on which the power series

$$\sum_{j=0}^{\infty} a_j(x-c)^j$$

converges is an interval centered about c. We call this set the *interval of convergence*. The series will converge absolutely and uniformly on compact subsets of the interval of convergence. The *radius* of the interval of convergence (called the *radius of convergence*) is defined to be half the length of the interval of convergence. Whether convergence holds at the endpoints of the interval will depend on the particular series being studied. Ad hoc methods must be used to check the endpoints. Let us use the notation $\mathcal{C}$ to denote the *open interval of convergence*.

It happens that, if a power series converges at either of the endpoints of its interval of convergence, then the convergence is uniform up to that endpoint. This is a consequence of Abel's partial summation test; details will be explored in the exercises.

On the interval of convergence $\mathcal{C}$, the power series defines a function f. Such a function is said to be *real analytic*. More precisely, we have:

Definition 13.2 A function f, with domain an open set $U \subset \mathbb{R}$ and range the real numbers, is called *real analytic* if, for each $c \in U$, the function f may be represented by a convergent power series on an interval of positive radius centered at c:

$$f(x) = \sum_{j=0}^{\infty} a_j(x-c)^j .$$

We need to know both the algebraic and the calculus properties of a real analytic function: is it continuous? differentiable? How does one add/subtract/multiply/divide two such functions?

Proposition 13.3 *Let*

$$\sum_{j=0}^{\infty} a_j(x-c)^j \quad \text{and} \quad \sum_{j=0}^{\infty} b_j(x-c)^j$$

be two power series each having common nontrivial interval of convergence $\mathcal{C}$ centered at c. Let $f(x)$ be the function defined by the first series and $g(x)$ the function defined by the second series. Then, on $\mathcal{C}$, it holds that

(1) $f(x) \pm g(x) = \sum_{j=0}^{\infty}(a_j \pm b_j)(x-c)^j$;

(2) $f(x) \cdot g(x) = \sum_{m=0}^{\infty} \sum_{j+k=m}(a_j \cdot b_k)(x-c)^m$.

Proof: Let

$$A_N = \sum_{j=0}^{N} a_j(x-c)^j \quad \text{and} \quad B_N = \sum_{j=0}^{N} b_j(x-c)^j$$

be, respectively, the Nth partial sums of the power series that define f and g. If C_N is the Nth partial sum of the series

$$\sum_{j=0}^{\infty}(a_j \pm b_j)(x-c)^j$$

then

$$
\begin{aligned}
f(x) \pm g(x) &= \lim_{N \to \infty} A_N \pm \lim_{N \to \infty} B_N = \lim_{N \to \infty} [A_N \pm B_N] \\
&= \lim_{N \to \infty} C_N = \sum_{j=0}^{\infty}(a_j \pm b_j)(x-c)^j .
\end{aligned}
$$

This proves **(1)**.

For **(2)**, let us restrict attention to x in a compact subset K of $\mathcal{C}$. Write

$$D_N = \sum_{m=0}^{N} \sum_{j+k=m}(a_j \cdot b_k)(x-c)^m \quad \text{and} \quad R_N = \sum_{j=N+1}^{\infty} b_j(x-c)^j .$$

Note that, since the series for g converges, we know that $|R_N(x)| \leq M$ for some positive M.

We have

$$
\begin{aligned}
D_N &= a_0 B_N + a_1(x-c)B_{N-1} + \cdots + a_N(x-c)^N B_0 \\
&= a_0(g(x) - R_N) + a_1(x-c)(g(x) - R_{N-1}) \\
&\quad + \cdots + a_N(x-c)^N(g(x) - R_0) \\
&= g(x)\sum_{j=0}^{N} a_j(x-c)^j \\
&\quad - [a_0 R_N + a_1(x-c)R_{N-1} + \cdots + a_N(x-c)^N R_0].
\end{aligned}
$$

Clearly,

$$g(x) \sum_{j=0}^{N} a_j (x - c)^j$$

converges to $g(x)f(x)$ as N approaches ∞. In order to show that $D_N \to g \cdot f$, it will thus suffice to show that

$$\left| a_0 R_N + a_1 (x - c) R_{N-1} + \cdots + a_N (x - c)^N R_0 \right|$$

converges to 0 as N approaches ∞. Fix x. Now we know that

$$\sum_{j=0}^{\infty} a_j (x - c)^j$$

is absolutely convergent so we may set

$$A = \sum_{j=0}^{\infty} |a_j| |x - c|^j .$$

Also $\sum_{j=0}^{\infty} b_j (x - c)^j$ is convergent. Therefore, given $\epsilon > 0$, we can find N_0 so that $N > N_0$ implies $|R_N| < \epsilon$. Thus we have

$$\left| a_0 R_N + a_1 (x - c) R_{N-1} + \cdots + a_N (x - c)^N R_0 \right|$$
$$\leq \quad \left| a_0 R_N + \cdots + a_{N-N_0} (x - c)^{N-N_0} R_{N_0} \right|$$
$$+ \left| a_{N-N_0+1} (x - c)^{N-N_0+1} R_{N_0-1} + \cdots + a_N (x - c)^N R_0 \right|$$
$$\leq \quad \sup_{M \geq N_0} R_M \cdot \left(\sum_{j=0}^{\infty} |a_j| |x - c|^j \right)$$
$$+ \left| a_{N-N_0+1} (x - c)^{N-N_0+1} R_{N_0-1} \cdots + a_N (x - c)^N R_0 \right|$$
$$\leq \quad \epsilon A + \left| a_{N-N_0+1} (x - c)^{N-N_0+1} R_{N_0-1} \cdots + a_N (x - c)^N R_0 \right| .$$

Thus

$$\left| a_0 R_N + a_1 (x - c) R_{N-1} + \cdots + a_N (x - c)^N R_0 \right|$$
$$\leq \quad \epsilon A + M \cdot \sum_{j=N-N_0+1}^{N} |a_j| |x - c|^j .$$

Since the series defining A converges, we find, on letting $N \to \infty$, that

$$\limsup_{N \to \infty} \left| a_0 R_N + a_1 (x - c) R_{N-1} + \cdots + a_N (x - c)^N R_0 \right| \leq \epsilon \cdot A .$$

Since $\epsilon > 0$ was arbitrary, we may conclude that

$$\lim_{N \to \infty} \left| a_0 R_N + a_1 (x - c) R_{N-1} + \cdots + a_N (x - c)^N R_0 \right| = 0 . \qquad \square$$

POINT OF CONFUSION 13.4 Observe that the form of the product of two power series provides some motivation for the form that the product of numerical series took in Theorem 7.50.

POINT OF CONFUSION 13.5 If we are going to manipulate or combine two power series, then we must assume that they are both defined for the same values of x. We may assume that they have the same interval of convergence, or we may assume that they are expanded about the same point c and work on the *intersection* of their intervals of convergence.

A Look Back

1. What is the radius of convergence of a power series?

2. When does a power series converge at the endpoints of the interval of convergence?

3. How do we add power series?

4. What other arithmetic operations can we perform on power series?

Exercises

1. Prove that the composition of two real analytic functions, when the composition makes sense, is also real analytic.

2. Prove that
$$\sin^2 x + \cos^2 x = 1$$
directly from the power series expansions.

3. Let $f(x) = \sum_{j=0}^{\infty} a_j x^j$ be a power series convergent on the interval $(-r, r)$ and let Z denote those points in the interval where f vanishes. Prove that if Z has an accumulation point in the interval then $f \equiv 0$. (**Hint:** If a is the accumulation point, expand f in a power series about a. What is the first nonvanishing term in that expansion?)

4. Prove that a function f on an interval I is real analytic if and only if f satisfies an estimate of the form
$$|f^{(x)}(k)| \leq C \cdot \frac{k!}{R^k}$$
for some $R > 0$ and x in a compact set K.

5. Show that the inverse of a (suitable) real analytic function is real analytic.

6. Show that the solution of the differential equation $y' + y = x$ will be real analytic.

7. Prove that $\cos 2x = \cos^2 x - \sin^2 x$ directly from the power series expansions.

8. Prove that $\sin 2x = 2 \sin x \cos x$ directly from the power series expansions.

* **9.** Prove the assertion from the text that, if a power series converges at an endpoint of the interval of convergence, then the convergence is uniform up to that endpoint.

*   **10.** Verify that the function

$$f(x) = \begin{cases} 0 & \text{if} \quad x = 0 \\ e^{-1/x^2} & \text{if} \quad x \neq 0 \end{cases}$$

is infinitely differentiable on all of $\mathbb{R}$ and that $f^{(k)}(0) = 0$ for every k. However, f is certainly not real analytic.

13.2   More on Power Series: Convergence Issues

Preliminary Remarks

In this section we learn Hadamard's elegant formula for the radius of convergence of a power series. And we learn more about the Taylor expansion. As previously noted, any smooth function has a (formal) Taylor expansion. The issue is whether it converges, and whether it converges back to the original function.

We now introduce the *Hadamard formula* for the radius of convergence of a power series.

Lemma 13.6 (Hadamard) *For the power series*

$$\sum_{j=0}^{\infty} a_j (x - c)^j \,,$$

define A and ρ by

$$A = \limsup_{n \to \infty} |a_n|^{1/n} \,,$$

$$\rho = \begin{cases} 0 & \text{if } A = \infty, \\ 1/A & \text{if } 0 < A < \infty, \\ \infty & \text{if } A = 0 \,, \end{cases}$$

then ρ is the radius of convergence of the power series about c.

Proof: Observing that

$$\limsup_{n \to \infty} |a_n (x - c)^n|^{1/n} = A|x - c| \,,$$

we see that the lemma is an immediate consequence of the Root Test.   □

Corollary 13.7 *The power series*

$$\sum_{j=0}^{\infty} a_j (x - c)^j$$

has radius of convergence ρ if and only if, when $0 < R < \rho$, there exists a constant $0 < C = C_R$ such that

$$|a_n| \le \frac{C}{R^n}.$$

POINT OF CONFUSION 13.8 The single most important attribute of a power series is its radius of convergence. It turns out that the radius of convergence is best understood in the context of the complex numbers. We cannot say much about that point in the present book, but some exercises will touch on the issue.

From the power series

$$\sum_{j=0}^{\infty} a_j (x - c)^j$$

it is natural to create the *derived series*

$$\sum_{j=1}^{\infty} j a_j (x - c)^{j-1}$$

using term-by-term differentiation.

Proposition 13.9 *The radius of convergence of the derived series is the same as the radius of convergence of the original power series.*

Proof: We observe that

$$\limsup_{n \to \infty} |n a_n|^{1/n} = \lim_{n \to \infty} n^{-1/n} \limsup_{n \to \infty} |n a_n|^{1/n}$$

$$= \limsup_{n \to \infty} |a_n|^{1/n}.$$

So the result follows from the Hadamard formula. $\square$

Proposition 13.10 *Let f be a real analytic function defined on an open interval I. Then f is continuous and has continuous, real analytic derivatives of all orders. In fact the derivatives of f are obtained by differentiating its series representation term by term.*

Proof: Since, for each $c \in I$, the function f may be represented by a convergent power series with positive radius of convergence, we see that, in a sufficiently small open interval about each $c \in I$, the function f is the uniform limit of a sequence of continuous functions: the partial sums of the power series representing f. It follows that f is continuous at c. Since the radius of convergence of the derived series is the same as that of the original series, it also follows that the derivatives of the partial sums converge uniformly on an open interval about c to a continuous function. It then follows from Theorem 12.17 that f is differentiable and its derivative is the function defined by the derived series. By induction, f has continuous derivatives of all orders at c.   □

We can now show that a real analytic function has a unique power series representation at any point.

Corollary 13.11 *If the function f is represented by a convergent power series on an interval of positive radius centered at c,*

$$f(x) = \sum_{j=0}^{\infty} a_j(x - c)^j \,,$$

then the coefficients of the power series are related to the derivatives of the function by

$$a_n = \frac{f^{(n)}(c)}{n!} \,.$$

Proof: This follows readily by differentiating both sides of the above equation n times, as we may by the proposition, and evaluating at $x = c$.   □

Finally, we note that integration of power series is as well-behaved as differentiation.

Proposition 13.12 *The power series*

$$\sum_{j=0}^{\infty} a_j(x - c)^j$$

and the series

$$\sum_{j=0}^{\infty} \frac{a_j}{j+1}(x - c)^{j+1}$$

obtained from term-by-term integration have the same radius of convergence, and the function F defined by

$$F(x) = \sum_{j=0}^{\infty} \frac{a_j}{j+1}(x - c)^{j+1}$$

on the common interval of convergence satisfies

$$F'(x) = \sum_{j=0}^{\infty} a_j(x - c)^j = f(x) \,.$$

Proof: The proof is left to the exercises.   □

We conclude this section with a consideration of Taylor series:

Theorem 13.13 (Taylor's Expansion) *For k a nonnegative integer, let f be a $k+1$ times continuously differentiable function on an open interval $I = (a - \epsilon, a + \epsilon)$. Then, for $x \in I$,*

$$f(x) = \sum_{j=0}^{k} f^{(j)}(a)\frac{(x-a)^j}{j!} + R_{k,a}(x),$$

where

$$R_{k,a}(x) = \int_{a}^{x} f^{(k+1)}(t)\frac{(x-t)^k}{k!}\,dt.$$

Proof: We apply integration by parts to the Fundamental Theorem of Calculus to obtain

$$
\begin{aligned}
f(x) &= f(a) + \int_{a}^{x} f'(t)\,dt \\
&= f(a) + \left(f'(t)\frac{(t-x)}{1!} \right)\Bigg|_{a}^{x} - \int_{a}^{x} f''(t)\frac{(t-x)}{1!}\,dt \\
&= f(a) + f'(a)\frac{(x-a)}{1!} + \int_{a}^{x} f''(t)\frac{x-t}{1!}\,dt.
\end{aligned}
$$

Notice that, when we performed the integration by parts, we used $t - x$ as an antiderivative for dt. This is of course legitimate, as a glance at the integration by parts theorem reveals. We have proved the theorem for the case $k = 1$. The result for higher values of k is obtained inductively by repeated applications of integration by parts. $\qquad\square$

Taylor's theorem allows us to associate with any infinitely differentiable function a formal expansion of the form

$$\sum_{j=0}^{\infty} a_j(x-a)^j.$$

However, there is no guarantee that this series will converge. Even if it does converge, it may not converge back to $f(x)$. An important example to keep in mind is the function

$$h(x) = \begin{cases} 0 & \text{if } x = 0 \\ e^{-1/x^2} & \text{if } x \neq 0. \end{cases}$$

This function is infinitely differentiable at every point of the real line (including the point 0—use l'Hôpital's Rule). However, all of its derivatives at $x = 0$ are equal to zero (this matter will be treated in the exercises). Therefore the formal Taylor series expansion of h about $a = 0$ is

$$\sum_{j=0}^{\infty} 0 \cdot (x-0)^j = 0.$$

We see that the formal Taylor series expansion for h converges to the zero function at every x, but not to the original function h itself.

In fact the theorem tells us that the Taylor expansion of a function f converges to f at a point x if and only if $R_{k,a}(x) \to 0$. In the exercises we shall explore the following more quantitative assertion:

An infinitely differentiable function f on an interval I has Taylor series expansion about $a \in I$ that converges back to f on a neighborhood J of a if and only if there are positive constants C, R such that, for every $x \in J$ and every k, it holds that

$$\left| f^{(k)}(x) \right| \le C \cdot \frac{k!}{R^k}.$$

The function h considered above should not be thought of as an isolated exception. For instance, we know from calculus that the function $f(x) = \sin x$ has Taylor expansion that converges to f at every x. But then, for ϵ small, the function $g_\epsilon(x) = f(x) + \epsilon \cdot h(x)$ has Taylor series that does *not* converge back to $g_\epsilon(x)$ for $x \ne 0$. Similar examples may be generated by using other real analytic functions in place of sine.

A Look Back

1. What estimate on the derivatives of a smooth function will imply that it is real analytic?

2. How does one prove Taylor's expansion?

3. How are the radius of convergence of a power series and the radius of convergence of its derived series related?

4. How are the radius of convergence of a power series and the radius of convergence of its integrated series related?

Exercises

1. Let f be an infinitely differentiable function on an interval I. If $a \in I$ and there are positive constants C, R such that, for every x in a neighborhood of a and every k, it holds that
   $$\left| f^{(k)}(x) \right| \le C \cdot \frac{k!}{R^k},$$
   then prove that the Taylor series of f about a converges to $f(x)$. (**Hint:** Estimate the error term.) What is the radius of convergence?

2. Let f be an infinitely differentiable function on an open interval I centered at a. Assume that the Taylor expansion of f about a converges to f at every point of I. Prove that there are constants C, R and a (possibly smaller) interval J centered at a such that, for each $x \in J$, it holds that
   $$\left| f^{(k)}(x) \right| \le C \cdot \frac{k!}{R^k}.$$

3. Give examples of power series, centered at 0, on the interval $(-1,1)$, which **(a)** converge only on $(-1,1)$, **(b)** converge only on $[-1,1)$, **(c)** converge only on $(-1,1]$, (d) converge only on $[-1,1]$.

4. Prove Corollary 13.11.

5. Prove Proposition 13.12.

6. The real analytic function $1/(1+x^2)$ is well defined on the entire real line. Yet its power series about 0 only converges on an interval of radius 1. Explain why. [**Hint:** Think in terms of the complex numbers.]

7. How does Exercise 6 differ for the function $1/(1-x^2)$?

*   **8.** Show that the function

$$h(x) = \begin{cases} 0 & \text{if} \quad x = 0 \\ e^{-1/x^2} & \text{if} \quad x \neq 0 \end{cases}$$

is infinitely differentiable (you will need to take special care at the origin, using l'Hôpital's Rule). What is its Taylor expansion at 0?

*   **9.** Assume that a power series converges at one of the endpoints of its interval of convergence. Use summation by parts to prove that the function defined by the power series is continuous on the half-closed interval including that endpoint.

*   **10.** The function defined by a power series may extend continuously to an endpoint of the interval of convergence without the series converging at that endpoint. Give an example.

*   **11.** Prove that, if a function on an interval I has derivatives of all orders which are positive at every point of I, then f is real analytic on I.

*   **12.** If $\{a_j\}_{j=0}^{\infty}$ is any sequence of real numbers then there is an infinitely differentiable function whose power series expansion about 0 is $\sum_j a_j x^j$. Explain why. [This is a theorem of E. Borel.] Of course the power series that you obtain here may very well not converge!

13.3   The Exponential and Trigonometric Functions

Preliminary Remarks

In the present section we give rigorous definitions of the exponential and trigonometric functions, and derive some of their most basic properties. We also take a look at the important constant π.

We begin by defining the exponential function:

Definition 13.14 The power series

$$\sum_{j=0}^{\infty} \frac{x^j}{j!}$$

converges, by the Ratio Test, for every real value of x. The function defined thereby is called the *exponential function* and is written $\exp(x)$.

POINT OF CONFUSION 13.15 The only other straightforward way to define the exponential function is that it is the solution of the differential equation $y' = y$. You can, if you wish, substitute the power series in Definition 13.14 into this differential equation to see that it is satisfied.

Proposition 13.16 *The function* $\exp(x)$ *satisfies*

$$\exp(a + b) = \exp(a) \cdot \exp(b)$$

for any complex numbers a and b.

Proof: We write the right-hand side as

$$\left(\sum_{j=0}^{\infty} \frac{a^j}{j!} \right) \cdot \left(\sum_{j=0}^{\infty} \frac{b^j}{j!} \right) .$$

Now convergent power series may be multiplied term by term. We find that the last line equals

$$\sum_{j=0}^{\infty} \left(\sum_{\ell=0}^{j} \frac{a^{j-\ell}}{(j-\ell)!} \cdot \frac{b^\ell}{\ell!} \right) . \tag{13.16.1}$$

However, the inner sum on the right side of this equation may be written as

$$\frac{1}{j!} \sum_{\ell=0}^{j} \frac{j!}{\ell!(j-\ell)!} a^{j-\ell} b^\ell = \frac{1}{j!}(a+b)^j .$$

Here we have used the Binomial Formula. It follows that line (13.16.1) equals $\exp(a+b)$. □

We set $e = \exp(1)$. This is consistent with our earlier treatment of the number e in Section 7.4. The proposition tells us that, for any positive integer k, we have

$$e^k = e \cdot e \cdots e = \exp(1) \cdot \exp(1) \cdots \exp(1) = \exp(k) .$$

If m is another positive integer then

$$(\exp(k/m))^m = \exp(k) = e^k ,$$

whence

$$\exp(k/m) = e^{k/m} .$$

We may extend this formula to *negative* rational exponents by using the fact that $\exp(a) \cdot \exp(-a) = 1$. Thus, for any rational number q,

$$\exp(q) = e^q .$$

Now note that the function exp is increasing and continuous. It follows (this fact is treated in the exercises) that if we set, for any $r \in \mathbb{R}$,

$$e^r = \sup\{e^q : q \in \mathbb{Q} \text{ and } q < r\}$$

(this is a *definition* of the expression e^r), then $e^x = \exp(x)$ for every real x. [You may find it useful to review the discussion of exponentiation in Sections 6.4, 7.4; the presentation here parallels those treatments.] We will adhere to custom and write e^x instead of $\exp(x)$.

Proposition 13.17 *The exponential function e^x, for $x \in \mathbb{R}$, satisfies*

(a) $e^x > 0$ *for all x;*

(b) $e^0 = 1$;

(c) $(e^x)' = e^x$;

(d) e^x *is strictly increasing;*

(e) *the graph of e^x is asymptotic to the negative x-axis;*

(f) *for each integer $N > 0$ there is a number c_N such that $e^x > c_N \cdot x^N$ when $x > 0$.*

Proof: The first three statements are obvious from the power series expansion for the exponential function.

If $s < t$ then the Mean Value Theorem tells us that there is a number ξ between s and t such that

$$e^t - e^s = (t - s) \cdot e^\xi > 0;$$

hence the exponential function is strictly increasing.

By inspecting the power series we see that $e^x > 1 + x$ hence e^x increases to $+\infty$. Since $e^x \cdot e^{-x} = 1$ we conclude that e^{-x} tends to 0 as $x \to +\infty$. Thus the graph of the exponential function is asymptotic to the negative x-axis.

Finally, by inspecting the power series for e^x, we see that the last assertion is true with $c_N = 1/N!$. $\qquad\square$

Now we turn to the trigonometric functions. The definition of the trigonometric functions that is found in calculus texts is unsatisfactory because it relies too heavily on a picture and because the continual need to subtract off superfluous multiples of 2π is clumsy. We have nevertheless used the trigonometric functions in earlier chapters to illustrate various concepts. It is time now to give a rigorous definition of the trigonometric functions that is independent of these earlier considerations.

Definition 13.18 The power series

$$\sum_{j=0}^{\infty} (-1)^j \frac{x^{2j+1}}{(2j+1)!}$$

converges at every point of the real line (by the Ratio Test). The function that it defines is called the *sine* function and is usually written $\sin x$.

The power series

$$\sum_{j=0}^{\infty} (-1)^j \frac{x^{2j}}{(2j)!}$$

converges at every point of the real line (by the Ratio Test). The function that it defines is called the *cosine* function and is usually written $\cos x$.

You may recall that the power series that we use to define the sine and cosine functions are precisely the Taylor series expansions for the functions sine and cosine that were derived in your calculus text. But now we *begin* with the power series and must derive the properties of sine and cosine that we need *from these series*.

In fact the most convenient way to achieve this goal is to proceed by way of the exponential function. [The point here is mainly one of convenience. It can be verified by direct manipulation of the power series that $\sin^2 x + \cos^2 x = 1$ and so forth, but the algebra is extremely unpleasant.] The formula in the next proposition is usually credited to Euler.

POINT OF CONFUSION 13.19 For Euler's result, we will use the complex numbers, in particular the special number i. Of course you know that $i^2 = -1$. More generally, you know that

$$(a + ib) \cdot (c + id) = (ac - bd) + i(bc + ad).$$

Refer to Section 5.2.

Proposition 13.20 *The exponential function and the functions sine and cosine are related by the formula (for x and y real and $i^2 = -1$)*

$$\exp(x + iy) = e^x \cdot (\cos y + i \sin y).$$

Proof: We shall verify the case $x = 0$ and leave the general case for the reader. Thus we are to prove that

$$e^{iy} = \cos y + i \sin y. \tag{13.20.1}$$

Writing out the power series for the exponential, we find that the left-hand side of (13.20.1) is

$$\sum_{j=0}^{\infty} \frac{(iy)^j}{j!}$$

and this equals

$$\left[1 - \frac{y^2}{2!} + \frac{y^4}{4!} - + \cdots\right] + i\left[\frac{y}{1!} - \frac{y^3}{3!} + \frac{y^5}{5!} - + \cdots\right].$$

Of course the two series on the right are the familiar power series for cosine and sine. Thus

$$e^{iy} = \cos y + i \sin y,$$

as desired. $\square$

In what follows, we think of the formula (13.20.1) as *defining* what we mean by e^{iy}. As a result,

$$e^{x+iy} = e^x \cdot e^{iy} = e^x \cdot (\cos y + i \sin y).$$

Notice that $e^{-iy} = \cos(-y) + i\sin(-y) = \cos y - i\sin y$ (we know that the sine function is odd and the cosine function even from their power series expansions). Then formula (13.20.1) tells us that

$$\cos y = \frac{e^{iy} + e^{-iy}}{2}$$

and

$$\sin y = \frac{e^{iy} - e^{-iy}}{2i}.$$

Now we may prove:

Proposition 13.21 *For every real x it holds that*

$$\sin^2 x + \cos^2 x = 1.$$

Proof: Simply substitute into the left side the formulas for the sine and cosine functions which were displayed before the proposition and simplify. $\square$

We list several other properties of the sine and cosine functions that may be proved by similar methods. The proofs are requested of you in the exercises.

Proposition 13.22 *The functions sine and cosine have the following properties:*

(a) $\sin(s + t) = \sin s \cos t + \cos s \sin t$;

(b) $\cos(s + t) = \cos s \cos t - \sin s \sin t$;

(c) $\cos(2s) = \cos^2 s - \sin^2 s$;

(d) $\sin(2s) = 2 \sin s \cos s$;

(e) $\sin(-s) = -\sin s$;

(f) $\cos(-s) = \cos s$;

(g) $\sin'(s) = \cos s$;

(h) $\cos'(s) = -\sin s$.

One important task to be performed in a course on the foundations of analysis is to define the number π and establish its basic properties. In a course on Euclidean geometry, the constant π is defined to be the ratio of the circumference of a circle to its diameter. Such a definition is not useful for our purposes (however, it *is* consistent with the definition about to be given here).

Observe that $\cos 0$ is the real part of e^{i0} which is 1. Thus if we set

$$\alpha = \inf\{x > 0 : \cos x = 0\}$$

then $\alpha > 0$ and, by the continuity of the cosine function, $\cos \alpha = 0$. We define $\pi = 2\alpha$.

Applying Proposition 13.21 to the number α yields that $\sin \alpha = \pm 1$. Since α is the *first* zero of cosine on the right half line, the cosine function must be positive on $(0, \alpha)$. But cosine is the derivative of sine. Thus the sine function is *increasing* on $(0, \alpha)$. Since $\sin 0$ is the imaginary part of e^{i0} which is 0, we conclude that $\sin \alpha > 0$ hence that $\sin \alpha = +1$.

Now we may apply parts **(c)** and **(d)** of Proposition 13.22 with $s = \alpha$ to conclude that $\sin \pi = 0$ and $\cos \pi = -1$. A similar calculation with $s = \pi$ shows that $\sin 2\pi = 0$ and $\cos 2\pi = 1$. Next we may use parts **(a)** and **(b)** of Proposition 13.22 to calculate that $\sin(x + 2\pi) = \sin x$ and $\cos(x + 2\pi) = \cos x$ for all x. In other words, the sine and cosine functions are 2π−periodic.

The business of calculating a decimal expansion for π would take us far afield. One approach would be to utilize the already-noted fact that the sine function is strictly increasing on the interval $[0, \pi/2]$ hence its inverse function

$$\mathrm{Sin}^{-1} : [0, 1] \to [0, \pi/2]$$

is well defined. Then one can determine (see Chapter 10) that

$$\left(\mathrm{Sin}^{-1}\right)'(x) = \frac{1}{\sqrt{1 - x^2}}\,.$$

By the Fundamental Theorem of Calculus,

$$\frac{\pi}{2} = \mathrm{Sin}^{-1}(1) = \int_0^1 \frac{1}{\sqrt{1 - x^2}}\, dx\,.$$

By approximating the integral by its Riemann sums, one obtains an approximation to $\pi/2$ and hence to π itself. This approach will be explored in more detail in the exercises.

Let us for now observe that

$$\begin{aligned}
\cos 2 &= 1 - \frac{2^2}{2!} + \frac{2^4}{4!} - \frac{2^6}{6!} + - \cdots \\
&= 1 - 2 + \frac{16}{24} - \frac{64}{720} + \cdots .
\end{aligned}$$

Since the series defining $\cos 2$ is an alternating series with terms that strictly decrease to zero in magnitude, we may conclude (following reasoning from Chapter 7) that the last line is less than the sum of the first three terms:

$$\cos 2 < -1 + \frac{2}{3} < 0 .$$

It follows that $\alpha = \pi/2 < 2$ hence $\pi < 4$. A similar calculation of $\cos(3/2)$ would allow us to conclude that $\pi > 3$.

A Look Back

1. How do we define the exponential function?
2. How do we define $\sin x$ and $\cos x$?
3. How is the exponential function related to sine and cosine?
4. How do we know that the exponential function is not a polynomial?

Exercises

1. Provide the details of the assertion preceding Proposition 13.17 to the effect that if we define, for any real r,

$$e^r = \sup\{e^q : q \in \mathbb{Q} \text{ and } q < r\},$$

   then $e^x = \exp(x)$ for every real x.

2. Prove the equality $\left(\mathrm{Sin}^{-1}\right)'(x) = 1/\sqrt{1 - x^2}$.

3. Find a formula for $\cos 4x$ directly from the power series expansions.

4. Find a formula for $\sin(x + \pi/2)$ directly from the power series expansions.

5. Use one of the methods described at the end of this section to calculate π to two decimal places.

6. Prove Proposition 13.22.

8. Prove that the trigonometric polynomials, that is to say, the functions of the form

$$p(x) = \sum_{j=-N}^{N} a_j \cos jx + b_j \sin jx ,$$

   are dense in the continuous functions on $[0, 2\pi]$ in the uniform topology.

9. Prove the general case of Proposition 13.20.

10. Prove that the logarithm function is characterized by the property that $\log(a \cdot b) = \log a + \log b$.

11. Find a formula for $\tan^4 x$ in terms of $\sin 2x$, $\sin 4x$, $\cos 2x$, and $\cos 4x$.

12. Refer to Exercise 1 for notation and terminology. Prove that the exponential function defined there is one-to-one and onto from the real line to the half-line.

13. Define $\log x = \int_1^x 1/t \, dt$. Using this definition of logarithm, derive these basic properties of the log function: **(a)** $\log(a \cdot b) = \log a + \log b$; **(b)** $\log a^b = b \cdot \log a$.

13.4   Logarithms and Powers of Real Numbers

Preliminary Remarks

The logarithm function is of interest because it is the inverse of the exponential function, but also because it is used to define entropy in physics. The logarithm is useful in understanding exponential functions in general. We study these matters in the present section.

Since the exponential function $\exp(x) = e^x$ is positive and strictly increasing, it is a one-to-one function from $\mathbb{R}$ to $(0, \infty)$. Thus it has a well-defined inverse function that we call the *natural logarithm*. We write this function as $\ln x$.

Proposition 13.23 *The natural logarithm function has the following properties:*

(a) $(\ln x)' = 1/x$;

(b) $\ln x$ *is strictly increasing*;

(c) $\ln(1) = 0$;

(d) $\ln e = 1$;

(e) *the graph of the natural logarithm function is asymptotic to the negative y axis;*

(f) $\ln(s \cdot t) = \ln s + \ln t$;

(g) $\ln(s/t) = \ln s - \ln t$.

Proof: These follow immediately from corresponding properties of the exponential function. For example, to verify part **(f)**, set $s = e^\sigma$ and $t = e^\tau$. Then

$$
\begin{aligned}
\ln(s \cdot t) &= \ln(e^\sigma \cdot e^\tau) \\
&= \ln(e^{\sigma + \tau}) \\
&= \sigma + \tau \\
&= \ln s + \ln t.
\end{aligned}
$$

The other parts of the proposition are proved similarly. □

Proposition 13.24 If a and b are positive real numbers then

$$a^b = e^{b \cdot \ln a} \, .$$

Proof: When b is an integer then the formula may be verified directly from the definition of logarithm. For $b = m/n$ a rational number the formula follows by our usual trick of passing to nth roots. For arbitrary b we use a limiting argument as in our discussions of exponentials in Sections 6.4 and 13.3.

Remark 13.25 We have discussed several different approaches to the exponentiation process. We proved the existence of nth roots, $n \in \mathbb{N}$, as an illustration of the completeness of the real numbers (by taking the supremum of a certain set). We treated rational exponents by composing the usual arithmetic process of taking mth powers with the process of taking nth roots. Then, in Sections 6.4 and 13.3, we passed to arbitrary powers by way of a limiting process.

Proposition 13.24 gives us a unified and direct way to treat all exponentials at once. This unified approach will prove (see the next proposition) to be particularly advantageous when we wish to perform calculus operations on exponential functions.

Proposition 13.26 Fix $a > 0$. The function $f(x) = a^x$ has the following properties:

(a) $(a^x)' = a^x \cdot \ln a$;

(b) $f(0) = 1$;

(c) if $0 < a < 1$ then f is decreasing and the graph of f is asymptotic to the positive x-axis;

(d) if $1 < a$ then f is increasing and the graph of f is asymptotic to the negative x-axis.

Proof: These properties follow immediately from corresponding properties of the function exp. □

The logarithm function arises, among other places, in the context of probability and in the study of entropy. The reason is that the logarithm function is uniquely determined by the way that it interacts with the operation of multiplication:

Theorem 13.27 Let $\phi(x)$ be a continuously differentiable function with domain the positive reals and which satisfies the identity

$$\phi(s \cdot t) = \phi(s) + \phi(t) \tag{13.27.1}$$

for all positive s and t. Then there is a constant $C > 0$ such that

$$\phi(x) = C \cdot \ln x$$

for all x.

Proof: Differentiate the equation (13.27.1) with respect to s to obtain

$$t \cdot \phi'(s \cdot t) = \phi'(s).$$

Now fix s and set $t = 1/s$ to conclude that

$$\phi'(1) \cdot \frac{1}{s} = \phi'(s).$$

We take the constant C to be $\phi'(1)$ and apply Proposition 13.23(a) to conclude that $\phi(s) = C \cdot \ln s + D$ for some constant D. But ϕ cannot satisfy (13.27.1) unless $D = 0$, so the theorem is proved. $\qquad\square$

Observe that the *natural logarithm function* is then the unique continuously differentiable function that satisfies the condition (13.27.1) and whose derivative at 1 equals 1. That is the reason that the natural logarithm function (rather than the common logarithm, or logarithm to the base ten) is singled out as the focus of our considerations in this section.

A Look Back

1. What is the definition of the natural logarithm function?

2. What is the definition of the common logarithm function?

3. How can we use the logarithm to define a^b for arbitrary positive a and any real b?

4. What is the derivative of the logarithm function?

Exercises

1. Prove Proposition 13.24 by following the suggested line of reasoning.

2. Prove Proposition 13.23, except for part (f).

3. Prove that condition (13.27.1) implies that $\phi(1) = 0$. Assume that ϕ is differentiable at $x = 1$ but make no other hypothesis about the smoothness of ϕ. Prove that condition (13.27.1) then implies that ϕ is differentiable at every $x > 0$.

4. Provide the details of the proof of Proposition 13.26.

5. Calculate

$$\lim_{j \to \infty} \frac{j^{j/2}}{j!}.$$

6. Give three distinct reasons why the natural logarithm function is not a polynomial.

7. At infinity, any nontrivial polynomial function dominates the natural logarithm function. Explain what this means, and prove it.

8. At infinity, any exponential function with base greater than 1 dominates any polynomial. Explain this statement and prove it.

*   **9.** Show that the hypothesis of Theorem 13.27 may be replaced with $f \in \mathrm{Lip}_\alpha([0, 2\pi])$, some $\alpha > 0$.

*   **10.** The *Lambert W function* is defined implicitly by the equation

$$z = W(z) \cdot e^{W(z)}.$$

It is a fact that any elementary transcendental function (sine, cosine, logarithm, exponential) may be expressed (with an elementary formula) in terms of the W function. Prove that this is so for the exponential function and the sine function.

*   **11.** Prove Euler's formula relating the exponential to sine and cosine *not* by using power series, but rather by using differential equations.

*   **12.** Refer to Exercise 10. Show that the Lambert W function is real analytic on its domain.

Appendix: Elementary Number Systems

Section A1.1. The Natural Numbers

Mathematics deals with a variety of number systems. The simplest number system is $\mathbb{N}$, the *natural numbers*. As we have already noted, this is just the set of positive integers $\{1, 2, 3, \ldots\}$. In a rigorous course of logic, the set $\mathbb{N}$ is constructed from the axioms of set theory. However, in this book we shall assume that you are familiar with the positive integers and their elementary properties.

The principal properties of $\mathbb{N}$ are as follows:

1. 1 is a natural number.

2. If x is a natural number then there is another natural number $\widehat{x}$ which is called the *successor* of x.

3. $1 \neq \widehat{x}$ for every natural number x.

4. If $\widehat{x} = \widehat{y}$ then $x = y$.

5. *(Principle of Induction)* If $\mathcal{P}$ is a property and if

   (a) 1 has the property $\mathcal{P}$;

   (b) whenever a natural number x has the property $\mathcal{P}$ it follows that $\widehat{x}$ also has the property $\mathcal{P}$;

   then all natural numbers have the property $\mathcal{P}$.

These rules, or *axioms*, are known as the Peano Axioms for the natural numbers (named after Giuseppe Peano (1858–1932) who developed them). We take it for granted that the usual set of positive integers satisfies these rules. Certainly 1 is in that set. Each positive integer has a "successor"—after 1 comes 2 and after 2 comes 3 and so forth. The number 1 is not the successor of any other positive integer. Two positive integers with the same successor must be the same. The last axiom is more challenging but makes good sense: if some property $\mathcal{P}(n)$ holds for $n = 1$ and if whenever it holds for n then it also holds for $n + 1$, then we may conclude that $\mathcal{P}$ holds for all positive integers.

We will spend the remainder of this section exploring Axiom 5, the Principle of Induction.

Example A1.1

Let us prove that, for each positive integer n, it holds that

$$1 + 2 + \cdots + n = \frac{n \cdot (n+1)}{2}.$$

We denote this equation by $\mathcal{P}(n)$, and follow the scheme of the Principle of Induction.

First, $\mathcal{P}(1)$ is true since then both the left and the right side of the equation equal 1. Now assume that $\mathcal{P}(n)$ is true for some natural number n. Our job is to show that it follows that $\mathcal{P}(n+1)$ is true.

Since $\mathcal{P}(n)$ is true, we know that

$$1 + 2 + \cdots + n = \frac{n \cdot (n+1)}{2}.$$

Let us add the quantity $n+1$ to both sides. Thus

$$1 + 2 + \cdots + n + (n+1) = \frac{n \cdot (n+1)}{2} + (n+1).$$

The right side of this new equality simplifies and we obtain

$$1 + 2 + \cdots + (n+1) = \frac{(n+1) \cdot ((n+1)+1)}{2}.$$

But this is just $\mathcal{P}(n+1) = \mathcal{P}(\widehat{n})$. *We have assumed $\mathcal{P}(n)$ and have proved $\mathcal{P}(\widehat{n})$*, just as the Principle of Induction requires.

Thus we may conclude that property $\mathcal{P}$ holds for all positive integers, as desired.   ∎

The formula that we derived in Example A1.1 was probably known to the ancient Greeks. However, a celebrated anecdote credits Karl Friedrich Gauss (1777–1855) with discovering the formula when he was nine years old. Gauss went on to become (along with Isaac Newton and Archimedes) one of the three greatest mathematicians of all time.

The formula from Example A1.1 gives a neat way to add up the integers from 1 to n for any n, without doing any work. Any time that we discover a new mathematical fact, there are generally several others hidden within it. The next example illustrates this point.

Example A1.2

The sum of the first m positive even integers is $m \cdot (m+1)$. To see this, note that the sum in question is

$$2 + 4 + 6 + \cdots + 2m = 2(1 + 2 + 3 + \cdots + m).$$

But, by the first example, the sum in parentheses on the right is equal to $m \cdot (m+1)/2$. It follows that

$$2 + 4 + 6 + \cdots + 2m = 2 \cdot \frac{m \cdot (m+1)}{2} = m \cdot (m+1). \qquad \square$$

Remark 13.28 *The second example could also be performed by induction (without using the result of the first example).*

Example A1.3

Now we will use induction incorrectly to prove a statement that is completely preposterous:

All horses are the same color.

There are finitely many horses in existence, so it is convenient for us to prove the slightly more technical statement

Any collection of k horses consists of horses which are all the same color.

Our statement $\mathcal{P}(k)$ is this last displayed statement.

Now $\mathcal{P}(1)$ is true: *one horse is the same color.* (Note: this is not a joke, and the error has not occurred yet.)

Suppose next that $\mathcal{P}(k)$ is true: we assume that any collection of k horses has the same color. Now consider a collection of $\widehat{k} = k + 1$ horses. Remove one horse from that collection. By our hypothesis, the remaining k horses have the same color.

Now replace the horse that we removed and remove a different horse. Again, the remaining k horses have the same color.

We keep repeating this process: remove each of the $k+1$ horses one by one and conclude that the remaining k horses have the same color. Therefore every horse in the collection is the same color as every other. So all $k+1$ horses have the same color. The statement $\mathcal{P}(k+1)$ is thus proved (assuming the truth of $\mathcal{P}(k)$) and the induction is complete.

Where is our error? It is nothing deep—just an oversight. The argument we have given is wrong when $\widehat{k} = k + 1 = 2$. For remove one horse from a set of two and the remaining (*one*) horse is the same color. Now replace the removed horse and remove the other horse. The remaining (*one*) horse is the same color. *So what?* We cannot conclude that the two horses are colored the same. Thus the induction breaks down at the outset; the reasoning is incorrect. ∎

Proposition A1.4

Let a and b be real numbers and n a natural number. Then

$$
\begin{aligned}
(a+b)^n \;=\; & a^n + \frac{n}{1}a^{n-1}b + \frac{n(n-1)}{2\cdot 1}a^{n-2}b^2 \\
& + \frac{(n(n-1)(n-2)}{3\cdot 2\cdot 1}a^{n-3}b^3 \\
& + \cdots + \frac{n(n-1)\cdots 2}{(n-1)(n-2)\cdots 2\cdot 1}ab^{n-1} + b^n.
\end{aligned}
$$

Proof: The case $n = 1$ being obvious, proceed by induction.   □

Example A1.5

The expression
$$
\frac{n(n-1)\cdots(n-k+1)}{k(k-1)\cdots 1}
$$
is often called the *kth binomial coefficient* and is denoted by the symbol
$$
\binom{n}{k}.
$$
Using the notation $m! = m \cdot (m-1) \cdot (m-2)\cdots 2\cdot 1$, for m a natural number, we may write the kth binomial coefficient as
$$
\binom{n}{k} = \frac{n!}{(n-k)!\cdot k!}.
$$ ∎

Section A1.2. The Integers

Now we will apply the notion of an equivalence class to *construct* the integers (both positive and negative). There is an important point of knowledge to be noted here. For the sake of having a reasonable place to begin our work, we took the natural numbers $\mathbb{N} = \{1, 2, 3, \dots\}$ as given. Since the natural numbers have been used for thousands of years to keep track of objects for barter, this is a plausible thing to do. Even people who know no mathematics accept the positive integers. However, the number zero and the negative numbers are a different matter. It was not until the fifteenth century that the concepts of zero and negative numbers started to take hold—for they do not correspond to explicit collections of objects (five fingers or ten shoes) but rather to *concepts* (zero books is the lack of books; minus 4 pens means that we owe someone four pens). After some practice we get used to negative numbers, but explaining in words what they mean is always a bit clumsy.

It is much more satisfying, from the point of view of logic, to *construct* the integers (including the negative whole numbers and zero) from what we already have, that is, from the natural numbers. We proceed as follows. Let $A = \mathbb{N} \times \mathbb{N}$, the set of ordered pairs of natural numbers. We define a relation $\mathcal{R}$ on A and A as follows:

$$(a, b) \text{ is related to } (a', b') \text{ if } a + b' = a' + b.$$

See Section 4.1 for the concept of equivalence relation.

Theorem A1.6

The relation $\mathcal{R}$ is an equivalence relation.

Proof: That (a, b) is related to (a, b) follows from the trivial identity $a+b = a+b$. Hence $\mathcal{R}$ is reflexive. Second, if (a, b) is related to (a', b') then $a + b' = a' + b$ hence $a' + b = a + b'$ (just reverse the equality) hence (a', b') is related to (a, b). So $\mathcal{R}$ is symmetric.

Finally, if (a, b) is related to (a', b') and (a', b') is related to (a'', b'') then we have

$$a + b' = a' + b \qquad \text{and} \qquad a' + b'' = a'' + b'.$$

Adding these equations gives

$$(a + b') + (a' + b'') = (a' + b) + (a'' + b').$$

Cancelling a' and b' from each side finally yields

$$a + b'' = a'' + b.$$

Thus (a, b) is related to (a'', b''). Therefore $\mathcal{R}$ is transitive. We conclude that $\mathcal{R}$ is an equivalence relation. $\qquad\qquad\square$

Now our job is to understand the equivalence classes which are induced by $\mathcal{R}$. [We will ultimately call this number system the integers $\mathbb{Z}$.] Let $(a, b) \in A$ and let $[(a, b)]$ be the corresponding equivalence class. If $b > a$ then we will denote this equivalence class by the integer $b - a$. For instance, the equivalence class $[(2, 7)]$ will be denoted by 5. Notice that if $a < b$ and $(a', b') \in [(a, b)]$ then $a + b' = a' + b$ hence $b' - a' = b - a$. Therefore the integer symbol that we choose to represent our equivalence class is *independent of which element of the equivalence class is used to compute it*.

If $(a, b) \in A$ and $b = a$ then we let the symbol 0 denote the equivalence class $[(a, b)]$. Notice that if (a', b') is any other element of $[(a, b)]$ then it must be that $a + b' = a' + b$ hence $b' = a'$; therefore this definition is unambiguous.

If $(a, b) \in A$ and $a > b$ then we will denote the equivalence class $[(a, b)]$ by the symbol $-(a - b)$. For instance, we will denote the equivalence class $[(7, 5)]$ by the symbol -2. Once again, if $a > b$ and if (a', b') is related to (a, b) then the equation $a + b' = a' + b$ guarantees that our choice of symbol to represent $[(a, b)]$ is unambiguous.

Thus we have given our equivalence classes names, and these names *look just like* the names that we usually give to integers: there are positive integers, and negative ones, and zero. But we want to see that these objects *behave* like integers. (As you read on, use the intuitive, non-rigorous mnemonic that the equivalence class $[(a, b)]$ stands for the integer $b - a$.)

First, do these new objects that we have constructed *add* correctly? Well, let $X = [(a, b)]$ and $Y = [(c, d)]$ be two equivalence classes. *Define* their sum to be $X + Y = [(a + c, b + d)]$. We must check that this is unambiguous. If $(\widetilde{a}, \widetilde{b})$ is related to (a, b) and $(\widetilde{c}, \widetilde{d})$ is related to (c, d) then of course we know that

$$a + \widetilde{b} = \widetilde{a} + b$$

and

$$c + \widetilde{d} = \widetilde{c} + d.$$

Adding these two equations gives

$$(a + c) + (\widetilde{b} + \widetilde{d}) = (\widetilde{a} + \widetilde{c}) + (b + d)$$

hence $(a + c, b + d)$ is related to $(\widetilde{a} + \widetilde{c}, \widetilde{b} + \widetilde{d})$. Thus, adding two of our equivalence classes gives another equivalence class, as it should.

Example A1.7

To add 5 and 3 we first note that 5 is the equivalence class $[(2, 7)]$ and 3 is the equivalence class $[(2, 5)]$. We add them componentwise and find that the sum is $[(2 + 2, 7 + 5)] = [(4, 12)]$. Which equivalence class is this answer? Looking back at our prescription for giving names to the equivalence classes, we see that this is the equivalence class that we called $12 - 4$ or 8. So we have rediscovered the fact that $5 + 3 = 8$. Check for yourself that, if we were to choose a different representative for 5—say $(6, 11)$—and a different representative for 3—say $(24, 27)$—then the same answer would result.

Now let us add 4 and -9. The first of these is the equivalence class $[(3, 7)]$ and the second is the equivalence class $[(13, 4)]$. The sum is therefore $[(16, 11)]$, and this is the equivalence class that we call $-(16 - 11)$ or -5. That is the answer that we would expect when we add 4 to -9.

Next, we add -12 and -5. Previous experience causes us to expect the answer to be -17. Now -12 is the equivalence class $[(19, 7)]$ and -5 is the equivalence class $[(7,2)]$. The sum is $[(26,9)]$, which is the equivalence class that we call -17.

Finally, we can see in practice that our method of addition is unambiguous. Let us redo the second example, calculating $4 + (-9)$, using $[(6,10)]$ as the equivalence class represented by 4 and $[(15,6)]$ as the equivalence class represented by -9. Then the sum is $[(21,16)]$, and this is still the equivalence class -5, as it should be. ∎

The assertion that the result of calculating a sum—no matter which representatives we choose for the equivalence classes—will give only one answer is called the "fact that addition is *well defined.*" In order for our definitions to make sense, it is essential that we check this property of well-definedness.

Remark A1.8

What is the point of this section? Everyone knows about negative numbers, so why go through this abstract construction? The reason is that, until one sees this construction, negative numbers are just imaginary objects—placeholders if you will—which are a useful notation but which do not exist. Now they *do* exist. They are a collection of equivalence classes of pairs of natural numbers. This collection is equipped with certain arithmetic operations, such as addition, subtraction, and multiplication. We now discuss these last two.

If $x = [(a,b)]$ and $y = [(c,d)]$ are integers, we define their *difference* to be the equivalence class $[(a+d, b+c)]$; we denote this difference by $x - y$.

Remark A1.9

We calculate $8 - 14$. Now $8 = [(1,9)]$ and $14 = [(3,17)]$. Therefore

$$8 - 14 = [(1+17, 9+3)] = [(18,12)] = -6,$$

as expected.

As a second example, we compute $(-4) - (-8)$. Now

$$-4 - (-8) = [(6,2)] - [(13,5)] = [(6+5, 2+13)] = [(11,15)] = 4.$$

Remark A1.10

When we first learn that $(-4) - (-8) = (-4) + 8 = 4$, the explanation is a bit mysterious: why is "minus a minus equal to a plus"? Now there is no longer any mystery: this property follows *from our construction* of the number system $\mathbb{Z}$. ∎

Finally, we turn to multiplication. If $x = [(a, b)]$ and $y = [(c, d)]$ are integers then we define their product by the formula

$$x \cdot y = [(a \cdot d + b \cdot c, a \cdot c + b \cdot d)].$$

This definition may be a surprise. Why did we not define $x \cdot y$ to be $[(a \cdot c, b \cdot d)]$? There are several reasons: first of all, the latter definition would give the wrong answer; moreover, it is not unambiguous (different representatives of x and y would give a different answer). If you recall that we think of $[(a, b)]$ as representing $b - a$ and $[(c, d)]$ as representing $d - c$ then the product should be the equivalence class that represents $(b - a) \cdot (d - c)$. That is the motivation behind our definition.

We proceed now to an example.

Example A1.11

We compute the product of -3 and -6. Now

$$(-3) \cdot (-6) = [(5, 2)] \cdot [(9, 3)] = [(5 \cdot 3 + 2 \cdot 9, 5 \cdot 9 + 2 \cdot 3)] = [(33, 51)] = 18,$$

which is the expected answer.

As a second example, we multiply -5 and 12. We have

$$-5 \cdot 12 = [(7, 2)] \cdot [(1, 13)] = [(7 \cdot 13 + 2 \cdot 1, 7 \cdot 1 + 2 \cdot 13)] = [(93, 33)] = -60.$$

Finally, we show that 0 times any integer A equals zero. Let $A = [(a, b)]$. Then

$$\begin{aligned}
0 \cdot A &= [(1, 1)] \cdot [(a, b)] = [(1 \cdot b + 1 \cdot a, 1 \cdot a + 1 \cdot b)] \\
&= [(a + b, a + b)] \\
&= 0.
\end{aligned}$$

∎

Remark A1.12

Notice that one of the pleasant byproducts of our construction of the integers is that we no longer have to give artificial explanations for why the product of two negative numbers is a positive number or why the product of a negative number and a positive number is negative. These properties instead follow automatically from our construction.

Of course we will not discuss division for integers; in general, division of one integer by another makes no sense *in the universe of the integers*.[1]

---

[1] Here the word "universe" is meant to mean the collection of all objects on which we are focusing our attention.

In the rest of this book we will follow the standard mathematical custom of denoting the set of all integers by the symbol $\mathbb{Z}$. We will write the integers not as equivalence classes, but in the usual way as $\cdots - 3, -2, -1, 0, 1, 2, 3, \ldots$. The equivalence classes are a device that we used to *construct* the integers. Now that we have the integers in hand, we may as well write them in the simple, familiar fashion.

In an exhaustive treatment of the construction of $\mathbb{Z}$, we would prove that addition and multiplication are commutative and associative, prove the distributive law, and so forth. But the purpose of this section is to demonstrate modes of logical thought rather than to be thorough.

Section A1.3. The Rational Numbers

In this section we use the integers, together with a construction using equivalence classes, to build the rational numbers. Let A be the set $\mathbb{Z} \times (\mathbb{Z} \setminus \{0\})$. Here the symbol $\setminus$ stands for "subtraction of sets": $\mathbb{Z} \setminus \{0\}$ denotes the set of all elements of $\mathbb{Z}$ *except* 0. In other words, A is the set of ordered pairs (a, b) of integers subject to the condition that $b \neq 0$. [*Think, intuitively and non-rigorously, of this ordered pair as "representing" the fraction a/b.*] We definitely want it to be the case that certain ordered pairs represent the same number. For instance,

The number $\frac{1}{2}$ should be the same number as $\frac{3}{6}$.

This example motivates our equivalence relation. Declare (a, b) to be related to (a', b') if $a \cdot b' = a' \cdot b$. [*Here we are thinking, intuitively and non-rigorously, that the fraction a/b should equal the fraction a'/b' precisely when $a \cdot b' = a' \cdot b$.*]

Is this an equivalence relation? Obviously the pair (a, b) is related to itself, since $a \cdot b = a \cdot b$. Also the relation is symmetric: if (a, b) and (a', b') are pairs and $a \cdot b' = a' \cdot b$ then $a' \cdot b = a \cdot b'$. Finally, if (a, b) is related to (a', b') and (a', b') is related to (a'', b'') then we have both

$$a \cdot b' = a' \cdot b \quad \text{and} \quad a' \cdot b'' = a'' \cdot b'.$$

Multiplying the left sides of these two equations together and the right sides together gives

$$(a \cdot b') \cdot (a' \cdot b'') = (a' \cdot b) \cdot (a'' \cdot b').$$

If $a' = 0$ then it follows immediately from the two equations preceding the last that both a and a'' must be zero. So the three pairs $(a, b), (a', b')$, and (a'', b'') are equivalent and there is nothing to prove. So we may assume that $a' \neq 0$. We know *a priori* that $b' \neq 0$; therefore we may cancel common terms in the last equation to obtain

$$a \cdot b'' = b \cdot a''.$$

Thus (a, b) is related to (a'', b''), and our relation is transitive.

The resulting collection of equivalence classes will be called the set of *rational numbers*, and we shall denote this set with the symbol $\mathbb{Q}$.

Example A1.13

The equivalence class $[(4, 12)]$ in the rational numbers contains all of the pairs $(4, 12), (1, 3), (-2, -6)$. (Of course it contains infinitely many other pairs as well.) This equivalence class represents the fraction $4/12$, which we sometimes also write as $1/3$ or $-2/(-6)$. ∎

If $[(a, b)]$ and $[(c, d)]$ are rational numbers then we define their *product* to be the rational number

$$[(a \cdot c, b \cdot d)].$$

This is well defined, for if (a, b) is related to $(\widetilde{a}, \widetilde{b})$ and (c, d) is related to $(\widetilde{c}, \widetilde{d})$ then we have the equations

$$a \cdot \widetilde{b} = \widetilde{a} \cdot b \quad \text{and} \quad c \cdot \widetilde{d} = \widetilde{c} \cdot d.$$

Multiplying together the left sides and the right sides we obtain

$$(a \cdot \widetilde{b}) \cdot (c \cdot \widetilde{d}) = (\widetilde{a} \cdot b) \cdot (\widetilde{c} \cdot d).$$

Rearranging, we have

$$(a \cdot c) \cdot (\widetilde{b} \cdot \widetilde{d}) = (\widetilde{a} \cdot \widetilde{c}) \cdot (b \cdot d).$$

But this says that the product of $[(a, b)]$ and $[(c, d)]$ (which is $[(ac, bd)]$) is related to the product of $[(\widetilde{a}, \widetilde{b})]$ and $[(\widetilde{c}, \widetilde{d})]$ (which is $(\widetilde{a}\widetilde{c}, \widetilde{b}\widetilde{d})$). So multiplication is unambiguous (i.e., well defined).

Example A1.14

The product of the two rational numbers $[(3, 8)]$ and $[(-2, 5)]$ is

$$[(3 \cdot (-2), 8 \cdot 5)] = [(-6, 40)] = [(-3, 20)].$$

This is what we expect: the product of $3/8$ and $-2/5$ is $-3/20$. ∎

If $q = [(a, b)]$ and $r = [(c, d)]$ are rational numbers and if r is not zero (that is, $[(c, d)]$ is not the equivalence class zero—in other words, $c \neq 0$) then we define the quotient q/r to be the equivalence class

$$[(ad, bc)].$$

We leave it to you to check that this operation is well defined.

Example A1.15

The quotient of the rational number $[(4, 7)]$ by the rational number $[(3, -2)]$ is, by definition, the rational number

$$[(4 \cdot (-2), 7 \cdot 3)] = [(-8, 21)].$$

This is what we expect: the quotient of $4/7$ by $-3/2$ is $-8/(21)$. ∎

How should we add two rational numbers? We could try declaring $[(a, b)] + [(c, d)]$ to be $[(a + c, b + d)]$, but this will not work (think about the way that we usually add fractions). Instead we define

$$[(a, b)] + [(c, d)] = [(a \cdot d + c \cdot b, b \cdot d)].$$

We turn to an example.

Example A1.16

The sum of the rational numbers $[(3, -14)]$ and $[(9, 4)]$ is given by

$$[(3 \cdot 4 + 9 \cdot (-14), (-14) \cdot 4)] = [(-114, -56)] = [(57, 28)].$$

This coincides with the usual way that we add fractions:

$$-\frac{3}{14} + \frac{9}{4} = \frac{57}{28}.$$ ∎

Notice that the equivalence class $[(0, 1)]$ is the rational number that we usually denote by 0. It is the additive identity, for if $[(a, b)]$ is another rational number then

$$[(0, 1)] + [(a, b)] = [(0 \cdot b + a \cdot 1, 1 \cdot b)] = [(a, b)].$$

A similar argument shows that $[(0, 1)]$ times any rational number $[(a, b)]$ gives $[(0, b)]$ or 0.

Of course the concept of subtraction is really just a special case of addition (that is, $x - y$ is the same thing as $x + (-y)$). So we shall say nothing further about subtraction.

In practice we will write rational numbers in the traditional fashion:

$$\frac{2}{5}, \quad \frac{-19}{3}, \quad \frac{22}{2}, \quad \frac{24}{4}, \quad \ldots.$$

In mathematics it is generally not wise to write rational numbers in mixed form, such as $2\frac{3}{5}$, because the juxtaposition of two numbers could easily be mistaken for multiplication. Instead we would write this quantity as the improper fraction $13/5$.

Table of Notation

Notation	Section	Definition
A, B	1.3	atomic statements
$\wedge$	1.3	and
$T,\ F$	1.3	true and false
$\vee$	1.3	or
$\sim$	1.4	not
$\Rightarrow$	1.5	implies
$\mathbb{R}$	1.5	real numbers
$\Leftrightarrow$	1.6	if and only if
$\forall$	1.7	for all
$\exists$	1.7	there exists
$\mathbb{N}$	2.2	the natural numbers
$\mathbb{Q}$	2.3	the rational numbers
$P(n)$	2.4	inductive statement
$S,\ T$	3.1	sets
$\mathbb{Z}$	3.2	the integers
$\in$	3.2	is an element of
$\subset$	3.2	is a subset of
$\emptyset$	3.2	empty set
$\cap$	3.2	intersection
$\cup$	3.2	union
$\setminus$	3.2	set-theoretic difference
^{c}S	3.2	complement of the set S
X	3.2	universal set
$S \times T$	3.4	set-theoretic product of S and T
$\mathcal{P}(S)$	3.4	power set of S
$\cup_{j=1}^{\infty} A_j$	3.5	countable union
$\cap_{j=1}^{\infty} B_j$	3.5	countable intersection
$\cup_{\alpha \in A} A_\alpha$	3.5	uncountable union
$\cap_{\alpha \in A} B_\alpha$	3.5	uncountable intersection
$\equiv$	3.5	is defined to be

Notation	Section	Definition
$\mathcal{R}$	4.1	a relation
$\sim$	4.1	are related
$<$	4.2	is less than
S	4.3	domain of a function
T	4.3	range of a function
f	4.3	a function
$f\colon S \to T$	4.3	a function from S to T
$f + g$	4.4	sum of functions
$f - g$	4.4	difference of functions
$f \cdot g$	4.4	product of functions
f/g	4.4	quotient of functions
$f \circ g$	4.4	composition of functions
f^{-1}	4.4	inverse of f
$\mathrm{card}(A)$	4.5	cardinality of A
$\mathrm{card}(A) < \mathrm{card}(B)$	4.5	cardinality of A less than cardinality of B
$\sup X$	5.1	supremum of X
$\mathrm{lub}\, X$	5.1	least upper bound of X
$\inf X$	5.1	infimum of X
$\mathrm{glb}\, X$	5.1	greatest lower bound of X
$|x|$	5.1	absolute value of x
$|x + y| \le |x| + |y|$	5.1	triangle inequality
$\mathcal{P}$	5.1	a cut
$\mathbb{C}$	5.2	the complex numbers
z	5.2	a complex number
i	5.2	the square root of -1
$\bar{z}$	5.2	complex conjugate
$|z|$	5.2	modulus of z
$e^{i\theta}$	5.2	complex exponential
$\{a_j\}$	6.1	a sequence
a_j	6.1	a sequence
a_{j_k}	6.2	a subsequence
$\liminf a_j$	6.3	limit infimum of a_j

Notation	Section	Definition
limsup a_j	6.3	limit supremum of a_j
a^j	6.4	a power sequence
e	6.4	Euler's number e
$\sum_{j=1}^{\infty} a_j$	7.1	a series
S_N	7.1	a partial sum
$\sum_{j=1}^{N} a_j$	7.1	a partial sum
$\sum_{j=1}^{\infty} (-1)^j b_j$	7.3	an alternating series
$j!$	7.4	j factorial
$\sum_{n=0}^{\infty} \sum_{j=0}^{n} a_j \cdot b_{n-j}$	7.5	the Cauchy product of series
(a,b)	8.1	open interval
$[a,b]$	8.1	closed interval
$[a,b)$	8.1	half-open interval
$(a,b]$	8.1	half-open interval
U	8.1	an open set
F	8.1	a closed set
∂S	8.1	boundary of S
^cS	8.1	complement of S
$\overline{S}$	8.2	closure of S
$\overset{\circ}{S}$	8.2	interior of S
$\{\mathcal{O}_\alpha\}$	8.3	an open cover
S_j	8.4	step in constructing the Cantor set
C	8.4	the Cantor set
$\lim_{E \ni x \to c} f(x)$	9.1	limit of f at c
ℓ	9.1	a limit
$f(L)$	9.3	image of the set L
$f^{-1}(W)$	9.2	inverse image of a set
$f(L)$	9.3	image of the set L
m	9.3	minimum for a function f

Notation	Section	Definition
M	9.3	maximum for a function f
$\lim_{x \to c^-} f(x)$	9.4	left limit of f at c
$\lim_{x \to c^+} f(x)$	9.4	right limit of f at c
$f'(x)$	10.1	derivative of f at x
df/dx	10.1	derivative of f
$\dot{f}$	10.1	derivative of f
$\text{Lip}_\alpha(I)$	10.3	space of Lipschitz functions
$C^{k,\alpha}(I)$	10.3	space of smooth functions of order k,α
$\mathcal{P}$	11.1	a partition
I_j	11.1	interval from the partition
Δ_j	11.1	length of I_j
$m(\mathcal{P})$	11.1	mesh of the partition
$\mathcal{R}(f, \mathcal{P})$	11.1	Riemann sum
$\int_a^b f(x)\,dx$	11.1	Riemann integral
$\int_b^a f(x)\ dx$	11.2	integral with reverse orientation
f_j	12.1	sequence of functions
$\{f_j\}$	12.1	sequence of functions
$\lim_{x \to c} f(x)$	12.2	limit of f as x approaches c
$\sum_{j=1}^\infty f_j(x)$	12.3	series of functions
$S_N(x)$	12.3	partial sum of a series of functions
$p(x)$	12.4	a polynomial
$\sum_{j=0}^\infty a_j(x - c)^j$	13.1	a power series
R_N	13.1	tail of the power series
ρ	13.2	radius of convergence
$f(x) = \sum_{j=0}^k f^{(j)}(a)\frac{(x-a)^j}{j!}$ $+R_{k,a}(x)$	13.2	Taylor expansion
$\exp(x)$	13.3	the exponential function
$\sin x$	13.3	the sine function
$\cos x$	13.3	the cosine function
$\text{Sin } x$	13.3	sine with restricted domain

Notation	Section	Definition
$\text{Cos}\,x$	13.3	cosine with restricted domain
$\ln x$	13.4	the natural logarithm function
$\binom{n}{k}$	A1.1	binomial coefficient
$[(a,b)]$	A1.2	an integer
$[(c,d)]$	A1.3	a rational number

GLOSSARY

Abel's convergence test   A test for convergence of series that is based on summation by parts.

absolutely convergent series   A series for which the absolute values of the terms form a convergent series.

absolute maximum   A number M is the absolute maximum for a function f if $f(x) \leq f(M)$ for every x.

absolute minimum   A number m is the absolute minimum for a function f if $f(x) \geq f(m)$ for every x.

absolute value   Given a real number x, its absolute value is the distance of x to 0.

accumulation point   A point x is an accumulation point of a set S if every neighborhood of x contains infinitely many distinct elements of S.

adding functions   If $f : \mathbb{R} \to \mathbb{R}$ and $g : \mathbb{R} \to \mathbb{R}$, then $(f+g)(x) \equiv f(x)+g(x)$.

algebraic number   A number that is the solution of a polynomial equation with integer coefficients.

alternating series   A series of real terms which alternate in sign.

alternating series test   If an alternating series has terms tending to zero then it converges.

"and"   The connective which is used for conjunction.

Archimedean Property   If a and b are positive real numbers then there is a positive integer n so that $na > b$.

319

Aristotelian logic   The logic developed by Aristotle of Athens and Macedon.

assumption   The antecedent of a mathematical thought. See *hypothesis*.

atomic statement   A sentence with a subject and a verb but no connectives.

axiom   An intuitively obvious statement on which our mathematical reasoning is based.

bijection   A one-to-one, onto function.

Binomial Expansion   The expansion, under multiplication, of the expression $(a+b)^n$.

Bolzano-Weierstrass Theorem   Every bounded sequence of real numbers has a convergent subsequence.

boundary of a set   The set of boundary points for the set.

boundary point   The point b is in the boundary of S if each neighborhood of b contains both points of S and points of the complement of S.

bounded above   A subset $S \subset \mathbb{R}$ is bounded above if there is a real number b such that $s \leq b$ for all $s \in$.

bounded below   A subset $S \subset \mathbb{R}$ is bounded below if there is a real number c such that $s \geq c$ for all $s \in$.

bounded sequence   A sequence a_j with the property that there is a number M so that $|a_j| \leq M$ for every j.

bounded set   A set S with the property that there is a number M with $|s| \leq M$ for every $s \in S$.

Cantor set   A compact set which is uncountable, has zero length, is perfect, is totally disconnected, and has many other unusual properties.

cardinality   Two sets have the same cardinality when there is a one-to-one correspondence between them.

cardinal number   The equivalence classes induced by the relation "same cardinality."

Cauchy Condensation Test   A series of decreasing, nonnegative terms converges if and only if its dyadically condensed series converges.

Cauchy criterion   A sequence a_j is said to be Cauchy if, for each $\epsilon > 0$, there is an $N > 0$ so that, if $j, k > N$, then $|a_j - a_k| < \epsilon$.

Cauchy criterion for a series   A series satisfies the Cauchy criterion if and only if the sequence of partial sums satisfies the Cauchy criterion for a sequence.

Cauchy product   A means for taking the product of two series.

Cauchy's Mean Value Theorem   A generalization of the Mean Value Theorem that allows the comparison of two functions.

Chain Rule   A rule for differentiating the composition of functions.

change of variable   A method for transforming an integral by subjecting the domain of integration to a one-to-one function.

closed set   The complement of an open set.

closure of a set   The set together with its boundary points.

codomain of a function   See *range of a function.*

codomain of a relation   See *range of a relation.*

common refinement of two partitions   The union of the two distinct partitions.

compact set   A set E is compact if every sequence in E contains a subsequence that converges to an element of E.

Comparison Test for Convergence   A series converges if it is majorized in absolute value by a convergent series.

Comparison Test for Divergence   A series diverges if it majorizes a divergent series.

complement of a set The set of points not in the set.

complete mathematical induction A form of induction in which one proves a statement $P(j)$ by showing that $P(1), P(2), \ldots P(n)$ implies $P(n+1)$.

complex conjugate Given a complex number $z = x + iy$, the conjugate is the number $\bar{z} = x - iy$.

complex numbers The set $\mathbb{C}$ of ordered pairs of real numbers equipped with certain operations of addition and multiplication.

composing functions If $f : \mathbb{R} \to \mathbb{R}$ and $g : \mathbb{R} \to \mathbb{R}$, then $f \circ g(x) \equiv f(g(x))$.

composition The composition of two functions is the succession of one function by the other.

conditionally convergent series A series which converges, but not absolutely.

conjunction The joining together of two statements.

connected set A set which cannot be separated by two disjoint open sets.

connectives The words which are used to connect logical statements. These are "and," "or," "not," "if-then," and "if and only if."

continuity at a point The function f is continuous at c if the limit of f at c equals the value of f at c. Equivalently, given $\epsilon > 0$, there is a $\delta > 0$ so that $|x - c| < \delta$ implies $|f(x) - f(c)| < \epsilon$.

continuous function A function which is continuous at each point c in the domain.

continuously differentiable function A function which has a derivative at every point, and so that the derivative function is continuous.

Continuum Hypothesis The hypothesis that there exists a set with cardinality strictly between that of the natural numbers and that of the real numbers.

contradiction A point of logic in which two assertions disagree.

contrapositive The contrapositive of the implication $\mathbf{A} \Rightarrow \mathbf{B}$ is $\sim \mathbf{B} \Rightarrow \backslash \mathbf{A}$.

convergence of a series   A series converges if and only if its sequence of partial sums converges.

convergence of a sequence (of numbers)   A sequence a_j with the property that there is a limiting element ℓ so that, for any $\epsilon > 0$, there is a positive integer N so that, if $j > N$, then $|a_j - \ell| < \epsilon$.

converse   For a statement "**A** implies **B**," the converse statement is "**B** implies **A**."

cosine function   The function $\cos x = \sum_{j=0}^{\infty} (-1)^j x^{2j}/(2j)!$.

countable set   Any set that has precisely the same cardinality as the natural numbers.

counterexample   An example which contradicts the flow of thought.

decreasing sequence   The sequence of real numbers a_j is decreasing if $a_1 \geq a_2 \geq a_3 \geq \cdots$.

Dedekind cut   A rational halfline that is bounded above in $\mathbb{Q}$. Used to construct the real numbers.

definition   The enunciation of the meaning of a mathematical term.

de Morgan's Laws   The identities

$$^c(A \cup B) = {}^cA \cap {}^cB$$

and

$$^c(A \cap B) = {}^cA \cup {}^cB.$$

Density Property   If $c < d$ are real numbers then there is a rational number q with $c < q < d$.

denumerable set   A set that is either empty, finite, or countable.

derivative   The limit $\lim_{t \to x}(f(x) - f(t))/(t - x)$ for a function f on an open interval.

derived power series   The series obtained by differentiating a power series term by term.

difference quotient   The quotient $(f(t) - f(x))/(t - x)$ for a function f on an open interval.

differentiable   A function that possesses the derivative at a point.

diophantine equation   A polynomial equation with integer coefficients for which we seek integer solutions.

direct proof   A proof that proceeds according to the statement being proved.

Dirichlet function   A function, taking only the values 0 and 1, which is highly discontinuous.

disconnected set   A set which can be separated by two disjoint open sets.

discontinuity of the first kind   A point at which a function f is discontinuous because the left and right limits at the point disagree.

discontinuity of the second kind   A point at which a function f is discontinuous because either the left limit or the right limit at the point does not exist.

disjoint sets   Sets $A_1, A_2, \ldots, A_k$ so that $A_1 \cap A_2 \cap \cdots \cap A_k = \emptyset$.

diverge to infinity   A sequence with elements that become arbitrarily large.

dividing functions   If $f : \mathbb{R} \to \mathbb{R}$ and $g : \mathbb{R} \to \mathbb{R}$, then $(f/g)(x) \equiv f(x)g(x)$ provided that $g(x)$ is never 0.

domain of a function   See *function*.

domain of a relation   For a relation $\mathcal{R}$ on sets S and T, the set of $s \in S$ such that there exists a $t \in T$ with $(s, t) \in \mathcal{R}$.

domain of integration   The interval over which the integration is performed.

dummy variable   A variable whose role in an argument or expression is formal. A dummy variable can be replaced by any other variable with no logical consequences.

elementary statement A sentence with a subject and a verb but no connectives.

element of A member of a given set.

empty set The set with no elements.

equality of sets Two sets are equal when they have the same elements.

equivalence classes The pairwise disjoint sets into which an equivalence relation partitions a set.

equivalence relation A relation on a set S that is *reflexive, symmetric,* and *transitive*. It partitions the set into equivalence classes.

exclusive "or" The use of "or" which disallows the possibility of both elements being true.

Euler's formula The identity $e^{iy} = \cos y + i \sin y$.

Euler's number This is the number $e = 10.71828\ldots$ which is known to be irrational, indeed transcendental.

even number A natural number is even if it is evenly divisible by 2.

exponential function The function $\exp(z) = \sum_{j=0}^{\infty} z^j/j!$.

extended set operations Operations in which the index set is larger than the set of natural numbers.

false An attribute of a sentence or statement. This attribute indicates the unacceptability or invalidity of the statement.

Fibonacci numbers The sequence $\{a_n\}$ of numbers beginning with $0, 1$ and so that $a_{n+2} = a_n + a_{n+1}$.

field A system of numbers equipped with operations of addition and multiplication and satisfying eleven natural axioms.

finite set A set that can be put in one-to-one correspondence with a set of the form $\{1, 2, \ldots, n\}$ for some positive integer n.

first-order logic The logic consisting of the elementary connectives, the universal quantifiers, an infinite string of variables, and finally parentheses.

"for all" The quantifier $\forall$ for making a statement about all objects of a certain kind.

function A *function* from a set A to a set B is a relation f on A and B such that, for each $a \in A$, there is one and only one pair $(a, b) \in f$. We call A the *domain* and B the *range* of the function.

Fundamental Theorem of Calculus A result relating the values of a function to the integral of its derivative: $f(x) - f(a) = \int_a^x f'(t)\,dt$.

geometric series This is a series of powers.

greatest lower bound The real number c is the greatest lower bound for the set $S \subset \mathbb{R}$ if b is a lower bound and if there is no lower bound that is greater than c.

higher derivatives The derivative of a derivative.

hypothesis The antecedent of a mathematical thought.

i The square root of -1 in the complex number system.

if An alternative phrase for converse implication.

"if and only if" The connective which is used for logical equivalence.

"if-then" The connective which is used for implication.

image of a function See *function*. The image of the function f is Image $f = \{b \in B : \exists a \in A \text{ such that } (a) = b\}$.

image of a relation For a relation $\mathcal{R}$ on sets S and T, the set of $t \in T$ such that there exists an $s \in S$ with $(s, t) \in \mathcal{R}$.

image of a set If f is a function then the image of E under f is the set $\{f(e) : e \in E\}$.

imaginary part Given a complex number $z = x + iy$, its imaginary part is y.

inclusive "or" The use of "or" which allows the possibility of both elements being true.

increasing sequence The sequence of real numbers a_j is increasing if $a_1 \leq a_2 \leq a_3 \leq \cdots$.

index set The set over which a collection of sets or objects is indexed.

inductive hypothesis The hypothesis of the main statement to be proved in an inductive argument.

infimum See *greatest lower bound.*

infinite set A set is infinite if it is not finite.

injective A function is injective if it is *one-to-one.*

integers The natural numbers, the negatives of the natural numbers, and zero.

integration by parts A device for integrating a product.

interior of a set The collection of interior points of the set.

interior point A point of the set S which has a neighborhood lying in S.

intermediate value theorem The result that says that a continuous function does not skip values.

intersection of sets The set of elements common to two or more given sets.

interval A subset of the reals that contains all its intermediate points.

interval of convergence of a power series An interval of the form $(c - \rho, c + \rho)$ on which the power series converges (uniformly on compact subsets of the interval).

inverse of a function If $f : \mathbb{R} \to \mathbb{R}$, then $g : \mathbb{R} \to \mathbb{R}$ is an inverse for f if $f \circ g(x) = x$ and $g \circ f(x) = x$ for all $x \in \mathbb{R}$.

irrational number A real number which is not rational.

isolated point of a set   A point of the set with a neighborhood containing no other point of the set.

k times continuously differentiable   A function that has k derivatives, each of which is continuous.

Lambert W function   A transcendental function W with the property that any of the standard transcendental functions (sine, cosine, exponential, logarithm) can be expressed in terms of W.

least upper bound   The real number b is the least upper bound for the set $S \subset \mathbb{R}$ if b is an upper bound and if there is no other upper bound that is less than b.

Least Upper Bound Property   The important defining property of the real numbers.

left limit   A limit of a function at a point c that is calculated with values of the function that are to the left of c.

lemma   An ancillary result that is used to prove a theorem.

l'Hôpital's Rule   A rule for calculating the limit of the quotient of two functions in terms of the quotient of the derivatives.

limit   The value ℓ that a function approaches at a point of or an accumulation point c of the domain. Equivalently, given $\epsilon > 0$, there is a $\delta > 0$ so that $|f(x) - \ell| < \epsilon$ whenever $|x - c| < \delta$.

limit infimum   The least limit of any subsequence of a given sequence.

limit supremum   The greatest limit of any subsequence of a given sequence.

Lipschitz function   A function that satisfies a condition of the form $|f(s) - f(t)| \leq C|s - t|$ or $|f(s) - f(t)| \leq |s - t|^{\alpha}$ for $0 < \alpha \leq 1$.

local extrema   Either a local maximum or a local minimum.

local maximum   The point x is a local maximum for the function f if $f(x) \geq f(t)$ for all t in a neighborhood of x.

local minimum   The point x is a local minimum for the function f if $f(x) \leq f(t)$ for all t in a neighborhood of x.

logically equivalent Two statements are logically equivalent if they have the same truth table.

logically independent Two statements are logically independent if neither one implies the other.

lower bound A real number c is a lower bound for a subset $S \subset \mathbb{R}$ if $s \geq c$ for all $s \in S$.

Mean Value Theorem If f is a continuous function on $[a, b]$, differentiable on the interior, then the slope of the segment connecting $(a, f(a))$ and $(b, f(b))$ equals the derivative of f at some interior point.

mesh of a partition The maximum length of any interval in the partition.

modulus The modulus of a complex number $z = x + iy$ is $|z| = \sqrt{x^2 + y^2}$.

modus ponendo ponens The most fundamental logical syllogism. It says that, if $\mathbf{A} \Rightarrow \mathbf{B}$ and $\mathbf{A}$, then we may conclude $\mathbf{B}$.

modus tollendo tollens A basic logical syllogism that is logically equivalent to *modus ponendo ponens*.

monotone sequence A sequence that is either increasing or decreasing.

monotonic function A function that is either monotonically increasing or monotonically decreasing.

monotonically decreasing function A function whose graph goes downhill when moving from left to right: $f(s) \geq f(t)$ when $s \leq t$.

monotonically increasing function A function whose graph goes uphill when moving from left to right: $f(s) \leq f(t)$ when $s \leq t$.

multiplying functions If $f : \mathbb{R} \to \mathbb{R}$ and $g : \mathbb{R} \to \mathbb{R}$, then $f \cdot g(x) \equiv f(x) \cdot g(x)$.

natural logarithm function The inverse function to the exponential function.

natural numbers The counting numbers $1, 2, 3, \ldots$.

necessary for   An alternative phrase for converse implication.

neighborhood of a point   An open set containing the point.

nested sets   Either sets A_j so that $A_1 \subset A_2 \subset \cdots$ or sets B_j so that $B_1 \supset B_2 \supset \cdots$.

Neumann series   A series of the form $1/(1 - \alpha) = \sum_{j=0}^{\infty} \alpha^j$ for $|\alpha| < 1$.

Newton quotient   The quotient $(f(t) - f(x))/(t - x)$ for a function f on an open interval.

non-terminating decimal expansion   A decimal expansion for a real number that has infinitely many nonzero digits.

"not"   The connective which is used for negation.

odd number   A natural number is odd if it has remainder 1 when divided by 2.

one-to-one function   A function that takes different values at different points of the domain.

only if   An alternative phrase for implication.

onto function   A function whose image equals its range.

open ball   The set of points at distance less than some $r > 0$ from a fixed point c.

open set   A set which contains a neighborhood of each of its points.

"or"   The connective which is used for disjunction.

ordered field   A field equipped with an order relation that is compatible with the field structure.

ordered pair   A pair (a, b) in which the terms appear in order.

order relation   A relation on a set S that induces an ordering on S.

ordinal arithmetic   The arithmetic of ordinal numbers.

ordinal number   In set theory, an ordinal number, or ordinal, is one generalization of the concept of a natural number that is used to describe a way to arrange a collection of objects in order, one after another.

pairwise disjoint sets   Sets $A_1, A_2, \ldots, A_k$ such that $A_j \cap A_k = \emptyset$ for every j, k.

partial ordering   Let $\mathcal{R}$ be a relation on a set S. We call $\mathcal{R}$ a *partial ordering* on S if it satisfies the following properties:

(a) For all $x \in S$, $(x, x) \in \mathcal{R}$;

(b) If $x, y \in S$ and both $(x, y) \in \mathcal{R}$ and $(y, x) \in \mathcal{R}$, then $x = y$;

(c) If $x, y, z \in S$ and both $(x, y) \in \mathcal{R}$ and $(y, z) \in \mathcal{R}$, then $(x, z) \in \mathcal{R}$.

partial sum of functions   The sum of the first N terms of a series of functions.

partial sum (of scalars)   The sum of the first N terms of a series of scalars.

partition of the interval $[a, b]$ A finite, ordered set of points $\mathcal{P} = \{x_0, x_1, x_2, \ldots, x_{k-1}, x_k\}$ such that

$$a = x_0 \leq x_1 \leq x_2 \leq \cdots \leq x_{k-1} \leq x_k = b.$$

Peano axioms   An axiom system for the natural numbers.

perfect set   A set which is closed and in which every point is an accumulation point.

pigeonhole principle   The result that, if $(n + 1)$ letters are delivered to n mailboxes, then one mailbox will contain two letters.

Pinching Principle   A criterion for convergence of a sequence that involves bounding it below by a convergent sequence and bounding it above by another convergent sequence with the same limit.

pointwise convergence of a sequence of functions   A sequence f_j of functions converges pointwise if $f_j(x)$ convergence for each x in the common domain.

polar form of a complex number   The polar form of a complex number z is $re^{i\theta}$, where r is the modulus of z and θ is the angle that the vector from 0 to

z subtends with the positive x-axis.

power series expanded about the point c A series of the form

$$\sum_{j=0}^{\infty} a_j (x - c)^j \, .$$

power set   The collection of all subsets of a given set.

Principle of Induction   A proof technique for establishing a statement $P(n)$ about the natural numbers.

product of sets   See *set-theoretic product*.

proof   An argument to establish the validity of a mathematical fact.

proof by contradiction   A proof that uses the law of the excluded middle.

proof by induction   A proof that is based on the Axiom of Induction in Zermelo-Frankel arithmetic.

proposition   A basic formulation of a mathematical truth.

propositional calculus   Sentential logic and elementary connectives.

Pythagorean theorem   The statement that, if a, b are the legs of a right triangle, and c is the hypotenuse, then $a^2 + b^2 = c^2$.

Pythagoras's theorem   The result that 2 does not have a rational square root.

quantifier   A logical device for making a quantitative statement. Our standard quantifiers are "for all" and "there exists."

radius of convergence of a power series   Half the length ρ of the interval of convergence.

range of a function   See *function*.

range of a relation   For a relation $\mathcal{R}$ on sets S and T, the range is the set T.

rational numbers   Numbers which may be represented as quotients of integers.

Ratio Test for Convergence   A series converges if the limit of the sequence of quotients of summands is less than 1.

Ratio Test for Divergence   A series diverges if the limit of the sequence of quotients of summands is greater than 1.

real analytic function   A function with a convergent power series expansion about each point of its domain.

real numbers   An ordered field $\mathbb{R}$ containing the rationals $\mathbb{Q}$ so that every nonempty subset with an upper bound has a least upper bound.

real part   Given a complex number $z = x + iy$, its real part is x.

rearrangement of a series   A new series obtained by permuting the summands of the original series.

reflexive   Let $\mathcal{R}$ be a relation on a set S. If $s \in S$, then $s\mathcal{R}s$.

relation   A relation on sets A and B is a subset of $A \times B$.

remainder term for the Taylor expansion   The term $R_{k,a}(x)$ in the Taylor expansion.

Riemann integrable   A function for which the Riemann integral exists.

Riemann integral   The limit of the Riemann sums.

Riemann sum   The approximate integral based on a partition.

right limit   A limit of a function at a point c that is calculated with values of the function to the right of c.

Rolle's Theorem   The special case of the Mean Value Theorem when $f(a) = f(b) = 0$.

Root Test for Convergence   A series converges if the limit of the nth roots of the nth terms is less than 1.

Root Test for Divergence   A series is divergent if the limit of the nth roots of the nth terms is greater than 1.

scalar   An element of either $\mathbb{R}$ or $\mathbb{C}$.

Schroeder-Bernstein Theorem   The result that says that if there is a one-to-one function from the set A to the set B and a one-to-one function from the set B to the set A then A and B have the same cardinality.

second-order logic   A logic more sophisticated than first-order logic in that it allows us to quantify over subsets of the domain of discourse M and also to consider functions F mapping $M \times M$ into M.

sequence of functions   A function from $\mathbb{N}$ into the set of functions on some space.

sequence (of scalars)   A function from $\mathbb{N}$ into $\mathbb{R}$ or $\mathbb{C}$ or a metric space. We often denote the sequence by a_j.

series of functions   An infinite sum of functions.

series (of scalars)   An infinite sum of scalars.

set   A collection of objects.

set-builder notation   The notation $S = \{x : c(x)\}$ for specifying a set S.

set formulation of induction   See *complete mathematical induction.*

set-theoretic difference   The set-theoretic difference $A \setminus B$ consists of those elements that lie in A but not in B.

set-theoretic isomorphism   A one-to-one, onto function.

set-theoretic product   If A and B are sets then their set-theoretic product is the set of ordered pairs (a, b) with $a \in A$ and $b \in B$.

simple ordering   Let $\mathcal{R}$ be a relation on a set S. We call $\mathcal{R}$ a simple ordering if it satisfies the following properties:

(a) If $x, y \in S$ and both $(x, y) \in \mathcal{R}$ and $(y, x) \in \mathcal{R}$, then $x = y$;

(b) If $x, y, z \in S$ and both $(x, y) \in \mathcal{R}$ and $(y, z) \in \mathcal{R}$, then $(x, z) \in \mathcal{R}$;

(c) If $x, y \in S$ are distinct, then either $(x, y) \in \mathcal{R}$ or $(y, x) \in \mathcal{R}$.

sine function   The function $\sin x = \sum_{j=0}^{\infty}(-1)^j x^{2j+1}/(2j+1)!$.

smaller cardinality   The set A has smaller cardinality than the set B if there is a one-to-one mapping of A to B but none from B to A .

strictly monotonically decreasing function   A function whose graph goes strictly downhill when moving from left to right: $f(s) > f(t)$ when $s < t$.

strictly monotonically increasing function   A function whose graph goes strictly uphill when moving from left to right: $f(s) < f(t)$ when $s < t$.

strict simple ordering   Let $\mathcal{R}$ be a relation on a set S. We call $\mathcal{R}$ a strict simple ordering if it satisfies the properties

(a) If $x, y \in S$ and $(x, y) \in \mathcal{R}$, then $(y, x) \notin \mathcal{R}$.

(b) If $x, y, z \in S$ and both $(x, y) \in \mathcal{R}$ and $(y, z) \in \mathcal{R}$, then $(x, z) \in \mathcal{R}$;

(c) If $x, y \in S$ are distinct, then either $(x, y) \in \mathcal{R}$ or $(y, x) \in \mathcal{R}$.

strong mathematical induction   See *complete mathematical induction.*

subfield   Given a field k, a subfield m is a subset of k which is also a field with the induced field structure.

subsequence   A sequence that is a subset of a given sequence with the elements occurring in the same order.

subset of   A subcollection of the members of a given set.

subtracting functions   If $f : \mathbb{R} \to \mathbb{R}$ and $g : \mathbb{R} \to \mathbb{R}$, then $(f - g)(x) \equiv f(x) - g(x)$.

successor   The natural number which follows a given natural number.

suffices for   An alternative phrase for implication.

summation by parts   A discrete analogue of integration by parts.

superset of   A supercollection of the members of a given set.

supremum   See *least upper bound.*

surjective A function is surjective if it is *onto*.

symmetric Let $\mathcal{R}$ be a relation on a set S. If $s, t \in S$ and $s\mathcal{R}t$, then $t\mathcal{R}s$.

Taylor's expansion The expansion $f(x) = \sum_{j=0}^{k} f^{(j)}(a)\frac{(x-a)^j}{j!} + R_{k,a}(x)$ for a given function f.

terminating decimal expansion A decimal expansion for a real number that has only finitely many nonzero digits.

theorem The formulation of a significant mathematical result.

"there exists" The quantifier $\exists$ for making an existence statement about some objects of a certain kind.

third-order logic Extends second-order logic to the extent that it allows us to treat sets of functions and also more abstract constructs.

totally disconnected set A set in which any two points can be separated by two disjoint open sets.

transcendental number A real number which is not algebraic.

transitive Let $\mathcal{R}$ be a relation on a set S. If $s, t, u \in S$ and $s\mathcal{R}t$ and $t\mathcal{R}u$, then $s\mathcal{R}u$.

triangle inequality The inequality

$$|a + b| \leq |a| + |b|$$

for real numbers a and b.

true An attribute of a sentence or statement. This attribute indicates the acceptability or validity of the statement.

truth table An array which shows the possible truth values of a statement.

uncountable set A set that does not have the same cardinality as the natural numbers.

undefinable terms Terms set forth at the outset of a mathematical theory. These terms cannot be defined, since there are no other terms preceding them.

uniform convergence of a sequence of functions   The sequence f_j of functions converges uniformly to a function f if, given $\epsilon > 0$, there is an $N > 0$ so that, if $j > N$, then $|f_j(x) - f(x)| < \epsilon$ for all x.

uniform convergence of a series of functions   A series of functions such that the sequence of partial sums converges uniformly.

uniformly Cauchy sequence of functions   A sequence of functions f_j with the property that, for $\epsilon > 0$, there is an $N > 0$ so that, if $j, k > N$, then $|f_j(x) - f_k(x)| < \epsilon$ for all x in the common domain.

uniformly continuous   A function f is uniformly continuous if, for each $\epsilon > 0$, there is a $\delta > 0$ so that $|f(s) - f(t)| < \epsilon$ whenever $|s - t| < \delta$.

union of sets   The collection of objects that lie in any one of a given collection of sets.

universal set   The set of which all other sets are a subset.

upper bound   A real number b is an upper bound for a subset $S \subset \mathbb{R}$ if $s \leq b$ for all $s \in S$.

Venn diagram   A pictorial device for showing relationships among sets.

Weierstrass Approximation Theorem   The result that any continuous function on $[0, 1]$ can be uniformly approximated by polynomials.

Weierstrass M-Test   A simple scalar test that guarantees the uniform convergence of a series of functions.

Weierstrass Nowhere Differentiable Function   A function that is continuous on $[0, 1]$ but that is not differentiable at any point of $[0, 1]$.

well defined   An operation on equivalence classes is well defined if the result is independent of the representatives chosen from the equivalence classes.

well ordering   A strict, simple ordering of a set S so that each nonempty subset has a least element.

Zero Test   If a series converges then its summands tend to zero.

Bibliography

[BOA1] R. P. Boas, *A Primer of Real Functions*, Carus Mathematical Monograph No. 13, John Wiley & Sons, Inc., New York, 1960.

[BUC] R. C. Buck, *Advanced Calculus*, 2d ed., McGraw-Hill Book Company, New York, 1969.

[FED] H. Federer, *Geometric Measure Theory*, Springer-Verlag , New York, 1969.

[HOF] K. Hoffman, *Analysis in Euclidean Space*, Prentice Hall, Inc., Englewood Cliff s, NJ, 1961. m

[KRA1] S. G. Krantz, *The Elements of Advanced Mathematics*, 3rd ed., CRC Press, Boca Raton, FL, 2012.

[KRA4] S. G. Krantz, *Handbook of Logic and Proof Techniques for Computer Scientists*, Birkhäuser, Boston, 2002.

[KRA5] S. G. Krantz, *Real Analysis and Foundations*, 4th ed., Taylor & Francis, Boca Raton, FL, 2017.

[KRA6] S. G. Krantz, *Function Theory of Several Complex Variables*, 2nd ed., American Mathematical Society, Providence, RI, 2001.

[KRP] S. G. Krantz and H. R. Parks, *A Primer of Real Analytic Functions*, 2nd ed., Birkhäuser Publishing, Boston, 2002.

[LOS] L. Loomis and S. Sternberg, *Advanced Calculus*, Addison-Wesley, Reading, MA, 1968.

[NIV] I. Niven, *Irrational Numbers*, Carus Mathematical Monograph No. 11, John Wiley & Sons, Inc., New York, 1951.

[ROY] H. Royden, *Real Analysis*, Macmillan, New York, 1961.

[RUD1] W. Rudin, *Principles of Mathematical Analysis*, 3rd ed., McGraw-Hill Book Company, New York, 1971.

[**RUD2**] W. Rudin, *Real and Complex Analysis*, McGraw-Hill Book Company, New York, 1966.

[**STRO**] K. Stromberg, *An Introduction to Classical Real Analysis*, Wadsworth Publishing, Inc., Belmont, CA, 1989.

Index

Printed in the United States
by Baker & Taylor Publisher Services